HUMAN PHYSIOLOGY EXAMINATION REVIEW

THIRD EDITION

HUMAN PHYSIOLOGY EXAMINATION REVIEW

THIRD EDITION

Robert S. Shepard, Ph.D.

Professor of Physiology
Wayne State University
School of Medicine
Detroit, Michigan

APPLETON-CENTURY-CROFTS/Norwalk, Connecticut

0-8385-3949-1

85 86 87 88 89 / 10 9 8 7 6 5 4 3 2 1

Prentice-Hall of Australia, Pty. Ltd., Sydney
Prentice-Hall Canada, Inc.
Prentice-Hall Hispanoamericana, S.A., Mexico
Prentice-Hall of India Private Limited, New Delhi
Prentice-Hall International, Inc., London
Prentice-Hall of Japan, Inc., Tokyo
Prentice-Hall of Southeast Asia (Pte.) Ltd., Singapore
Whitehall Books Ltd., Wellington, New Zealand
Editora Prentice-Hall do Brasil Ltda., Rio de Janeiro

Library of Congress Cataloging in Publication Data

Shepard, Robert Stanley, 1927-
    Human physiology examination review.

    Includes bibliographies and index.
    1. Human physiology—Examinations, questions, etc.
I. Title. [DNLM: 1. Physiology—examination questions.
QT 18 S547h]
QP40.S48 1985   612'.0076   85-5982
ISBN 0-8385-3949-1

Dedicated to Azro Rodgers Shepard 1897–1972

Contents

Preface

The examination question serves a multitude of functions. It is used as (1) a whip to enforce conformity and increase effort, (2) a test of the acquisition of facts, (3) a test of problem-solving skills, (4) a feedback system that indicates progress and recommends further work, and (5) a predictor of competency.

The following types of questions are common on machine-scored examinations such as Part I of the National Board Examination on Physiology, as well as the FLEX examination:

1. Type A: Single answer multiple choice
   a. Stem followed by 4 to 5 possible matches
   b. 4 to 5 possible matches followed by the stem
2. Type K: Multiple answer multiple choice (see Chapter 1, questions 8 and 9 for an example of a modification of this approach)
   a. Stem followed by 4 numbered possible matches
3. Type C: Comparison (see Chapter 4, question 2)
   a. 4 possibilities precede the stem
4. Type B: Matching (see Chapter 10, question 1 for a modification of this type)
   a. Similar to 1b in format, but the same A through E selections are used for a number of different questions
5. Sets (see Chapter 7, questions 41 through 45 for an example of a modification of this approach)
   a. Type G: A descriptive header followed by two or more type A related questions (see Chapter 7, question 1)
   b. Type N: Similar to the above but with type K questions

Other types of questions include:

6. Statement–Reason (see Chapter 1, questions 18 through 21)
7. Quantitative comparisons (see Chapter 4, question 3 for a modification of this approach)
8. True–False
9. Fill in the blanks (see Chapter 4, question 1)
10. List or discuss (see Chapter 29, questions 12 and 19 through 23)

I have incorporated many of the comparison, quantitative comparison, and true–false approach questions into a type A format. I have used these modifications because I wanted (1) a book that, unlike the National Board and most other machine-graded examinations, was organized by subject, rather than by question format and

(2) a series of questions that did not abruptly change from one format to a markedly different format and then back to the first format.

Probably the question that gives the student the greatest number of problems is item 6, Statement–Reason. Here, he or she is being asked if a certain event (Statement: norepinephrine increases coronary blood flow, for example) is caused by a certain mechanism (Reason: norepinephrine stimulates alpha adrenergic receptors). In this case both the "Statement" and "Reason," when read independently, are correct, but there is no cause and effect relationship. Since physiology is concerned with mechanisms of action, this type of question is a very important one. Unfortunately a few students read too much into this type of question. For example, the following should be marked correct:

Statement

_____ An increased aortic pressure causes a decreased heart rate because

Reason

an increased aortic pressure causes a reflex increased stimulation of parasympathetic neurons to the heart.

Those that justify marking it incorrect because it is not the stimulation of parasympathetic neurons that is the mechanism, but rather the release of acetylcholine by the vagal neurons, are reading more into the question than is there.

I have written this book with three types of reader in mind: the student who is attempting to learn physiology for the first time, the student who is reviewing physiology after having completed formal training in the discipline, and the teacher who is assembling an examination. My intent in this edition has been to provide comprehensive, effective coverage of the field. In order to achieve this goal without markedly increasing the size of the book, I have introduced a number of information-concentrated items. Many of the approximately 900 questions suggest large numbers of additional questions that could be asked. For example, question 1 in Chapter 10 could be divided into eight separate multiple choice questions, and question 21 in Chapter 9, a single question on a diagram or record, could be expanded into a number of additional questions.

For many of those beginning the study of physiology, the question–answer format is an important learning tool. For those who find parts of the written material in this book inadequate for their needs, I have provided references where more complete discussions of specific physiologic principles are found. These reference texts, however, contain diverse physiologic perspectives. For example, one author emphasizes that the transmembrane potential is the result of an electrogenic pump, while another maintains that it is due to the differential permeability of the cell membrane. Read three different texts and you may find three different formulae for the equilibrium potential for K^+:

$$E_K = 61.5 \log \frac{(K^+_o)}{(K^+_i)}, \text{ mammal at 37° C;}$$

$$E_K = -58 \log \frac{(K)_i}{(K)_o} \text{ mv, frog at 27° C;}$$

$$E_K = \frac{(2.303 \text{ RT})}{zF} \log \frac{(K^+_i)}{(K^+_o)}, \text{ mammal at a temperature of T.}$$

To the mature physiologist, many of these differences are either inconsequential or valid pieces of a complex set of mechanisms. Consider, for example, the first two formulae given above. Their apparent differences concern the facts that (1) the experimental conditions were different, and thus the "constants" (61.5 and 58) were different, and (2) the mathematical approaches were different [log $(K^+_o/K^+_i) = -$ log (K_i/K_o)]. On the other hand, the difference between the first two formulae and the third is more fundamental. The first two are based on the concept that since the inside of the resting cell membrane is negatively charged, the *resting transmembrane potential should be regarded as negative*, and from this reference point, E_K becomes *negative*. The third formula represents a less traditional approach, in which the outside of the membrane is the reference point, and therefore the *resting transmembrane potential and E_K are called positive*.

Although this lack of uniformity can represent an important barrier for some students, it also represents part of the strength of physiology. Physiology is a science that has its roots in the radicalism of Michael Servetus (he was burned at the stake in 1553), the genius of Leonardo da Vinci, and the perseverance and experimental insights of William Harvey, William Beaumont, Claude Bernard, Otto Loewi, Corneille Heymans, and many others.

Most of the questions included in this text are of the types contained in Part I of the National Board Examination on Physiology and in the FLEX examination. I have classified each as follows:

Major Classes:
C : Circulation
D : Digestion
E : Endocrinology
G: Growth and aging
H : Heart
K : Kidney and urinary systems
M: Muscle, skeletal and smooth
N : Nervous system and excitability
R : Respiration
S : Sexual reproduction
W: Work and metabolism

Subgroups:
a : abnormal
b : blood
c : chemistry
d : diffusion
e : electrical
f : flow of a gas or liquid
g : gases
h : heat
i : ionic environment
l : liver
m: mathematical calculations
n : neurons
o : omega (resistance to flow)
p : pressure of a gas or liquid
q : volume of a gas or liquid
s : sensation
u : urine
v : vibrations (sounds, etc.)

Type of Question:
1: simple recall (recognition)
2: simple problem solving (use of data to reach a conclusion)
3: complex problem solving

By using this system the reader can select questions that meet his own particular goals. If he wishes, for example, to develop his problem-solving ability, he can limit his selection of questions to those containing a 2 or 3 in the last column of the classification. In reviewing the performance of students on the National Board Examinations, for instance, I find that those examined do more poorly on questions involving mathematical calculations than on questions involving simple recall. Thus, if your goal is to pass this examination and you are weak in mathematical calculations, you would be well advised to review those questions containing an "m" in the classification. If, on the other hand, you are interested in abnormalities of the heart (murmurs, nodal rhythm, etc.), select questions containing both an "H" and an "a" in the classification.

My attempt to classify each question as a recall or problem-solving item is based on the assumption that the reader is just beginning his study of physiology. A question that involves reasoning for a beginner may be answered by memory by an examinee who has completed a course in physiology. The recall process in a multiple choice question is (1) read the question's stem and possible answers, (2) recall appropriate fact(s), and (3) answer the question. Problem solving involves (1) question, (2) recall, (3) interpretation, and (4) conclusions and answer. One examinee might go through the following steps:

1.	question:	hyperventilation causes (A) alkalosis, (B) acidosis?
2.	recall:	hyperventilation causes the elimination of CO_2, which is an acid-forming gas
3.	interpretation:	the elimination of an acid-forming gas causes alkalosis
4.	answer:	hyperventilation causes alkalosis

For most examinees the answer might be simple recall.

Let me also say a few words about the art of taking examinations. In using this book, I presume your goal is not so much to get the right answer as to learn the material. Therefore, don't let a lucky guess deceive you. In taking any examination, don't be misled by extraneous factors. (1) Don't look for an answer pattern. The answers are frequently determined by a table of random numbers or some other randomizing technique. (2) Regard each question as an honest attempt to assess your skills. The student who is looking for trick questions is usually his own worst enemy. (3) Don't read into questions material that is not there. For example, where the question states: "On the basis of the following data calculate . . .," you are not being asked to use *all* the data but merely to select the relevant data to reach a certain conclusion. (4) Work in the context of the question. In other words, if you are asked to do a calculation and the answers from which you are to select are all to two-place accuracy, don't calculate to the fourth place. If there is no penalty for guessing, never leave an answer blank. (5) Organize your time. In a timed examination it is usually important to complete the examination. Don't spend 50% of your time on a question worth 1% of the total score. Conquer that tendency to be compulsive before it is too late.

Important Notes to the Reader

References

A numbered list of reference books follows. Accompanying each question is a number combination that identifies the reference source and the page or pages where information pertinent to the question may be found. The first number corresponds to the textbook in the list, and the second number refers to the page of that textbook. For example (3:25) refers to the third book in the list, page 25; Ganong's *Review of Medical Physiology*.

Please note that in the case of reference book number two, Brobeck's *Best and Taylor's Physiological Basis of Medical Practice*, (2:5-102) refers, for example, to this second book in the list, section 5, page 102.

Special Coding of Questions

Accompanying each question, located beneath the reference source, is a letter and number combination that classifies each question according to subject matter and type, as explained in the Preface. Review the explanation given there, as the classification system will be helpful to you in your studies.

For example, **mic 2** identifies the question material as involving mathematical calculations, ionic environment, and chemistry and as being of the simple problem-solving type.

Location of Answers

In the text, the answers directly follow the questions. Each answer is indented, set in small type, and, when more than one choice is explained, the correct answer choice is circled.

Although readers will utilize the text differently in meeting their own study needs, it is advisable to cover the answer until you have attempted to answer the question.

List of References

1. Berne RM, Levy MN: *Physiology*. St. Louis, Mosby, 1983
2. Brobeck JR (ed): *Best and Taylor's Physiological Basis of Medical Practice*, 10th ed. Baltimore, Williams and Wilkins, 1979

3. Ganong WF: *Review of Medical Physiology*, 11th ed. Los Altos, California, Lange, 1983
4. Guyton AC: *Textbook of Medical Physiology*, 6th ed. Philadelphia, Saunders, 1981
5. Mountcastle VB (ed): *Medical Physiology*, 14th ed. Boston, Little, Brown, 1980

PART ONE
INTRODUCTION

1
Basic Principles

1

*(1:474; 2:3-94;
3:468; 4:210)*
Wmpf 1

Which of the following mathematical relationships is correct?

A. 1 mm Hg = 1330 dynes/cm^2
B. 1 calorie = 1330 dynes/cm^2
C. 1 watt = 10^{-7} joules
D. 1 erg = 1 dyne cm^{-1}
E. 1 poise = 2.39×10^{-8} calories

Answer and Explanation
(A.) Pressure is measured in both mm Hg and dynes/cm^2. In answers B through E we are dealing with an illogical mixture of units (i.e., we are comparing apples and potatoes).
B. The calorie is a unit of work and the dyne/cm^2 is a unit of pressure.
C. The watt is a unit of power and the joule is a unit of work.
D. The erg is a unit of work and the dyne/cm is a unit of tension. One dyne cm^{-1} = 1 dyne/cm.
E. The poise is a unit of viscosity and the calorie is a unit of work.

2

(3:15)
mic 2

Five grams of CaCl$_2$ were added to 200 ml of water. The atomic weight of calcium is 40 g and that of chlorine is 35.5 g. What is the *concentration* of this solution?

A. 2.5%
B. 3.0%
C. 10 g of CaCl$_2$/liter
D. 15 g of CaCl$_2$/liter
E. 20 g of CaCl$_2$/liter

Answer and Explanation
A. Concentration of CaCl$_2$

$$= \frac{5 \text{ g}}{200 \text{ ml}} \times \frac{5}{5} = \frac{25 \text{ g}}{1000 \text{ ml}} = 25 \text{ g/liter}$$

$$25 \text{ g}/1000 \text{ g} = 2.5 \text{ g}/100 \text{ g} = 2.5\%.$$

3

(2:5-20; 3:15; 5:4)
mic 2

The concentration of $CaCl_2$ reported for the previous question is equivalent to which of the following?

A. less than 200 mM/liter
B. between 200 and 220 mM/liter
C. between 220 and 230 mM/liter
D. between 230 and 240 mM/liter
E. more than 240 mM/liter

Answer and Explanation
C. Concentration of $CaCl_2$

$$= \frac{25 \text{ g/liter}}{\text{molecular wt}} = \frac{25 \text{ g/liter}}{(40 + 35.5 + 35.5)(\text{g/mol})}$$

$$= 0.225 \text{ mol/liter} = 225 \text{ mM/liter}.$$

4

(2:5-7; 3:15; 4:41;
5:1159)
mic 2

The *concentration of Ca* in the solution discussed in the previous question is

A. less than 225 mEq/liter
B. 225 mEq/liter
C. 300 mEq/liter
D. 440 mEq/liter
E. over 440 mEq/liter

Answer and Explanation
E. Concentration = (molar concentration) × (valence)
   = (225 mM/liter) × (2 mEq/mM)
   = 450 mEq/liter.

5

(2:5-12; 3:25;
4:448; 5:1784)
mic 2

An aqueous solution contains 0.1 *mM of H^+* and 0.1 mM of Cl^- per liter. On the basis of these data, one can conclude that the solution has

A. a pH of less than 3 and is acid
B. a pH of less than 3 and is basic
C. a pH of 4 and is acid
D. a pH of 4 and is basic
E. a concentration of OH^- of more than 0.1 mM/liter

Answer and Explanation
C. pH = −log (H^+)

(H^+) = 0.1 mM/liter = 10^{-4} mol/liter = −log (10^{-4}) = 4.

Since the solution has a pH of less than 7, it can be characterized as acid (i.e., it has a concentration of H^+, which is greater than 10^{-7} mol/liter).

E. $(H^+) \cdot (OH^-) = 10^{-14}$

$(OH^-) = (10^{-14})/(H^+) = (10^{-14})/(10^{-4}) = 10^{-10} = 0.0000000001$ mol/liter.

6

(2:5-12; 3:25;
4:448)
mic 3

Solution 1 has a pH of 7.4, *solution 2* a pOH of 4, *solution 3* a hydrogen ion concentration of 10^{-6} M, and *solution 4* a hydroxide ion concentration of 10^{-5} M. Arrange the solutions in order of increasing *acidity*:

A. solutions 1, 2, 3, 4
B. solutions 1, 3, 4, 2

C. solutions 2, 4, 3, 1
D. solutions 2, 4, 1, 3
E. solutions 3, 1, 4, 2

Answer and Explanation
D. • Solution 1: pH = 7.4
• Solution 2: pH = 10; pOH = 4
• Solution 3: pH = 6;$(H^+) = 10^{-6}$ mol/liter
• Solution 4: pH = 9; $(OH^-) = 10^{-5}$ mol/liter
• The least acid pH is 10 (solution 2). The most acid pH is 6 (solution 3).

7

(1:31; 2:5-7; 3:16; 4:41; 5:1159)
ic 1

Intracellular fluid characteristically contains

A. 95 mEq of Cl^-/liter
B. 80 mEq of Ca^{++}/liter
C. 100 mEq of HCO_3^-/liter
D. 70 mEq of Na^+/liter
E. 155 mEq of K^+/liter

Answer and Explanation
E. The intracellular environment contains approximately 190 mEq of cations and 190 mEq of anions per liter. The major cation is K^+ (157 mEq/liter), and the major anions are phosphate (113 mEq/liter) and proteins (74 mEq/liter). Cl^-, Ca^{++}, HCO_3^-, and Na^+ are all found in higher concentration in the blood plasma and other intercellular fluids than in the intracellular fluid.

8

(1:824; 2:5-5; 3:16; 4:391; 5:1150)
qic 1

Check each of the following statements concerning *intracellular fluid* that is true:

1. it contains over 50% of the body water
2. it has a higher osmotic pressure than extracellular fluid
3. it has a higher concentration of organic anions than extracellular fluid

Which one of the following best summarizes your conclusions?

A. statement 1 is true
B. statement 2 is true
C. statements 1 and 2 are true
D. statements 1 and 3 are true
E. statements 2 and 3 are true

Answer and Explanation
D. The intracellular fluid contains about 55% of the body's water and over four times the concentration of protein anions than does either plasma or interstitial fluid.

9

(1:8; 2:1-18; 3:16; 4:41; 5:10)
dc 2

Check each of the following statements concerning the *diffusion* of water that is true:

1. the net flux of water is from an area of low solute concentration to an area of high solute concentration
2. requires a semipermeable membrane
3. is an energy-requiring process

(Continued)

Which one of the following best summarizes your conclusions?

A. statement 1 is true
B. statement 2 is true
C. statement 3 is true
D. statements 2 and 3 are true
E. statements 1, 2, and 3 are true

Answer and Explanation
A. Diffusion is a random movement of particles that may occur either through semipermeable membranes or in their absence and does not require a source of energy.

10

(1:13; 2:1-27; 3:17; 4:47; 5:1153)
Cipm 3

Nine grams of NaCl are added to 1 liter of water. At $37°C$, this solution exerts an osmotic pressure of 7.6 atmospheres. How much osmotic pressure would be exerted by a solution containing 20 g of glucose per liter of water under similar circumstances? The molecular weight for glucose is 180 g and that for NaCl is 58.5 g.

A. less than 3 atmos
B. 5.6 atmos
C. 7.6 atmos
D. 10 atmos
E. more than 15 atmos

Answer and Explanation
A. The osmotic pressure exerted by a solution is directly related to the molar concentration of particles in that solution.

NaCl: $\dfrac{(9 \text{ g/liter})}{(58.5 \text{ g/mol})} = 0.15$ mol/liter.

Glucose: $\dfrac{(20 \text{ g/liter})}{(180 \text{ g/mol})} = 0.11$ mol/liter.

Since each particle of NaCl ionizes to two particles (Na^+ and Cl^-) when placed in solution, the NaCl solution contains 0.30 mol of particles per liter. If we assume negligible ionization for glucose, the osmotic pressure of the glucose solution will be:

$\dfrac{(0.11 \text{ mol/liter})}{(0.30 \text{ mol/liter})} \times (7.6 \text{ atmos}) = 2.8$ atmos.

11

(1:823; 2:5-3; 3:13; 1:391; 5:1149)
dqm 2

The following data were obtained from a patient by injecting an indicator and determining its concentration in the blood:

a.	total body water (indicator = DHO = heavy water):	42 liters
b.	extracellular water, fast (indicator = thiocyanate):	11 liters
c.	extracellular water, slow (indicator = thiocyanate):	19 liters
d.	plasma water (indicator Evans blue):	3 liters
e.	blood cell volume (calculated from hematocrit):	5 liters

A. What was the patient's intracellular water volume?
B. What was the patient's interstitial water volume?

Answer and Explanation
A. *Intracellular volume* = a − c = 42 − 19 = 23 liters. Line b represents that part of the *extracellular water* that is freely accessible to the indicator. It equals interstitial volume + plasma volume. Line c includes b (*fast extracellular volume*) + transcellular volume (in the joints and cerebral ventricles) + cartilage and bone water.
B. *Interstitial volume* = b − d = 11 − 3 = 8 liters.

12

One liter of *isotonic saline* was injected intravenously into a 70-kg woman. The injection had no effect on capillary hydrostatic pressure. After 15 minutes the (check each correct answer)

1. extracellular water would increase by more than 900 ml
2. interstitial water would increase by more than 900 ml
3. intracellular water would increase by more than 900 ml
4. plasma water would be increased by more than 900 ml
5. total body water would increase by more than 900 ml

Answer and Explanation
1, 5. The solute (Na^+) would be distributed in the plasma and interstitial space. Therefore the extracellular and total body water would each be increased by about 1 liter and the intracellular volume would remain unchanged. Since the plasma volume represents about 27% of the fast extracellular space, its volume would increase by about 270 ml. Since the interstitial volume represents 73%, its volume would increase by about 730 ml.

13

If 1 month later, 1 liter of plasma was given intravenously to the woman in question 12 and once again the capillary hydrostatic pressure stayed constant, what would be the effect on fluid volume 15 minutes after the injection (choose one or more correct answers from possibilities 1 through 5 in question 12).:

Answer and Explanation
1, 4, 5. The *colloids* in the plasma do not readily move into the interstitial space and therefore the plasma volume 15 minutes after the injection will be increased by about 1 liter and the interstitial and intracellular space will be unchanged.

14

If 2 months later, 1 liter of an *isosmolal solution of urea* was injected intravenously to the woman in question 12, and once again the capillary hydrostatic pressure stayed constant, what would be the effect on fluid volume 15 minutes after the injection (choose one or more correct answers from possibilities 1 through 5 in question 12).

Answer and Explanation
5. Since urea freely passes the capillary and cell membrane barrier it will be distributed throughout the extracellular and intracellular spaces. The intracellular volume will increase 670 ml if none of the water enters cartilage, bone, or transcellular water.

15

Is the urea solution in question 14 (1) hypertonic, (2) hypotonic, or (3) isotonic? Why is it important to distinguish between *permeant* and *impermeant* particles in determining *tonicity*?

Answer and Explanation
2. A red blood cell placed in isosmolar urea will swell. This is because urea diffuses into the cell and brings water with it. It is the impermeants (particles that do not permeate or move through cell membranes) that determine tonicity and the impermeants plus the permeants that determine *osmolarity*.

16

Two chambers (H and I) of equal volume are separated by a membrane (M) that is impermeable to *protein ions* ($Prot^-$), but permeable to K^+ and Cl^-. Chambers H and I initially contain water plus the same concentration of cations and anions. The initial ionic contents of the chambers are as follows.

(Continued)

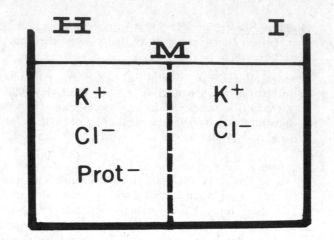

This situation would result in

A. a net movement of Cl^- into chamber I
B. a net movement of K^+ into chamber I
C. a net movement of water into chamber I
D. all of the above are true
E. $[K^+_H][Cl^-_H] = [K^+_I][Cl^-_I]$

Answer and Explanation

A. Cl^- will move into H because there is a higher *concentration* of Cl^- in chamber I.
B. Some K^+ will follow the negatively charged chloride into H.
C. Since the net flux of solute is into H, the net flux of water will also be into H.
(E.) The *Donnan effect* is that the product of the concentration of diffusible ions on both sides of a semipermeable membrane are equal. Blood plasma and *intracellular fluid*, like chamber H, have diffusible ions and important levels of protein ions. The perivascular and the *interstitial fluid*, like chamber I, have diffusable ions, but very few protein ions.

17

(1:19; 2:1-18; 3:19; 4:840; 5:21)
Wcd 2

The following diagram shows the rate of entry of various molecules into a cell:

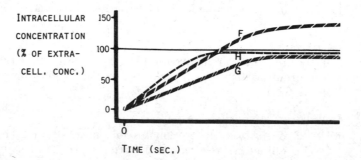

Choose the correct statement.

A. If curves F and G represent molecules that enter only by *diffusion* through *pores*, then curve F represents the diffusion of a larger molecule than curve G.
B. If curves G and H represent molecules that enter only through the lipid portion of the cell membrane then curve H represents a less lipid soluble molecule than curve G.
C. If curve F represents an ion being moved by *facilitated diffusion* then curve G represents the movement of the same ion after *ATPase* inhibition by *ouabain*.
D. If curve H represents a molecule being moved by facilitated diffusion then curve G

represents the movement of the same molecule after the addition of a molecule of similar chemical structure.

E. Curve F represents the diffusion of a negatively charged particle into the cytoplasm.

Answer and Explanation

A. The rate of diffusion through pores is inversely related to the size of the molecule.

B. The rate of diffusion through lipid is directly proportional to the degree of *lipid solubility*.

C. Facilitated diffusion is not an *energy-requiring phenomenon*.

(D.) The rate of facilitated diffusion of a molecule will be inversely related to the concentration of other molecules competing for the *transport* system.

E. Since the cytoplasmic side of the cell membrane is *negatively charged,* curve F could represent the diffusion of a positively charged ion into the cell. Negatively charged ions that diffuse through the cell membrane would obtain an intracellular concentration which is less than their extracellular concentration. For example HCO_3^- is formed in the cell from CO_2 produced by the cell, but has a lower intracellular concentration than its extracellular concentration.

Questions 18 through 21 consist of two main parts: a statement and a reason for the statement. Choose the best answer for each of these questions from the options A through E below and mark your choice in the space provided.

A. both the statement and the reason are true and are related as cause and effect (i.e., as indicated)
B. both the statement and the reason are true but are not related as cause and effect
C. the statement is true but the reason is false
D. the statement is false but the reason is an accepted fact or principle
E. both the statement and the reason are false

18

(1:5; 2:1-19; 3:19; 4:45; 5:6)
d 1

Statement
_____ Particle size is the major determinant for the penetration of a cell membrane by small nonpolar and polar molecules because

Reason
the cell membrane behaves as if it has small pores with a diameter of 0.4 to 0.8 nm.

Answer and Explanation

D. The *cell membrane* behaves as if it has many different channels for penetration. Many of the polar molecules, as well as inorganic ions, may pass through the membrane only via the hypothesized *pores*, and therefore their penetration will be determined by particle size. *Lipid solubility* is the major determinant for the rate of penetration of the cell membrane by many nonpolar molecules. For example, water traverses the membrane more rapidly than hydrated K^+ because the former has a smaller diameter. CO_2 traverses the membrane more rapidly than O_2 because of its higher lipid solubility.

19

(1:18; 2:1-20; 3:19; 4:43; 5:10)
d 2

Statement
_____ Facilitated diffusion is a process that has the property of competitive inhibition and saturation because

Reason
facilitated diffusion requires a pump mechanism for which there is only a limited amount of energy.

Answer and Explanation

C. *Facilitated diffusion* is a type of *carrier-mediated* transport that occurs along a concentration gradient and does not require an energy source. It demonstrates *saturation*, because there are a limited number of carriers. This type of transport is responsible for the speed at which glucose traverses the membrane.

20

(1:486; 2:1-12; 3:3; 4:16; 5:677)
Wc 1

Statement
_____ Mitochondria are the major source of energy in the cell because

Reason
mitochondria contain the enzymes responsible for oxidative phosphorylation.

Answer and Explanation
A. The *oxidative phosphorylation* of 2 mol of pyruvic acid that occurs during the citric acid cycle results in 13 times more energy for the cell than the anaerobic catabolism of 1 mol of glucose during glycolysis.

21

(1:905; 2:1-7; 3:7; 4:17; 5:699)
Wc 1

Statement
_____ The *nucleus* serves to regulate the synthesis of proteins by the cell because

Reason
the nucleus contains the DNA that serves as a template for the synthesis of RNA.

Answer and Explanation
A. *Deoxyribonucleic acid* is that part of the genetic material that helps form messenger *ribonucleic acid.*

22

(2:1-11; 3:41; 4:20; 5:1133)
Wc 1

The *lysosome*

A. produces secretory granules that store hormones and enzymes
B. contains a variety of enzymes responsible for the digestion of bacteria and worn out native cellular components
C. synthesizes proteins such as hormones that are secreted by the cell
D. synthesizes steroids in steroid-secreting cells
E. is a site to detoxification in some cells

Answer and Explanation
B.

23

(2:1-7; 3:7; 4:32)
Wc 1

Select from possibilities A through E in question 22 the function or functions of each of the following:

1. Agranular *endoplasmic reticulum:*   _____
2. *Ribosome* on endoplasmic reticulum: _____
3. *Golgi complex:*   _____

Answer and Explanation
1. D, E.
2. C.
3. A.

PART TWO
PERIPHERAL NERVOUS CONTROL

2
Organization of the Peripheral Nerves

1

(1:72; 2:1-3; 3:31; 4:116; 5:153)
Nn 1

Check each of the following statements about *myelin* that is true:

1. it is found in greater concentration in the white matter of the spinal cord than in the gray matter
2. it usually ensheaths the axon hillock
3. it usually forms an uninterrupted coating around axons

Which one of the following best summarizes your conclusions?

A. statement 1 is true
B. statement 2 is true
C. statement 3 is true
D. statements 2 and 3 are true
E. statements 1, 2, and 3 are true

Answer and Explanation

(A.) Myelinated axons are the major constituent of the white matter of the cord. Cell bodies, dendrites, and unmyelinated axons, on the other hand, are the major constituents of the gray matter. It is in the gray matter of the cord and brain that synaptic connections are made, and it is in the white matter that most of the central nervous system conduction occurs. It is the presence or absence of myelin that permits us to distinguish grossly these two functionally different areas.

B. The axon hillock, the axon endings, the dendrites, and the cell body in a myelinated neuron contain no myelin.

C. In myelinated axons, myelin is not found at the nodes of Ranvier.

2

(1:317; 2:9-59; 3:174; 4:116; 5:895)
Nn 1

Which of the following is an example of an *unmyelinated neuron*?

A. anterior horn cell
B. sensory neuron from the Golgi tendon organ
C. preganglionic parasympathetic neuron
D. preganglionic sympathetic neuron
E. postganglionic parasympathetic neuron

Answer and Explanation

E. Postganglionic sympathetic and parasympathetic neurons, and neurons carrying the sensation of slow pain, are small-diameter (less than 1.3 microns), slow-conducting (less

(Continued)

than 2.3 m/sec) neurons. They are less susceptible to asphyxia than other neurons and are usually unmyelinated.

3

(1:157; 2:1-41; 3:102; 4:614; 5:391)
NsMa 3

A disease of unknown etiology has affected the right leg and thigh in such a way that you suspect that the *unmyelinated neurons* in that part are not conducting impulses. Which of the following test results on the affected part would give you evidence that your hypothesis is correct?

result 1:   increased period of latency in the withdrawal reflex in response to pain
result 2:   inability to perceive pain
result 3:   increased period of latency in the knee jerk
result 4:   paralysis of skeletal muscle

A. result 1
B. results 2 and 3
C. result 4
D. results 1, 2, and 3
E. none of the above

Answer and Explanation
A. There are two types of pain fibers: (1) large-diameter, myelinated type A neurons and (2) small-diameter, unmyelinated dorsal root C fibers. The complete absence of the latter would not affect the period of latency. It might, on the other hand, decrease the duration of sensation and eliminate the dull lingering pain.
B, D. The afferent limb of the knee jerk reflex contains proprioceptor neurons (i.e., neurons from stretch receptors). The efferent limb of the reflex contains somatic efferent neurons. Both of these limbs are large-diameter, myelinated type A neurons.
(E.)

4

(1:49; 2:1-31; 3:38; 4:116; 5:61)
Nei 1

A neuron that exhibits *saltatory conduction* differs from one that does not, in that the former

A. produces its action potential through a Na^+ influx
B. produces its action potential through a K^+ influx
C. exhibits continuous conduction
D. conducts from one node of Ranvier to the next
E. conducts impulses more slowly

Answer and Explanation
A. all neurons apparently produce their action potentials by a Na^+ influx.
B. None do this.
C, (D.) Saltatory (L. *saltare*, to leap) conduction is a process whereby an action potential at one node produces one at a neighboring node. In unmyelinated neurons there is no leaping of action potentials but rather a continuous conduction.
E. Saltatory conduction results in a more rapid movement of the action potential through the neuron.

5

(1:46; 2:1-41; 3:38; 4:117; 5:70)
Nnef 1

The *speed* at which a myelinated axon conducts an action potential is directly related to

A. the diameter of the dendrites
B. the diameter of the axon
C. the length of the axon
D. the amount of axonal branching
E. the quantity of acetylcholine initiating the action potential

Answer and Explanation
A, (B.) The fastest conducting unmyelinated neurons have the largest diameter. Myelinated neurons conduct more rapidly than unmyelinated neurons with a similar axon diameter.

6

(1:46; 2:1-41; 3:41; 4:117; 5:72)
Nnes 2

An investigator (1) cuts a sensory nerve, (2) stimulates its central end with a single supermaximal stimulus, and (3) records electrical activity several centimeters central to the point of stimulation. What sort of *electrical activity* does he or she find?

A. minimal, since neurons cannot produce or conduct action potentials if their axons have been severed
B. a single spike potential
C. multiple spike potentials, since reverberating circuits are characteristic of most nerves
D. multiple spike potentials, because unmyelinated neurons conduct more rapidly than myelinated neurons
E. multiple spike potentials, because myelinated neurons conduct more rapidly than unmyelinated neurons

Answer and Explanation
E. Most nerves are collections of myelinated and unmyelinated neurons of different diameters. Since a supermaximal stimulus is, by definition, a signal sufficiently strong to excite all the neurons in the nerve, and since the velocity of conduction is directly related to the diameter of the neuron and the presence of a myelin sheath, it will take different periods of time for each neuron to send its signal from the point of origin to the recording electrode. This results in multiple action potentials arriving at different points in time. The reverberating circuit is also an important mechanism for producing multiple action potentials but is a characteristic of the central nervous system and not concerned with generating potentials in sensory nerves.

7

(1:35; 2:1-38; 3:36; 4:117; 5:54)
Nnes 2

An investigator (1) dissects out a single, afferent, unmyelinated neuron, (2) applies a single *subthreshold stimulus* to it, and (3) records the response to stimulation near the cathode of the stimulating electrode. What type of electrical activity does he or she find?

A. none, since a neuron does not respond to subthreshold stimuli
B. an increase in the transmembrane potential
C. a decrease in the transmembrane potential
D. a spike potential that is not propagated
E. a spike potential that is propagated for less than five internodes

Answer and Explanation
A. A neuron does not produce spike potentials in response to a subthreshold stimulus, but it does show other electrical changes.
B, (C.) If the resting transmembrane potential changes from −90 to −80 mv in response to a subthreshold stimulus, this change is called a decrease in the transmembrane potential. For many, this is a confusing perspective, but vocabulary is an essential part of any science, and this is an important part of the vocabulary of physiology.
D, E. The subthreshold stimulus produces a local excitatory state that makes the cell more excitable. The electrical pattern for the local excitatory state is never spiked, does not initiate activity in another innervated structure, and is therefore never referred to as a spike or action potential.

8

(1:37; 2:1-55; 3:36; 4:118; 5:54)
Nne 2

In an experiment, the axon of a motor neuron is stimulated with a single *superthreshold stimulus*. In this experiment one expects to record

A. no action potential in that part of the axon central to the site of stimulation
B. no action potential in that part of the axon peripheral to the site of stimulation
C. an action potential in all parts of the axon
D. an action potential in all parts of the axon and cell body
E. an action potential throughout the motor neuron and in the neurons innervating the motor neuron

Answer and Explanation

A, B, Ⓒ. An axon can conduct either toward or away from its cell body. In a motor neuron, a stimulus conducted toward the central nervous system is called an *antidromic impulse*, and one moving away from the central nervous system is called *orthodromic*. In the case of a sensory neuron, conduction away from the cord or brain is called antidromic. The significance of an antidromic stimulus is discussed under *axon reflex*.
D. The cell body does not exhibit action potentials. In this case, it produces an excitatory postsynaptic potential.
E. Motor neurons do not cause action potentials in the neurons that innervate them. The synapse between motor and internuncial neurons permits only impulse transmission to the motor neuron.

9

(1:90; 2:9-55; 3:105; 4:597; 5:413)
Nn 1

Check each of the following that contains *visceral afferent neurons*:

1. dorsal roots of spinal nerves
2. vagus nerves
3. phrenic nerves
4. splanchnic nerves

Which one of the following best summarizes your conclusions?

A. statement 1 is correct
B. statement 4 is correct
C. statements 2, 3, and 4 are correct
D. statements 1 and 4 are correct
E. statements 1, 2, 3, and 4 are correct

Answer and Explanation

E. Both afferent and efferent neurons are found side by side in most of the peripheral nerves. Areas where this is not characteristic include the dorsal and ventral roots of the spinal nerves, the gray rami communicantes, and certain cranial nerves (i.e., I, II, and VIII). Vagus nerves contain sensory neurons from the lungs, trachea, esophagus, aortic arch, aortic bodies, and most of the abdominal viscera. Phrenic nerves contain afferent neurons from the pericardium.

10

(1:314; 2:9-59; 3:174; 4:710; 5:894)
Nn 1

The cervical vagus nerve contains

A. afferent neurons from mechanoreceptors in the heart and pulmonary vessels
B. afferent neurons from mechanoreceptors and chemoreceptors in the aorta
C. all of the above
D. postganglionic parasympathetic neurons from the medulla oblongata
E. afferent neurons from mechanoreceptors and chemoreceptors in and near the carotid sinus

Answer and Explanation
C. The *cardiopulmonary* and *aortic arch* baroreceptors and the *aortic bodies* send their signals up the *vagi* (tenth cranial nerve).
D. The vagus contains *descending preganglionic parasympathetic* neurons. In the dog, it also contains ascending postganglionic sympathetic neurons.
E. They send their signals up the *ninth cranial nerve*.

11

(1:80; 2:1-43; 3:88; 4:577; 5:336)
sNne 2

Information concerning the intensity of a stimulus is a function of (check each correct answer):

1. the amplitude of the action potential in a sensory neuron
2. the frequency of action potentials in a single neuron
3. the number of sensory neurons activated

Which one of the following best summarizes your conclusions?

A. statement 1 is correct
B. statement 2 is correct
C. statement 3 is correct
D. statements 1 and 3 are correct
E. statements 2 and 3 are correct

Answer and Explanation
A. The relationship between the height of the action potential in a single neuron and the stimulus is an *all-or-none* relationship. One does not change the character of the action potential by increasing the intensity of stimulation above threshold. On the other hand, the amplitude of the action potential is modified by changes in the environment.
E. Within limits, as one increases the intensity of a stimulus, more sensory neurons will be stimulated, and each will conduct impulses at a progressively greater *frequency*.

12

(1:84; 2:1-77; 3:53; 4:133; 5:680)
NnM 2

If a single threshold stimulus is applied to a *somatic efferent* neuron, there will result

A. a twitch contraction of a single muscle fiber
B. a twitch contraction of a group of muscle fibers
C. a tetanic contraction of a single muscle fiber
D. a tetanic contraction of a group of muscle fibers
E. an increase in irritability of the fibers that neuron innervates

Answer and Explanation
A. Stimulation of a motor neuron to skeletal muscle results in the contraction of all of the muscle fibers that neuron innervates.
B. In skeletal muscle, a single stimulus to one of its motor neurons causes a single contraction in each of the fibers in that particular *motor unit*.
C, D. In skeletal muscle, but not all other types of muscle, a maintained contraction (i.e., a tetanic contraction) normally occurs only in response to a series of stimuli. Muscle *rigor and contracture* are not classified as contractions.
E. Autonomic neurons frequently act on cardiac and smooth muscle by increasing or decreasing their irritability. The role of the somatic efferent neuron, on the other hand, is not to change irritability but to initiate contraction.

13

*(1:450; 2:3-47;
3:55; 4:144; 5:969)*
NnMHe 1

The stimulation of *one* cell by another is frequently brought about by the first cell releasing acetylcholine. In which of the following is acetylcholine not the *intercellular transmitter* substance?

A. the stimulation of *one* ventricular cell by another
B. the stimulation of a skeletal muscle fiber by a somatic efferent neuron
C. the stimulation of smooth muscle by a postganglionic parasympathetic neuron
D. the stimulation of a postganglionic parasympathetic neuron by a preganglionic parasympathetic neuron
E. the stimulation of the adrenal medulla by a preganglionic sympathetic neuron

Answer and Explanation
A. In cardiac cells, unlike skeletal muscle cells, there is so little electrical resistance between certain cells that the action potential is transmitted from one cell to the next. This intercellular transmission occurs at the gap junction of the intercalated disk of cardiac muscle and is also noted in some smooth muscle. Motor neurons, on the other hand, stimulate, facilitate, or inhibit activity by the release of one of three transmitter substances: acetylcholine, norepinephrine, or dopamine.

14

*(1:226; 2:9-71;
3:97; 4:635; 5:397)*
NnM 1

The usual reflex response to the electrical stimulation of a pain fiber from the hand is

A. facilitation of the motor neurons to the ipsilateral extensor muscles for the forearm
B. inhibition of the motor neurons to the ipsilateral flexor muscles for the forearm
C. facilitation of the motor neurons to the ipsilateral extensor and inhibition of the motor neurons to the ipsilateral flexor muscles for the forearm
D. facilitation of the motor neurons to the contralateral flexor muscles for the forearm
E. facilitation of the motor neurons to the contralateral extensor muscles for the forearm

Answer and Explanation
E. One responds reflexly to pain by ipsilateral (same side as pain) flexion and contralateral (opposite side) extension. The antagonists to these reactions are inhibited.

15

*(1:760; 3:350;
4:832; 5:1324)*
NnMa 2

Which of the following reflexes in the adult disappear in the absence of functional connections between the *spinal cord* and the brain?

A. vomiting reflex
B. sweating reflex
C. withdrawal reflex
D. erection of the penis
E. all of the above disappear

Answer and Explanation
A. Neuron connections essential for the vomiting reflex are located in the cervical cord and the medulla of the brain. The other listed reflexes have sufficient synaptic connections in the cord to permit them to continue functioning after the loss of brain function. These reflexes may be considerably modified, however, since the normally functioning brain acts to inhibit and facilitate reflex action in the cord.

16

(1:77; 2:8-3; 3:84; 4:589; 5:329)
sNn 2

Check each of the following statements about *specialized receptors* that is true:

1. there are only five types (i.e., those for touch, vision, olfaction, hearing, and taste)
2. each responds to only one modality
3. under normal circumstances, they are responsible for all sensory input into the central nervous system

Which one of the following best summarizes your conclusions?

A. all of the above statements are false
B. all of the above statements are true
C. statement 2 is true
D. statements 1 and 2 are true
E. statements 2 and 3 are true

Answer and Explanation

A. Many of the current concepts of sensation are overly simplistic and data back to the philosophers of the pre-Christian era. Some of the receptor groups not included in the five listed in statement 1 include the acceleration-sensing system in the utricle and semicircular canals, the proprioceptors of skeletal muscle, tendons, joints, and the osmoreceptors of the hypothalamus. Although the various receptors may have a lower threshold for one particular modality than the others, they certainly cannot be characterized as being insensitive to other types of stimuli. Pressure on the eyeball can, for example, cause the sensation of light. Finally, it would be overly simplistic to state that all sensation is due to the stimulation of specialized receptors. The free nerve endings in the skin are not associated with specialized receptors and probably also serve as important sources of sensory information to the central nervous system. There is evidence that some types of sensation are due to the pattern of impulses arriving at the central nervous system. For example, the stimulation of one type of receptor in the eye can give the sensation of red; another type can give the sensation of green; the stimulation of both together can give the sensation of yellow.

17

(1:77; 2:8-3; 3:84; 4:588; 5:329)
sN 2

An individual's manual dexterity is most dependent on the functioning of his

A. enteroceptors
B. exteroceptors
C. nociceptors
D. proprioceptors
E. teleceptors

Answer and Explanation

A. *Enteroceptors* (interoceptors) respond to internal stimuli such as an irritation of the intestine and are associated with the sensation of pain, hunger, etc.
B. *Exteroceptors* respond to the near at hand external environment (touch, heat, etc.).
C. *Nociceptors* respond to potentially damaging stimuli (pain, nausea, etc.). They may be either extero- or interoreceptors.
D. *Proprioceptors* are located in muscles and joints and tell us about the position of our body in space. Some of the other receptors also may contribute to our manual dexterity.
E. *Teleceptors* respond to stimuli that are at a distance (vision, hearing, etc.). None of the attempts to classify the sensory organs has been entirely satisfactory. For example, heat receptors in the skin act as teleceptors when they are stimulated by the heat of the sun and as exteroceptors when touching warm water.

18

(1:221; 2:9-73; 3:97; 4:634; 5:201)
MNn 2

Internuncial neurons (i.e., interneurons)

A. are an essential part of the withdrawal (nociceptive) reflex
B. are an essential part of the stretch (myotatic) reflex
C. are an essential part of all reflexes
D. are always excitatory
E. are always inhibitory

Answer and Explanation
A. The withdrawal reflex, like most reflexes in the body, is multisynaptic (i.e., involves at least one internuncial neuron).
B. The knee jerk and other stretch reflexes are monosynaptic.
D, E. They may be either inhibitory or excitatory.

19

(1:213; 2:9-78; 3:92; 4:628; 5:703)
sMN 2

Muscle spindles are specialized receptors for stretch that contain *annulospiral endings*. Which of the following situations would cause the stimulation of the annulospiral endings in the triceps muscle of the arm (check each correct answer)?

1. contraction of the extrafusal fibers of the triceps
2. contraction of the intrafusal fibers of the triceps
3. contraction of antagonistic muscles
4. contraction of synergistic muscles

Which one of the following best summarizes your conclusions?

A. all of the above are correct
B. all of the above are incorrect
C. statements 1 and 3 are correct
D. statements 2 and 3 are correct
E. statements 1 and 4 are correct

Answer and Explanation
A, B, C. The extrafusal fibers are parallel to the annulospiral endings, and therefore their contraction would decrease any stretch of the endings (i.e., prevent their stimulation). Synergistic muscles would have a similar action.
D. The intrafusal fibers are in series with the endings, and therefore their contraction would cause the stretch of the endings. Stretching the entire biceps muscle, by the contraction of antagonistic muscles or other means, would also stretch the endings.

20

(1:214; 2:9-78; 3:94; 4:633; 5:714)
sMN 2

The *Golgi tendon organ* differs in function from the annulospiral endings in that the tendon organ (check each correct answer):

1. is more sensitive to the distending force
2. causes a reflex inhibition of homonymous alpha efferent neurons
3. causes a reflex inhibition of ipsilateral, heteronymous, antagonistic, alpha neurons

Which one of the following best summarizes your conclusions?

A. statement 1 is correct
B. statement 2 is correct
C. statements 1 and 2 are correct
D. statements 1 and 3 are correct
E. statements 1, 2, and 3 are correct

Answer and Explanation

A. A threshold distending force for the annulospiral endings is about 2 g, whereas threshold for the Golgi tendon organ is about 200 g.

B. The tendon organ seems to protect the muscle from a reflex contraction that might do damage. It performs this function through a multisynaptic reflex that inhibits motor neurons to the homonymous muscle.

Check each of the following statements about the *gamma efferent neuron* that is true:

1. it is an A group motor neuron with a diameter smaller than that for alpha efferent neurons
2. it innervates intrafusal fibers
3. it innervates muscle fibers that stretch the annulospiral endings

21

*(1:213; 2:9-79;
3:94; 4:628; 5:711)*
NnMs 1

Which one of the following best summarizes your conclusions?

A. all of the above are true
B. all of the above are false
C. statements 1 and 2 are true
D. statement 2 is true
E. statements 2 and 3 are true

Answer and Explanation

A. The stimulation of gamma efferent neurons causes a contraction of intrafusal fibers and a stretch of the annulospiral endings in the muscle spindle, which causes a reflex stimulation of alpha efferent neurons, which causes the stimulation of extrafusal fibers and, therefore, an increase in muscle tone.

22

*(1:213; 2:9-79;
3:95; 4:628; 5:770)*
NmMas 2

During an operation, a patient has several of his lumbar, posterior roots destroyed. After the operation the patient experiences a difficulty in standing erect and a decrease in *muscle tone*. The reason for these symptoms is that the operation

A. has destroyed some of the somatic efferent neurons to the leg and thigh
B. has destroyed sympathetic neurons to the leg and thigh
C. has destroyed parasympathetic neurons to the leg and thigh
D. has destroyed neurons carrying impulses from the Krause end-bulbs
E. has destroyed neurons from the annulospiral endings

Answer and Explanation

A, B, C. Destruction of the dorsal roots does not cause the destruction of motor neurons.
D. The Krause end-bulbs are cold receptors. They are not involved in postural reflexes.
E. The annulospiral endings are stimulated by muscle stretch and produce a reflex stimulation of alpha efferent neurons and, therefore, a contraction of the homonymous muscle. These stretch reflexes help maintain a high level of muscle tone and are an important part of our postural reflex system.

23

*(1:213; 2:9-79;
3:162; 4:628; 5:791)*
NnMa 3

As a result of a war injury, the *midbrain* in a patient is severed between the inferior and superior colliculi. This resulted in a marked increase in extensor muscle tone and an extensor muscle *hyperreflexia*. Both of these symptoms can be relieved by transection of the dorsal roots. What might be responsible for these symptoms?

A. generalized loss of facilitation
B. generalized loss of inhibition
C. decreased stretch of annulospiral endings in muscle
D. loss of inhibition delivered to gamma efferent neurons
E. decreased alpha efferent motor activity

(Continued)

Answer and Explanation

A, B. The symptoms reported above can result from a loss of inhibition, but the fact that they can be relieved by transection of the sensory roots indicates a more specific mechanism.

C. A decreased stretch would result in decreased tone and hyporeflexia.

D. Increased stimulation of gamma efferent neurons (1) facilitates more contraction of intrafusal muscle fibers, (2) facilitates more stretch of annulospiral endings, (3) facilitates more reflex stimulation of alpha efferent neurons, (4) brings about increased muscle tone (extrafusal fibers).

E. This would cause a decrease in muscle tone.

24

(1:214; 2:9-78; 3:94; 4:633; 5:714)
NnMs 2

A normal, healthy woman is attempting to lift a heavy load, and her muscles "give out." After this experience, there is no apparent problem in neuromuscular function. What is the most likely mechanism responsible for the abrupt cessation of skeletal muscle *contraction*?

A. the activation of stretch receptors in the Golgi tendon organ
B. the activation of stretch receptors in the annulospiral endings
C. skeletal muscle ischemia
D. the inactivation of stretch receptors in the Golgi tendon organ
E. the inactivation of stretch receptors in the annulospiral endings

Answer and Explanation

A. The receptors in the Golgi tendon organ have a higher threshold for stretch than those in the annulospiral endings and are, therefore, activated only in response to possibly dangerous distention. This causes a reflex inhibition of the homonymous motor neurons to the extrafusal fibers and, therefore, a cessation of contraction.

B. The receptors in the annulospiral endings facilitate further contraction of the extrafusal fibers. This facilitation can be overwhelmed by inhibition initiated by the Golgi tendon organ.

C. An inadequate blood flow is an unlikely mechanism in a short-term contraction in a healthy subject.

25

(1:53; 2:9-61; 3:72; 4:713; 5:177)
NnMca 2

A patient comes to your office complaining of muscle weakness. You administer *neostigmine* intramuscularly and the muscle weakness disappears. What is the mechanism of action of the neostigmine?

A. it blocks the action of acetylcholine
B. it blocks the action of norepinephrine
C. it interferes with the action of amine oxidase
D. it interferes with the action of acetylcholine esterase
E. it interferes with the action of carbonic anhydrase

Answer and Explanation

A. Acetylcholine (ACh) initiates skeletal muscle contraction; blocking its action would make the muscles weaker.

B. Norepinephrine does not play an important role in controlling skeletal muscle contraction.

C. Amine oxidase is an enzyme that facilitates the destruction of catecholamines (norepinephrine, etc.).

D. *Acetylcholine esterase* is the enzyme that catalyzes the destruction of acetylcholine. By interfering with the body's natural destroyer of ACh you can exaggerate ACh's action. In the disease *myasthenia gravis*, neostigmine causes an increased muscle strength. In a healthy individual, on the other hand, there is no noticeable change in strength in response to neostigmine.

E. Carbonic anhydrase catalyzes the formation of H_2CO_3 from CO_2 and water.

26

(1:550; 2:3-112; 3:480; 4:717; 5:90)
NHRMa 2

Decentralization is a process in which all the nerves going to and coming from an organ are cut. This separation of an organ from the central nervous system results in

A. an increased frequency of contraction in the case of the heart
B. a contraction that is less forceful in the case of the diaphragm
C. hypertrophy in the case of skeletal muscle
D. a decreased sensitivity to acetylcholine in the case of the salivary glands
E. inability to produce concentrated urine in the case of the kidney

Answer and Explanation

A. The heart, unlike skeletal muscle, continues to contract in the absence of any connection with the central nervous system. In human beings, a decentralization of the heart, produced by drugs or by surgery, results in the heart rate changing from about 70 to about 110 beats per minute. Apparently, in human beings at rest, the cardiac nerves have an over-all effect of decreasing the rate of firing of the cardiac *pacemaker*. In cats, on the other hand, the cardiac nerves have an opposite action.

B. The diaphragm, like all other skeletal muscles, stops contracting when it is decentralized.

C. All skeletal muscle slowly atrophies in response to decentralization.

D. Salivary glands, as well as most other organs, become *hypersensitive* to acetylcholine after decentralization. This is probably because, after decentralization, the quantity of acetylcholine esterase in the organ decreases.

E. Nerves are far less important in the control of the kidney than hormones such as aldosterone, parathormone, and antidiuretic hormone (ADH). ADH facilitates the production of concentrated urine.

Questions 27 through 29 consist of two main parts: a statement and a reason for the statement. Choose the best answer for each of these questions from the options A through E below and mark your choice in the space provided.

A. both the statement and the reason are true and are related as cause and effect (i.e., as indicated)
B. both the statement and the reason are true but are not related as cause and effect
C. the statement is true but the reason is false
D. the statement is false but the reason is an accepted fact or principle
E. both the statement and the reason are false

27

(1:50; 2:1-56; 3:64; 4:565; 5:203)
Nn 1

Statement
_____ In a four-neuron reflex arc the efferent neuron does not stimulate an internuncial neuron because

Reason
the efferent neuron does not release a transmitter substance.

Answer and Explanation

C. The *motor (efferent) neuron* apparently does not release a *neurotransmitter* in the central nervous system, but it does release a transmitter at its axon endings. For example, preganglionic autonomic, postganglionic parasympathetic, and somatic motor neurons all release the neurotransmitter acetylcholine. Most postganglionic sympathetic motor neurons release norepinephrine.

28

*(1:50; 2:9-73; 3:65;
4:576; 5:188)*
Nn 2

Statement

_____ A seven-neuron *reflex arc* requires more time than a three-neuron arc because

Reason

in a seven-neuron arc there is a total synaptic delay in excess of 3 msec, and in a three-neuron arc the synaptic delay is less than 2 msec.

Answer and Explanation

A. The signal is delayed by about 0.5 msec at each *synapse*. In most cases, this is apparently the result of the time it takes to release the transmitter, have it diffuse a short distance, and have it activate the next neuron in the arc. *Reaction times* are characteristically shorter for simple acts (for example, pushing a button every time you see a light flash) than for complex acts (pushing a button every time you see a red light preceded by a yellow light), because there are fewer synapses involved in the simple act. In a number of cases, it is possible, through training, to decrease your reaction time. This is apparently done by decreasing the number of synapses involved in a reaction. For example, initially a reaction may have the spinal cord, brain stem, and forebrain as essential parts of the reflex arc. With training, one can sometimes bypass the forebrain and brain stem as part of the reaction, and now the reaction seems to become automatic. This is possibly part of what happens when one learns to ski or ride a bicycle.

29

*(1:37; 2:1-43; 3:36;
4:118; 5:48)*
NMH 1

Statement

_____ We know that an *all-or-none* relationship exists between a stimulus and the gastrocnemius muscle because

Reason

the muscle produces the same force of contraction in response to a threshold stimulus and a superthreshold stimulus.

Answer and Explanation

E. There is not an all-or-none relationship between any skeletal muscle (i.e., the organ, not the fiber) and its stimulus. A *superthreshold stimulus* produces a more forceful contraction in skeletal muscle than does a threshold stimulus. On the other hand, a superthreshold stimulus and a threshold stimulus produce a contraction of the same force when applied to a single muscle fiber. The all-or-none law applies to the relationship between the stimulus and the character of contraction and/or the character of the action potential in the *skeletal muscle* cell, the *axon*, and the *motor unit*. It does not apply to the skeletal muscle or the nerve. In a neuron, a subthreshold stimulus produces no action potential. A superthreshold and a threshold stimulus each produce an action potential of the same *amplitude* and *velocity*. In a nerve, on the other hand, a threshold stimulus may cause action potentials in 10 neurons, and a superthreshold stimulus may cause action potentials in 50 neurons.

3
Transmembrane Potential

(1:25; 2:1-34; 3:21; 4:111; 5:113)
Neid 3

1 Under a particular set of experimental conditions, a mammalian axon was found to be permeable to only one ion, K^+. The equilibrium potential for K^+ in this axon was –97 mv. On the basis of these data, we will expect to find

A. that the K^+ concentration inside the cell is 5 to 10 times that outside the cell when the resting potential is –97 mv

B. approximately the same concentration of K^+ inside and outside the cell when the resting potential is –97 mv

C. when the resting potential is –97 mv, an intracellular concentration of negatively charged ions that is more than 0.5% higher than the intracellular concentration of positively charged ions

D. an increase in the transmembrane potential if we increase the permeability of the membrane to Na^+

E. a resting transmembrane potential of –97 mv

Answer and Explanation

A, B. The *equilibrium potential* for K^+ (E_K) is determined by the *activity* gradient for K^+ that exists between the extracellular and intracellular fluids. It is the transmembrane potential that would occur when the net flux of K^+ across the membrane equals 0 if the cell membrane were permeable only to one ion, K^+. We can estimate E_K by using the *Nernst equation* and assuming that the concentration gradient for K^+, $(K^+_o)/(K^+_i)$, is approximately equal to the activity gradient for K^+:

$$E_K = 61.5 \log \frac{(K^+_o)}{(K^+_i)} = -97 \text{ mv.}$$

If the concentration gradient for K^+ were 0.1, it would mean that K^+ was 10 times more concentrated inside the cell (K^+_i) than outside (K^+_o) and that E_K would equal –61.5 mv, since the log of 0.1 is –1. Since E_K equals –97 mv, the concentration of K^+ inside the cell must be more than 10 times that outside the cell, and therefore statements A and B must be wrong.

C. When the resting potential is –97 mv, there is a separation of charge across the membrane, but we should still consider the cytoplasm and extracellular fluid electrically neutral, since this small separation of charge in the immediate region of the membrane does not measurably move the *anion-to-cation concentration gradient* away from 1 in either the extracellular or the intracellular fluid.

D. An increase in *sodium conductance* (i.e., permeability) will decrease the transmembrane potential (i.e., move it toward E_{Na}, which is approximately + 66 mv).

Ⓔ

2

*(1:38; 2:1-47; 3:39;
4:111; 5:45)*
Neid 2

In the generation of the initial phase of an action potential, there is a movement of the *transmembrane potential* (check each correct answer):

1. toward the equilibrium potential for Na^+
2. toward the equilibrium potential for K^+
3. toward the equilibrium potential for Cl^-

Which one of the following best summarizes your conclusions?

A. statement 1 is correct
B. statement 2 is correct
C. statement 3 is correct
D. statements 1 and 2 are correct
E. statements 1, 2, and 3 are correct

Answer and Explanation

A. The change of the transmembrane potential from −90 mv to +40 mv is due to an increased Na^+ conductance.

3

*(1:23; 2:1-33; 3:16;
4:398; 5:14)*
Neid 2

Check each of the following statements that is true. In a *resting skeletal muscle cell*:

1. the electrical and chemical gradients for Na^+ are directed into the cell
2. the electrical and chemical gradients for Cl^- are in opposite directions
3. K^+ efflux is impeded by the electrical gradient

Which one of the following best summarizes your conclusions?

A. statement 1 is true
B. statement 2 is true
C. statement 3 is true
D. statements 1 and 3 are true
E. statements 1, 2, and 3 are true

Answer and Explanation

E. Na^+ and Cl^- are more concentrated in the intercellular environment than in the intracellular environment, and thus their chemical gradients are into the cell. The opposite is true of K^+. The inside of the cell membrane is negative in reference to the outside, and thus the electrical gradient for positively charged ions is into the cell and the gradient for negatively charged ions is out of the cell.

4

*(1:31; 2:1-34; 3:19;
4:53; 5:14)*
Neid 2

Select the two mechanisms most likely to be responsible for the production of the *resting transmembrane potential* in an axon or skeletal muscle fiber.

1. a pump that extrudes fewer positively charged particles from the cell than it brings into the cell
2. a pump that extrudes more K^+ than it brings in Na^+
3. a pump that extrudes more Na^+ than it brings in K^+
4. a membrane that is more permeable to Na^+ than to K^+
5. a membrane that is more permeable to K^+ than to Na^+

Answer and Explanation

3, 5. The *electrogenic pump* and *differential permeability* of the plasma membrane are both responsible for the resting transmembrane potential. Consistent with the pump hypothesis is the generally accepted conclusion that an active transport system extrudes 3 Na^+ from the cell for every 2 K^+ carried into the cell. Consistent with the permeability hypothesis is the conclusion that the plasma membrane during its resting potential is more permeable to K^+ than to Na^+. For example, the permeability coefficient for a resting skeletal muscle plasma membrane has been calculated to be 10^{-4} cm/sec for Cl^-, 10^{-8} cm/sec for K^+, and 10^{-10} cm/sec for Na^+. If such a cell had a Na^+–K^+ pump that carried approximately equal quantities of Na^+ out of and K^+ into the cell, the cell would lose more positively charged particles through diffusion than it would gain. It would therefore tend to develop a negative transmembrane potential if there were no compensating movement of other charged particles.

5

(1:27; 2:5-5; 3:17; 4:366; 5:16)
Ndei 2

The following concentrations are noted inside (C_i) and outside (C_o) a cell:

	Inside	Outside
Na^+	10 mM	115 mM
K^+	120 mM	5 mM
Cl^-	? mM	120 mM
Protein$^-$	? mM	0 mM

If the cell volume remains constant and the cell membrane is freely permeable to K^+ and Cl^- but impermeable to Na^+ and protein$^-$, what would the *concentration of Cl^-* be inside the cell?

A. 5 mM
B. 10 mM
C. 40 mM
D. 110 mM
E. 130 mM

Answer and Explanation

A. *Donnan and Gibbs* demonstrated that in the presence of nondiffusible ions, the diffusible ions distribute themselves so that the product of their concentrations on one side of the membrane is equal to the product of their concentrations on the other side of the membrane:

$$(K^+_i) \times (Cl^-_i) = (K^+_o) \times (Cl^+_o)$$

$$(120 \text{ mM}) \times (Cl^-_i) = (5 \text{ mM}) \times (120 \text{ mM})$$

$$(Cl^-_i) = 5 \text{ mM}.$$

6

(1:13; 2:1-27; 3:16; 4:397; 5:1154)
ipmd 2

Using the data from question 5, determine what *concentration of protein* would be necessary for osmotic balance.

A. 10 mM
B. 25 mM
C. 85 mM
D. 105 mM
E. 120 mM

Answer and Explanation

D. Osmotic balance requires the same molar concentration of particles on both sides of the cell membrane. Therefore:

$$Na^+_i + K^+_i + Cl^-_i + Protein^-_i = Na^+_o + K^+_o + Cl^-_o$$

$$10 \text{ mM} + 120 \text{ mM} + 5 \text{ mM} + Pr^-_i = 115 \text{ mM} + 5 \text{ mM} + 120 \text{ mM}$$

$$Pr^-_i = 240 \text{ mM} - 135 \text{ mM} = 105 \text{ mM}.$$

7

(1:35; 2:1-34; 3:16; 4:44; 5:58)
NMide 2

If the *permeability* of a resting skeletal muscle cell to K$^+$ is increased while the permeability of the cell to Na$^+$ stays constant

A. the transmembrane potential would decrease
B. the cell would become more excitable
C. the cell would become more excitable because of a decrease in the transmembrane potential
D. the transmembrane potential would increase
E. the transmembrane potential would not change

Answer and Explanation

A, B, C. A decrease in the transmembrane potential can increase excitability, but there is no decrease in potential.

D. The equilibrium potential for K$^+$ is approximately −97 mv when the transmembrane potential is −90 mv. Therefore, an increased permeability for K$^+$ would result in a K$^+$ efflux and a movement of the transmembrane potential toward −97 mv. This, by convention, is called an increase in the transmembrane potential.

8

(1:38; 2:1-36; 3:35; 4:111; 5:64)
Nide 3

In an experiment, an investigator (1) isolates an axon and places it is in *Ringer's solution,* (2) inserts an intracellular electrode through the cell membrane, (3) measures the resting transmembrane potential, (4) stimulates the axon, and (5) measures its action potential (*see Record A*). The investigator then replaces the Ringer's solution with one of a different electrolyte content and again records the axon's resting transmembrane potential and action potential (*see Record B*). On the basis of these data how do you think the second solution differs from the Ringer's solution?

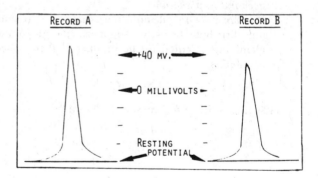

The second solution contains

A. a lower concentration of Na$^+$
B. a lower concentration of K$^+$
C. a higher concentration of Na$^+$
D. a higher concentration of K$^+$
E. a higher concentration of K$^+$ and a lower concentration of Na$^+$

Answer and Explanation

A. *Record B* has a reduced overshoot (30 mv *vs.* 60 mv). This is due to a reduced concentration of Na$^+$ in the extracellular environment.
B. This would cause a more negative resting potential.
C. This would cause an increased overshoot.
D. This would cause a less negative resting potential.
E. This would cause a less negative resting potential and a decreased overshoot.

9

(3:41; 4:119;
5:1519)
Neid 1

A decrease in the extracellular concentration of *calcium* causes (check each correct statement):

1. decreased membrane stability
2. hyperpolarization
3. decreased excitability

Which one of the following best summarizes your conclusions?

A. statement 1 is correct
B. statement 2 is correct
C. statements 1 and 2 are correct
D. statements 1, 2, and 3 are correct
E. statements 2 and 3 are correct

Answer and Explanation

A. Calcium is a structural part of the cell membrane and tends to stabilize it and therefore decrease its excitability. It also tends to increase contractility. Hypocalcemia has the opposite actions.

10

(1:43; 2:1-49; 3:59;
4:144; 5:123)
Mdei 2

The following records were obtained (I) before and (II) after a change in the external environment of the uterus.

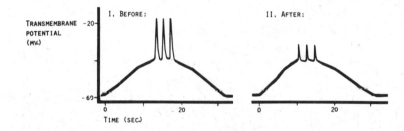

What was the change? The extracellular concentration of

A. Na^+ was increased
B. Ca^{++} was increased
C. K^+ was increased
D. K^+ was decreased
E. Ca^{++} was decreased

Answer and Explanation

C, D. In these records the slow *sine wave* on which the action potentials are superimposed is characteristic of many smooth muscles. It, like the resting potential of skeletal muscle, is increased (i.e., becomes more negative) by decreases in extracellular K^+ and decreased by increases in extracellular K^+. Note that this potential has not changed in these records. Apparently this sine wave also serves as a *pacemaker potential*. Note that as the sine wave becomes less negative it eventually initiates the development of spike potentials.

E. Smooth muscle, unlike other muscles, produces action potentials that do not necessarily have an *overshoot*. These potentials are due to the influx of both Na^+ and Ca^{++} and therefore are decreased in amplitude when the extracellular concentration of either Na^+ or Ca^{++} is decreased. In order to answer this question correctly you must recognize that the sine wave potential has not changed, but the spike potentials have decreased in amplitude.

11

*(1:449; 2:1-48;
3:40; 4:109; 5:65)*
Neid 1

In excitable cells, *repolarization* is most closely associated with which of the following events?

A. Na^+ efflux
B. Na^+ influx
C. K^+ efflux
D. K^+ influx
E. decreased excitability

Answer and Explanation

A, B. In excitable cells, during repolarization, there may be some Na^+ diffusion, but the event responsible for the rapid repolarization of cells as different as neurons and cardiac cells, and always found during the repolarization of these cells, is a K^+ efflux (i.e., movement of K^+ from the cell into the intercellular space). Most of the Na^+ gained by the cell during depolarization is moved from the cell by the Na^+ pump after repolarization.
Ⓒ **D, E.** The repolarization process is responsible for a return of excitability.

12

*(1:397; 2:1-50;
3:41; 4:109; 5:20)*
Nnei 2

The following *action potential* is recorded from a somatic efferent neuron:

At what interval would the O_2 consumption in milliliters of O_2 most exceed the resting level?

A. interval f-g
B. interval g-h
C. interval h-i
D. interval i-j
E. interval j-k

Answer and Explanation

A, B, C, D. During these periods, the neuron is relying primarily on anaerobic sources of energy.
Ⓔ During this period, the neuron is repaying its O_2 debt. In addition, the energy-requiring $Na - K$ pump is active. A rock at the top of a cliff in some respects is similar to a neuron in the resting state. It takes relatively little energy to push the rock down the cliff or to initiate the action potential. It takes a great deal of energy to carry the rock back up or to move intracellular Na^+ out of and extracellular K^+ into the cytoplasm. In other words, during interval j-k, recovery is going on. During the intervals that precede it, the major process is diffusion (i.e., diffusion of Na^+ during f-h and diffusion of K^+ during h-j). The process of diffusion, unlike active transport, does not require the expenditure of energy.

13

(1:39; 2:1-45; 3:41; 4:111; 5:65)
Nndei 2

In the *action potential* illustrated in question 12

A. interval g-h is caused by an active transport of Na^+ into the cell
B. interval g-h is caused by a diffusion of Na^+ into the cell
C. interval g-h is caused by pinocytosis
D. interval h-i is caused by active transport of K^+
E. interval h-i is caused by active transport of Na^+

Answer and Explanation

A. The influx of Na^+ during this period is due to (1) an increased Na^+ *conductance* and (2) the *electrochemical gradient.*
B. *Diffusion* is a movement of particles along a concentration gradient. It is not energy requiring.
C. *Pinocytosis* is a process of small vessicle formation whereby a small part of the external environment is brought into the cell. It is apparently important in bringing certain large molecules into the cell but plays little or no role in Na^+ transport.
D, E. Interval h-i is caused by a diffusion of K^+ out of the cell. Most of the *active transport* of Na^+ and K^+ occurs after period h-i.

14

(1:35; 2:1-37; 3:37; 4:109; 5:46)
Nne 2

In the *transmembrane potential* illustrated in question 12

A. point f represents the threshold potential or firing level for the cell
B. the movement of the transmembrane potential from point f toward point g is called depolarization
C. the movement of the transmembrane potential from point f toward point g is associated with a decreased excitability
D. the cell is refractory during interval j-k
E. all of the above statements are true

Answer and Explanation

B. The movement of the potential from point f to g, as well as the movement from g to h, is an example of *depolarization.* Period f-g is caused by a local excitatory state during which there is a progressive increase in *excitability.* At point g the *threshold* for the cell is reached, the rate of depolarization increases, and the cell becomes *refractory* (i.e., it loses its excitability). The cell regains its excitability during period i-j.

15

(1:52; 2:1-56; 3:81; 4:139; 5:157)
Mne 1

The *end-plate potential* of skeletal muscle is best characterized as

A. a local reversal of charge originating at the end plate
B. a reversal of charge originating at the end plate and propagated throughout the cell
C. a decrease in the transmembrane potential that is propagated throughout the cell
D. a local decrease in the transmembrane potential that is caused by an increased permeability to Na^+ and K^+
E. a local decrease in the transmembrane potential that is associated with little or no increase in Na^+ conductance

Answer and Explanation

A, B. Under normal conditions, the end-plate potential will change from −90 mv to −10 mv. This is not a reversal of charge but a decrease in the transmembrane potential.
C. The end-plate potential, unlike the action potential, is a local (i.e., nonpropagated) event.
D, E. The depolarization of the end plate differs from the action potential found elsewhere in the muscle membrane in that the former is due to increased Na^+ and K^+ conductance and the latter is due to increased Na^+ conductance alone.

16

(1:52; 2:1-56; 3:81; 4:139; 5:157)
MNe 1

The end plate of a normally innervated skeletal muscle cell can be distinguished from the rest of the cell membrane in that the *end plate*

A. will initiate a contraction in response to the local application of acetylcholine
B. will depolarize when exposed to an excess of extracellular K^+
C. will depolarize in response to an excess of extracellular Ca^{++}
D. has all of the above characteristics
E. has none of the above characteristics

Answer and Explanation
A.
B. Both membranes are *permeable* to K^+ prior to stimulation and will depolarize in response to hyperkalemia.
C. Neither membrane is permeable to Ca^{++} prior to stimulation and neither will depolarize in response to hypercalcemia.

17

(1:58; 2:1-88; 3:81; 4:118; 5:188)
Nne 1

The *excitatory postsynaptic potential* (EPSP) differs from the end-plate potential in that the EPSP is

A. propagated
B. a reversal of charge
C. not associated with an increased permeability to Na^+
D. not decreased by curare
E. initiated by acetylcholine

Answer and Explanation
A, B, C. Both EPSP and the end-plate potential are nonpropagated depolarizations caused in part by an increased permeability of a part of the cell (end plate, dendrites, cell body) to Na^+.
D. E. In some cases the EPSP is initiated by acetylcholine, in other cases probably by something else. Under physiologic conditions, the end-plate potential is apparently always initiated by acetylcholine.

18

(1:58; 2:1-88; 3:65; 4:118; 5:188)
Nne 1

The *inhibitory postsynaptic potential*

A. probably occurs only in the brain
B. is a local depolarization caused by an increase in $g_{Ca^{++}}$
C. is a local hyperpolarization caused by a decrease in $g_{Ca^{++}}$
D. is a local depolarization caused by an increase in g_{Cl^-}
E. is a local hyperpolarization caused by an increase in g_{Cl^-}

Answer and Explanation
A. They have been seen in spinal motor neurons.
E. An increased g_{Cl^-} causes a diffusion of Cl^- to the cytoplasmic side of the cell membrane.

19

(1:52; 2:1-56; 3:82; 4:139; 5:157)
Nne 1

Miniature end plate potentials (MEPPs) are found at the postjunctional membrane of the neuromuscular junction. Their amplitude is usually between 0.5 and 1.0 mv. The MEPP

A. is due to the leakage of small quantities of acetylcholine from vesicles in the quiescent somatic efferent neuron
B. is decreased in amplitude by antiacetylcholine esterases (neostigmine for example)
C. is well characterized by both of the above statements

D. cause a series of weak twitches
E. cause a series of weak tetanic contractions

Answer and Explanation

Ⓐ

20

(1:64; 2:1-57; 3:77; 4:568; 5:214)
Nnec 2

Gamma-amino butyric acid (GABA) is believed to be one of the inhibitory transmitters in the spinal cord and brain. A likely mechanism of action for GABA is to

A. increase the permeability of a neuron to Cl^-
B. increase the permeability of a neuron to Na^+
C. increase the permeability of the cell to Ca^{++}
D. decrease the permeability of the cell to Ca^{++}
E. increase the production of acetylcholine esterase

Answer and Explanation

Ⓐ This serves to stabilize the membrane potential at the equilibrium potential for Cl^-.
B, C. This would depolarize the membrane.
D. This would have no effect on excitability, since the cell already has a low Ca^{++} permeability.

21

(1:52; 2:1-56; 3:77; 4:139; 5:157)
Mnea 2

Curare in therapeutic doses

A. decreases the amplitude of the skeletal muscle end-plate potential
B. prevents propagation of an action potential in skeletal muscle
C. prevents propagation of an action potential in skeletal muscle and cardiac muscle
D. enhances the action of acetylcholine esterase
E. enhances the action of catecholamines (epinephrine, etc.)

Answer and Explanation

Ⓐ Curare blocks the action of acetylcholine (ACh) at the neuromuscular junction. If sufficient quantities of curare are given, the amplitude of the end-plate potential becomes so small that it is incapable of initiating an action potential.
B, C. The rest of the muscle is still excitable and will produce propagated reversals of charge (i.e., action potentials) in response to electrical stimulation.

22

(1:59; 2:1-37; 3:68; 4:577; 5:192)
Nne 2

In an experiment it is noted that (1) the *stimulation of sensory neuron A* causes the activation of 80 motor neurons, (2) the stimulation of *sensory neuron B* causes the activation of 120 motor neurons, and (3) the stimulation of *sensory neurons A and B* at the same time causes the activation of 250 motor neurons. Apparently, the stimulation of *motor neurons A and B* at the same time results in

A. occlusion
B. temporal summation
C. summation in what was previously a subliminal fringe
D. occlusion in what was previously a subliminal fringe
E. a procedural artifact

Answer and Explanation

A. *Occlusion* is a situation in which the response to *A* plus the response to *B* is the stimulation of a greater number of motor neurons than the response to the simultaneous stimulation of *A and B*.

(Continued)

B. Temporal *summation* is a phenomenon in which two or more subthreshold stimuli arriving at different times cause stimulation. The phenomenon described above is more likely to be spatial summation.

C. Apparently, each sensory neuron is stimulating some motor neurons and increasing the irritability of others. Those motor neurons that, in response to stimulation of sensory neurons, have an increased *excitability* are said to be in the subliminal fringe.

23

(1:77; 2:8-8; 3:86; 4:590; 5:330)
Nse 2

Check each of the following that is characteristic of the *generator potential* of the pacinian corpuscle:

1. it exhibits an all-or-none relationship with its stimulus
2. it is a propagated phenomenon
3. like the action potential, it exhibits an overshoot

Which one of the following best summarizes your conclusions?

A. all of the above are correct
B. none of the above are correct
C. statement 1 is correct
D. statement 3 is correct
E. statements 1 and 2 are correct

Answer and Explanation

A, B. The generator potential, or receptor potential, like the excitatory postsynaptic potential, and the end-plate potential, (1) is a graded response (i.e., not all-or-none), which (2) can initiate a propagated potential but is not itself propagated. It (3) will not exhibit an overshoot. The *overshoot* occurs when there is a marked increased permeability to Na^+ only not associated with a similar increased permeability to K^+. The overshoot is characteristic of the response of axons and striated muscle cells. Here, the increased Na^+ *conductance* is marked and precedes the increased K^+ conductance.

24

(1:78; 2:8-140; 3:86; 4:594; 5:335)
Nse 3

In an experiment, an investigator (1) records the resting transmembrane potential for a stretch receptor, (2) stretches the receptor while recording the transmembrane potential, and (3) releases the stretch. The following record is obtained from this series of experiments:

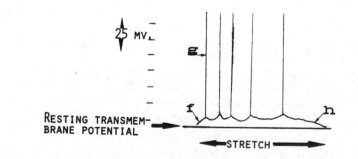

Which of the following interpretations of the record are correct?

A. the initial response to stretch (i.e., at f) is the production of an action potential
B. the initial response to stretch is hyperpolarization
C. stretch produces at g a high-amplitude generator potential
D. stretch produces at f a hyperpolarization and at g an action potential
E. adaptation occurs at h

Answer and Explanation
A. What is seen at f is a generator potential (i.e., a nonpropagated depolarization).
B. The phenomenon at f is depolarization.
C. The phenomenon at g is not a generator potential, but the first of a volley of action potentials.
D, (E.) A decrease in a receptor potential during stimulation is called *adaptation*. The progressive decrease in the frequency of action potentials as the stretch is maintained is also characterized as adaptation.

25

(1:44; 2:1-38; 3:36; 4:118; 5:46)
NMe 3

A 2-volt, square-wave stimulus was found to cause a contraction in a mammalian gastrocnemius muscle when the stimulus was applied to the muscle for 0.6 msec but not when it was applied for a shorter duration. In addition, it was noted that a 1-volt, square-wave stimulus with a duration of 100 msec was the lowest voltage to cause a contraction. On the basis of these experiments, we can conclude that the muscle had

A. a rheobase of 1 volt
B. a chronaxie of 0.6 msec
C. both of the above
D. a utilization time of 100 msec
E. a rheobase of 2 volts and a chronaxie of 0.6 msec

Answer and Explanation
(C.) The *rheobase* for an excitable structure is the current or voltage necessary to produce a contraction or action potential when that current or voltage is applied for an interval that is equal to or exceeds the utilization time. We can assume that the utilization time is more than 0.6 msec and less than 100 msec. At body temperature, it should be well below 10 msec. Therefore, the rheobase should equal 1 volt. The *chronaxie* is the minimum period during which twice the rheobase (2×1 volt) must be applied in order to produce a contraction and/or action potential.
D. The *utilization time* is the minimum duration that the weakest possible stimulus must be applied in order to obtain an action potential and/or a contraction. There is no evidence presented in this problem from which you could conclude that the utilization time is 100 msec.

Questions 26 through 30 consist of two main parts: a statement and a reason for the statement. Choose the best answer for each of these questions from the options A through E below and mark your choice in the space provided.

A. both the statement and the reason are true and are related as cause and effect (i.e., as indicated)
B. both the statement and the reason are true but are not related as cause and effect
C. the statement is true but the reason is false
D. the statement is false but the reason is an accepted fact or principle
E. both the statement and the reason are false

26

(1:38; 2:1-36; 3:23; 4:106; 5:14)
Neid 2

Statement
_____ If the permeability of an axon membrane to K^+ and Cl^- is 0, the transmembrane potential will equal 30 to 70 mv because

Reason
the equilibrium potential for Na^+ is between 30 and 70 mv.

Answer and Explanation
A. Apparently we can satisfactorily calculate the *transmembrane potential* (E_m) of an axon if

(Continued)

we know the *permeability coefficients* (P), intracellular concentrations, and extracellular concentrations of each of these three ions (Na^+, K^+, and Cl^-). The equation for this calculation is called the *Goldman equation*:

$$E_m = 61.5 \log \frac{P_K(K^+_o) + P_{Na}(Na^+_o) + P_{Cl}(Cl^-_i)}{P_K(K^+_i) + P_{Na}(Na^+_i) + P_{Cl}(Cl^-_o)}.$$

Since P_K and P_{Cl} both equal zero:

$$E_m = 61.5 \log \frac{(Na^+_o)}{(Na^+_i)} = E_{Na}.$$

27

(1:38; 2:1-36; 3:21; 4:109; 5:64)
Neid 1

Statement

_____ If one changes the transmembrane potential of an axon from –90 to –30 mv for 5 msec, a current will move into the axon because

Reason

this change in the transmembrane potential decreases Na^+ conductance (g_{Na^+}).

Answer and Explanation

C. A change in the *transmembrane potential* from –90 mv to –30 mv opens Na^+ *gates* (an increase in g_{Na^+}), and the predominant current through the membrane becomes an inward Na^+ current (I_{Na^+}).

28

(1:38; 2:1-47; 3:40; 4:112; 5:63)
Neid 1

Statement

_____ Most investigators believe that Na^+ channels in the cell membrane of an axon can be blocked without blocking K^+ channels because

Reason

when they expose an axon to *tetrodotoxin* and then depolarize the axon, I_{Na^+} is lost, but I_{K^+} is retained.

Answer and Explanation

A. I_{K^+} is an outward current that is initiated by *depolarization* and causes repolarization. It, unlike I_{Na^+}, is little affected by changes in the concentration of extracellular Na^+.

29

(1:23; 2:1-34; 3:39; 4:945; 5:61)
Neid 2

Statement

_____ The resting transmembrane potential (E_s) of an axon is more sensitive to *hypernatremia* than *hyperkalemia* because

Reason

the cell membrane has a higher g_{Na^+} than g_{K^+} during the resting state.

Answer and Explanation

E. E_s is more sensitive to the extracellular concentration of K^+ because in the stable or resting state, the membrane is more permeable to K^+. In other words, during E_s: $g_{K^+} > g_{Na^+}$.

30

(1:35; 2:1-34; 3:450; 4:945; 5:14)
Neid 2

Statement

_____ Hyperkalemia causes hyperpolarization because

Reason

hyperkalemia causes a decreased concentration of K^+ in the cytoplasm.

Answer and Explanation

E. *Hyperkalemia* causes *depolarization* because it changes the concentration gradient for K^+ across the cell membrane. As a result, the net movement of K^+ out of the resting cell decreases, and the inside of the cell membrane becomes less negative. In the terms of the

Nernst equation, we can say that hyperkalemia makes the *equilibrium potential* for K^+ (E_{K+}) less negative and therefore makes the membrane potential (E_m) less negative:

$$61.5 \log \frac{(\text{increased } K^+_o)}{(K^+_i)} = \text{less negative } E_K = \text{less negative } E_m.$$

A less negative E_m means a decreased E_m. It is important to realize that the ratio of (K^+_o/K^+_i) is always less than 1 and that the log of a number below 1 is negative. In other words, the *log to the base 10* of a ratio of 1 equals 0, a ratio of 0.1 equals -1, and a ratio of 0.01 equals -2.

4
Autonomic Nervous System

(1:314; 2:9-54;
3:174; 4:710; 5:896)
Nn 1

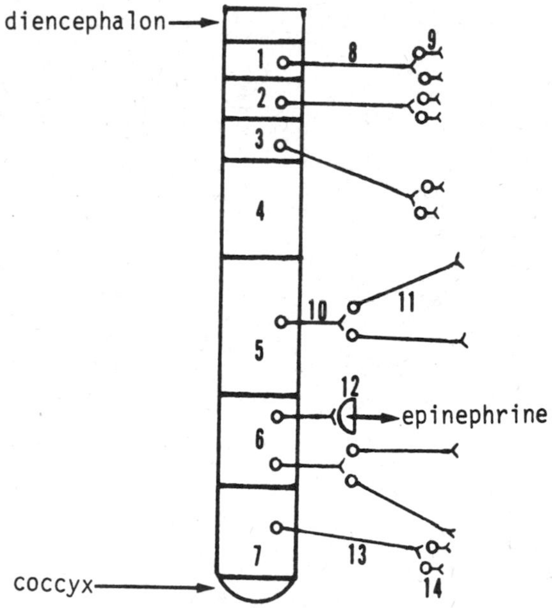

1   The anatomic relationships that exist for the motor division of the autonomic nervous system are represented in the following diagram:

Use the numbers in this diagram to answer the following questions.

- What are the three divisions of the *brain stem* from which autonomic neurons originate (diagram labels 1, 2, 3)?
- No autonomic neurons originate from what part of the *spinal cord* (diagram label 4)?
- Autonomic neurons originate from what parts of the spinal cord (diagram labels 5, 6, 7)?
- What autonomic neuron originates from the brain stem (diagram label 8)?
- What autonomic neurons does it innervate (diagram label 9)?
- What autonomic neuron originates from the thoracic and lumbar cord (diagram label 10)?
- What autonomic neurons does it innervate (diagram label 11)?

- What gland is innervated by preganglionic sympathetic neurons (diagram label 12)?
- What autonomic motor neurons have functional connections in the sacral cord (diagram labels 13, 14)?

Answer and Explanation

1. midbrain
2. pons
3. medulla oblongata
4. cervical spinal cord (also usually L-4 through S-1 and caudal to S-4)
5. thoracic
6. lumbar
7. sacral spinal cord

8. preganglionic parasympathetic neuron
9. postganglionic parasympathetic neurons
10. preganglionic sympathetic neuron
11. postganglionic sympathetic neurons
12. adrenal medulla
13. preganglionic parasympathetic neuron
14. postganglionic parasympathetic neurons

2

(1:318; 2:9-54; 3:176; 4:715; 5:896)
Nn 1

Phrases 1 through 4 each represent the first part of a sentence:

1. Stimulation of postganglionic parasympathetic neurons causes
2. Stimulation of postganglionic sympathetic neurons causes
3. Stimulation of either postganglionic (A) parasympathetic or (B) sympathetic neurons causes
4. Stimulation of neither postganglionic (A) parasympathetic nor (B) sympathetic neurons causes

Phrases A through U each represent the second part of a sentence. Write the one best match from 1 through 4 in the space provided in A through U.

A. _____ all of these neurons to release acetylcholine.
B. _____ all of these neurons to release norepinephrine.
C. _____ an increased heart rate and/or stroke volume.
D. _____ an increased conduction time in the A-V node of the heart.
E. _____ vasoconstriction in the gastrocnemius muscle.
F. _____ vasodilation in the gastrocnemius muscle.
G. _____ release of epinephrine from the adrenal medulla.
H. _____ glycogenolysis by the liver.
I. _____ renin secretion.
J. _____ vasopressin secretion by the pituitary gland.
K. _____ a positive inotropic and chronotropic response by the intestine.
L. _____ relaxation of the internal anal sphincter.
M. _____ contraction of the detrusor muscle of the urinary bladder.
N. _____ secretion by the salivary glands.
O. _____ secretion by the sweat glands.
P. _____ contraction of the radial muscle of the iris.
Q. _____ contraction of the circular muscle of the iris.
R. _____ contraction of the ciliary muscle of the eye.
S. _____ contraction of the extrinsic muscles of the eye.
T. _____ erection of the penis.
U. _____ ejaculation in the male.

Answer and Explanation

A.1.
B.4. Not all postganglionic sympathetic neurons are *adrenergic* (i.e., release norepinephrine). Some are *cholinergic*.
C.2.
D.1.
E.2. There are no parasympathetic neurons in the appendages.

(Continued)

F.2. Postganglionic, cholinergic, sympathetic fibers cause vasodilation and adrenergic, sympathetic fibers cause vasoconstriction.

G.4. The *medulla* is innervated by preganglionic, sympathetic fibers. It, like most postganglionic, sympathetic fibers, produces norepinephrine, but, unlike these fibers, converts the norepinephrine to epinephrine (by methylation).

H.2.

I.2.

J.4. *Vasopressin* (also called antidiuretic hormone) secretion is controlled by internuncial neurons originating in the hypothalamus of the brain.

K.1. Acetylcholine has a positive *chronotropic* and *inotropic* action on the intestine and a negative action on the heart.

L.1. Acetylcholine promotes GI motility by increasing peristaltic activity and dilating *sphincters.*

M.1.

N.3. Sympathetic fibers cause the release of a *saliva* with a high concentration of *mucin.* Parasympathetic fibers cause the release of a saliva with a high concentration of salivary *amylase.*

O.2. These are postganglionic, cholinergic fibers.

P.2. This causes pupillary dilation (*mydriasis*).

Q.1. This causes pupillary constriction (*miosis*).

R.1. This causes *accommodation* for near vision, an event that is associated with miosis.

S.4. These are all striated muscle. They are stimulated by *somatic efferent neurons.*

T.1.

U.2.

3

(1:325; 2:9-65; 3:181; 4:710; 5:917)
NHC 1

Which of the following structures is an important *reflex center* for autonomic nervous activity?

A. cerebellum
B. basal ganglia
C. both the cerebellum and basal ganglia
D. choroid plexus
E. the hypothalamus

Answer and Explanation

A, B, C. One cannot say that these areas play no role in the control of the autonomic nervous system, but there can be little doubt that (1) they play a much more important role in the integration of skeletal muscle activity and (2) they play a much less important role than the hypothalamus in the integration of ANS activity.

D. Cerbrospinal fluid is formed by the choroid plexus.

E. The hypothalamus is a part of the diencephalon (see figure for question 1). Through its control over sympathetic and parasympathetic neurons, it plays an important role (1) in temperature regulation and (2) in initiating the cardiovascular responses found during exercise, fear, and rage.

During an operation, a surgeon cuts all the neurons leading into the right and left *sympathetic chain* of ganglia. What are the immediate consequences of this operation?

(1:318; 2:9-173; 3:482; 4:249; 5:909)
NHCDa 2

A. an inability of the arterioles of the arm to constrict
B. an inability of the arterioles of the arm to constrict in response to a decrease in pressure in the carotid sinus
C. an inability of the heart to speed up in response to a decrease in pressure in the carotid sinus

D. an inability of the arterioles to constrict or the heart to speed up in response to a decrease in pressure in the carotid sinus

E. a disappearance of intestinal peristalsis

Answer and Explanation

A. Arterioles may constrict in response to stretch and changes in their environment in the absence of a nervous system.

B. There are no parasympathetic neurons in the arm, forearm, thigh, and leg; the only efferent neurons available for constricting and dilating their arterioles are the sympathetic neurons. Since all sympathetic neurons pass into the sympathetic chain, bilateral sympathectomy eliminates the reflex control of the blood vessels of the appendages.

C. One can cause a speeding up of the heart by either a stimulation of cardiac sympathetic neurons or an inhibition of cardiac parasympathetic neurons in the vagus nerve. The cranial parasympathetic neurons are left intact after this operation.

D, E. Peristalsis is due to a type of nerve net in the intestine which continues to function in the absence of any connections with the central nervous system.

5

(1:546; 2:9-34; 3:175; 4:437; 5:1339)
Nn 2

How would a patient respond to a *bilateral transection of the vagus nerves in the lower half of the neck?*

A. a decreased ability to release renin (in response to a decreased right atrial pressure) due to the section of sensory neurons

B. a tendency to aspirate food and other material due to section of motor neurons to the extrinsic muscles of the larynx

C. both of the above are correct

D. loss of the cephalic phase of gastric secretion due to the section of postganglionic parasympathetic neurons

E. loss of intestinal peristalsis due to the section of preganglionic parasympathetic neurons

Answer and Explanation

C. The cervical vagi carry sensory neurons from the cardiopulmonary receptors (important in the regulation of blood volume and sympathetic tone to the cardiovascular system), aortic receptors, lungs, etc. to the brain. Some of the motor neurons that pass down the vagi form the recurrent laryngeal nerves. They carry impulses up the neck to the striated muscles of the larynx.

D. The parasympathetic fibers in the neck originate from the brain stem and synapse with postganglionic neurons in the heart, alimentary tract, etc. In other words they are all preganglionic neurons.

E. The postganglionic neurons in the intestine continue to cause peristalsis even after the destruction of the preganglionic fibers which innervate them.

6

(1:319; 2:9-60; 3:176; 4:711; 5:899)
NCHD 2

Parasympathetic and sympathetic neurons exert approximately equal but opposite direct actions on

A. the arterioles of the arm

B. the force of ventricular contraction

C. the pancreas

D. all of the above

E. none of the above

Answer and Explanation

E. *Parasympathetic neurons* do not innervate structures in the arm, forearm, thigh, or leg. Arteriolar dilation in these appendages can be brought about by the stimulation of

(Continued)

postganglionic, cholinergic sympathetic neurons or by the inhibition of adrenergic sympathetic neurons. Parasympathetic neurons, unlike sympathetic neurons, exert little direct influence on the ventricles. They do play a major role in the control of the sinoatrial node, atrioventricular node, and atrial myocardium. The pancreas, on the other hand, receives an important parasympathetic innervation, but little innervation from sympathetic neurons.

7

(1:318; 2:9-62; 3:177; 4:712; 5:898)
NEn 1

Which of the following usually has norepinephrine as its major secretion?

A. the adrenal medulla
B. the adrenal cortex
C. postganglionic sympathetic neurons to the radial muscles of the iris
D. postganglionic sympathetic neurons to sweat glands
E. postganglionic parasympathetic neurons to circular muscles of the iris

Answer and Explanation

A. The adrenal medulla, unlike other producers of catecholamines in the body, has the capacity to methylate most of the norepinephrine it produces and, in this manner, convert it to epinephrine. The adrenal medulla usually has a secretion of about 80% epinephrine and 20% norepinephrine. During hypoxia or asphyxia, however, norepinephrine can constitute over 50% of the secretion.
B. The adrenal cortex is an important producer of steroids (mineralocorticoids, glucocorticoids, androgens, and estrogens), but not of catecholamines.
C. Adrenergic sympathetic neurons to the iris secrete norepinephrine.
D. Human sweat glands and some arterioles are innervated by postganglionic, cholinergic, sympathetic neurons.
E. All postganglionic parasympathetic neurons release acetylcholine, not catecholamines.

8

(1:1060; 2:9-62; 3:291; 4:567; 5:212)
Nbcn 1

Which of the following statements best characterizes *dopamine*. It

A. is a catecholamine precursor of norepinephrine
B. (like norepinephrine) usually is at a higher concentration in the blood than epinephrine
C. (like epinephrine and norepinephrine) has a half-life in the blood of about 2 minutes
D. (like epinephrine and norepinephrine) is excreted in the urine as 3-methoxy-4-hydroxymandelic acid, conjugated metanephrine, and normetanephrine
E. is well characterized by all of the above statements

Answer and Explanation

B. In the relaxed subject the concentration of *catecholamines* in the plasma is norepinephrine (300 pg/ml) greater than dopamine (200 pg/ml) greater than epinephrine (30 pg/ml).

9

(1:318; 2:9-64; 3:291; 4:714; 5:907)
NCH 1

Epinephrine affects smooth muscle tone in the blood vessels and the character of the heart's contraction in the following ways:

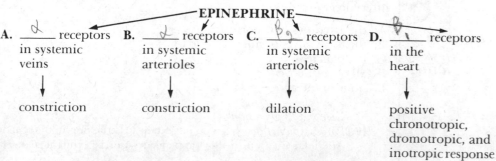

EPINEPHRINE

A. ____ α ____ receptors in systemic veins
↓
constriction

B. ____ α ____ receptors in systemic arterioles
↓
constriction

C. ____ β₂ ____ receptors in systemic arterioles
↓
dilation

D. ____ β₁ ____ receptors in the heart
↓
positive chronotropic, dromotropic, and inotropic response

A. What type of *adrenergic receptor* does it stimulate in the veins to produce venous constriction?

B. What type of receptor does it stimulate in the arterioles to produce arteriolar constriction?

C. What type of receptor does it stimulate in the arterioles to decrease arteriolar constriction?

D. What type of receptor does it stimulate in the heart to produce an increased *frequency* of contraction, *velocity* of conduction, or *force* of contraction?

Answer and Explanation

A. alpha

B. alpha

C. beta$_2$

D. beta$_1$

Where must norepinephrine be applied to produce an increased heart rate?

(1:319; 2:3-50; 3:176; 4:168; 5:1055)
NH 1

Answer and Explanation

It must be applied to the *beta$_1$ receptors of the sinoatrial node*. (When norepinephrine is applied to the SA node, it normally increases the atrial and ventricular rate. When it is applied to cardiac fibers elsewhere in the heart, it has a positive dromotropic and inotropic action on those fibers but seldom a chronotropic action.)

11

All systemic arterioles are not equally *sensitive to epinephrine*. Which one of the following arterioles is the least sensitive to epinephrine?

(1:319; 2:3-210; 3:496; 4:347; 5:1095)
NCf 2

A. skeletal muscle arteriole

B. cerebral arteriole

C. cutaneous arteriole

D. afferent renal arteriole

E. intestinal arteriole

Answer and Explanation

B. Intra-arterial injections of epinephrine into the cerebral artery produce less marked changes in blood flow than do similar injections into any other systemic artery. Two major factors in controlling cerebral blood flow are the concentration of CO_2 and H^+ in the cerebral blood. Both are potent dilators of the cerebral arterioles.

12

Atropine is a drug that blocks the action of acetylcholine on smooth muscles, glands, and the heart. Which of the following actions would you expect atropine to have?

(1:320; 2:9-120; 3:176; 4:715; 5:899)
NMHCu 2

A. loss of control over the diaphragm

B. cardiac asystole

C. pupillary constriction

D. decreased bronchial secretions

E. failure of skeletal muscle vasoconstriction in response to the stimulation of sympathetic efferent neurons

Answer and Explanation

A. Atropine does not interfere with the stimulation of the diaphragm or other skeletal muscles.

B. Acetylcholine modifies the intrinsic activity of the heart. Since it does not initiate contractions of the heart, blocking its action will not stop the heart from contracting.

(Continued)

C. Acetylcholine causes pupillary constriction. Blocking its action would cause pupillary dilation.

D. Atropine-like agents are used prior to a general anesthesia to dry the mucous membranes of the respiratory tract. In preventing the accumulation of mucus in the tract, they reduce the occurrence of atelectasis.

E. All postganglionic vasoconstrictor neurons are sympathetic fibers that release norepinephrine. Atropine does not block the action of norepinephrine, nor does it block the action of acetylcholine on the postganglionic neuron.

13

(1:532; 2:9-60; 3:480; 4:717; 5:908)
NMHC 2

Tetraethylammonium chloride (TEAC) is a drug that *blocks* the action of acetylcholine on postganglionic autonomic neurons. Which of the following actions would you expect TEAC to have in a resting subject?

A. a decrease in the heart rate
B. an increase in the heart rate
C. disappearance of intestinal peristalsis
D. an increase in the arterial blood pressure
E. prevention of skeletal muscle vasodilation in response to a local hypoxia

Answer and Explanation

A, **B.** Although the heart rate of a resting subject may be controlled by a combination of sympathetic tone, which tends to speed the heart, and parasympathetic tone, which tends to slow the heart, the over-all effect of the autonomic nervous system on heart rate in the resting subject is one of slowing it down. Therefore, when TEAC blocks the influence of the central nervous system on both sympathetic and parasympathetic neurons, the heart rate increases.

C. Peristalsis requires functioning groups of intrinsic neurons, but does not require signals from the central nervous system.

D. The arterial blood pressure (P_a) is approximately equal to the cardiac output (I) times the peripheral resistance (Ω): P_a (mm Hg) = (I ml/min) $\times$ (Ω mm Hg/ml/min). Loss of autonomic tone causes some minor changes in cardiac output and marked arteriolar vasodilation (decreased peripheral resistance). In other words, the loss of adrenergic sympathetic tone to the arterioles causes a decreased arterial pressure.

E. Hypoxia can produce arteriolar vasodilation in the absence of a nervous innervation.

14

(1:319; 2:9-69; 3:291; 4:714; 5:1055)
NnCp 3

A cat is (1) anesthetized, (2) injected intravenously with 2.5 μg of *epinephrine* per kilogram of body weight, (3) injected intravenously with 15 mg of dibenamine per kilogram of body weight, and (4) injected 20 minutes later with 2.5 μg of epinephrine per kilogram of body weight. The first injection of epinephrine caused an increase in arterial pressure. The second injection of epinephrine caused a decrease in arterial pressure. What is the most likely explanation of these results?

A. dibenamine blocks the vasoconstrictor action of epinephrine
B. dibenamine blocks the vasodilator action of epinephrine
C. dibenamine blocks the action of acetylcholine on smooth muscle and the heart
D. dibenamine is a ganglionic blocking agent
E. dibenamine catalyzes the conversion of epinephrine to norepinephrine

Answer and Explanation

A. Epinephrine apparently stimulates at least two types of receptor systems in arterioles. One type, the alpha receptor system, produces vasoconstriction (increased smooth muscle tone) when activated. The other type, the beta receptor system, produces vasodilation (decreased smooth muscle tone) when activated. Dibenamine blocks the vasoconstrictor action (i.e., blocks alpha receptors). In other words, it unmasks the vasodilator action of epinephrine.

B. It is the vasodilator action of epinephrine that causes the decreased arterial pressure.

C. Acetylcholine causes arteriolar dilation and decreased arterial pressure.

D, E. Norepinephrine elevates the arterial pressure more than the same quantity of epinephrine does.

15

(1:1064; 2:9-64; 3:291; 4:714; 5:1055)
NnCp 3

Drugs that block the vasoconstrictor action of epinephrine in skeletal muscle are called *alpha receptor blocking agents*. Drugs that block the vasodilator action of epinephrine in skeletal muscle are called *beta blocking agents*. Norepinephrine can be best characterized as having which of the following actions?

A. it has the same action on all skeletal muscle arterioles as epinephrine

B. it produces a less profound stimulation of the beta receptors of arterioles than epinephrine and therefore produces a less profound increase in arterial pressure

C. it produces a less profound stimulation of the beta receptors of arterioles than epinephrine and therefore produces a more profound increase in arterial pressure

D. it produces a less profound stimulation of the alpha receptors of arterioles than epinephrine and therefore produces a less profound increase in arterial pressure

E. it produces a less profound stimulation of the alpha receptors of arterioles than epinephrine and therefore produces a more profound increase in arterial pressure

Answer and Explanation

A. Norepinephrine produces a general vasoconstriction in skeletal muscle. Under most conditions, epinephrine produces vasodilation in skeletal muscle.

B. Norepinephrine is a general vasoconstrictor and would, therefore, be expected to produce a more profound increase in arterial pressure.

Ⓒ.

16

(1:1064; 2:9-64; 3:508; 4:714; 5:1055)
NnHC 2

A patient who has been receiving an *alpha blocking agent* (phenoxybenzamine) has complained of dizziness and fainting while walking up a single flight of stairs. During this exercise, his

A. cardiac output would be similar to that before the exercise

B. heart rate would increase

C. response is well characterized by both of the above

D. systemic resistance would fall to a lesser extent than before the drug was given

E. venous tone would increase to the same extent as before the drug was given

Answer and Explanation

B. Under these conditions (A) the cardiac output and (B) heart rate will increase (beta receptors in the heart are intact), (D) the decrease in systemic resistance will be exaggerated, because the patient has lost the constriction of the renal, visceral, and cutaneous arterioles that usually occurs during exercise, and the increased venous tone during exercise will be lost. There will also be a loss of some vasodilator reflexes. This loss is less extensive than the *vasoconstrictor* loss and therefore less important. The dizziness and fainting during the exercise is apparently due to the low systemic resistance which causes an arterial hypotension and brain ischemia.

(Continued)

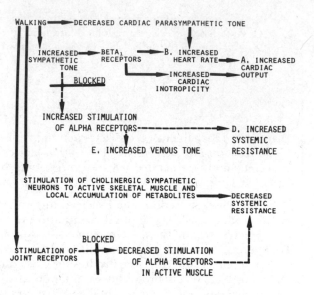

17

*(1:1064; 2:9-64;
3:291; 4:714;
5:1055)*
NnCH 3

Dichloroisoproterenol (DCI) is classified as a *beta adrenergic blocking agent*. It prevents the vasodilator and cardioaccelerator actions of epinephrine but does not prevent the vasoconstrictor action of either epinephrine or norepinephrine. These data are consistent with which of the following conclusions?

1. epinephrine produces vasodilation by stimulating beta receptors
2. epinephrine produces vasodilation by stimulating alpha receptors
3. epinephrine produces vasoconstriction by stimulating alpha receptors
4. epinephrine produces cardioacceleration by stimulating beta receptors
5. norepinephrine produces vasoconstriction by stimulating alpha receptors

Which of the following best summarizes your conclusions?

A. 1, 2, 3, 4, and 5 are correct
B. 1, 3, 4, and 5 are correct
C. 2, 3, 4, and 5 are correct
D. 3 and 4 are correct
E. 1 is correct

    Answer and Explanation

B. *Epinephrine*   →   *alpha receptors   →   vasoconstriction

               →   **beta$_1$ receptors   →   vasodilation

               →   **beta$_1$ receptors   →   cardiac acceleration ↑ HR

    Norepinephrine   →   **beta$_1$ receptors   →   cardiac acceleration ↑ HR

               →   *alpha receptors   →   vasoconstriction

    *Blocked by dibenamine, **Blocked by DCI*

18

*(1:1064; 2:9-64;
3:291; 4:714;
5:1055)*
NnHC 3

Norepinephrine and epinephrine both increase the heart rate and produce vasoconstriction in the skin. On the other hand, epinephrine produces vasodilation in skeletal muscle where norepinephrine produces vasoconstriction. These observations, as well as those presented in previous questions, are consistent with the view that

A. norepinephrine stimulates only alpha receptors
B. norepinephrine stimulates only beta receptors
C. norepinephrine stimulates alpha receptors of the heart and beta receptors of the arterioles

D. norepinephrine stimulates alpha receptors of the arterioles and beta receptors of the heart, but not beta receptors of the arterioles

E. epinephrine stimulates only beta receptors

Answer and Explanation

B_1
HLA

D. The data are consistent with the view that there are at least two types of beta receptors. The so-called beta$_1$ receptors are found in the heart, liver, and adipose tissue, and the beta$_2$ receptors are found in many smooth muscles (the smooth muscle of the arterioles, for example) and glands.

Epinephrine → *alpha receptors → vasoconstriction
→ **beta$_2$ receptors → vasodilation
→ **beta$_1$ receptors → cardiac acceleration
Norepinephrine → **beta$_1$ receptors → cardiac acceleration
→ *alpha receptors → vasoconstriction

*Blocked by dibenamine, **Blocked by DCI

19

(1:1064; 2:9-64; 3:176; 4:714; 5:1055)
NcCH 3

On the basis of the *adrenergic receptor* concepts developed in the preceding questions, we can conclude that:

1. epinephrine produces a more marked increase in arterial pressure than norepinephrine
2. norepinephrine produces a more marked increase in arterial pressure than epinephrine
3. norepinephrine has a weaker beta$_2$ stimulator action than epinephrine
4. epinephrine produces a more extensive vasoconstriction than norepinephrine
5. arterioles that have both alpha and beta receptors will dilate in response to norepinephrine

Which of the following best characterizes your conclusions?

A. conclusions 1, 4, and 5 are correct
B. conclusions 1, 3, 4, and 5 are correct
C. conclusions 2 and 3 are correct
D. conclusions 2, 3, and 5 are correct
E. conclusions 4 and 5 are correct

Answer and Explanation

C. Epinephrine → alpha receptors → vasoconstriction → increased arterial pressure
→ beta$_2$ receptors → vasodilation → decreased arterial pressure
Norepinephrine → alpha receptors → vasoconstriction → increased arterial pressure

Therefore, conclusions 1, 4, and 5 are not valid, and conclusions 2 and 3 are valid.

Questions 20 through 22 consist of two main parts: a statement and a reason for the statement. Choose the best answer for each of these questions from the options A through E below and mark your choice in the space provided.

A. both the statement and the reason are true and are related as cause and effect (i.e., as indicated)
B. both the statement and the reason are true but are not related as cause and effect
C. the statement is true but the reason is false
D. the statement is false but the reason is an accepted fact or principle
E. both the statement and the reason are false

20

(1:314; 2:3-172; 3:479; 4:712; 5:1052)
NC 2

Statement

_____ Atropine blocks the vasodilator action of postganglionic vasodilator neurons to the gastrocnemius muscle because

Reason

atropine blocks the action of all postganglionic parasympathetic neurons on smooth muscle.

Answer and Explanation

B. Since *vasodilator neurons* to the appendages are all apparently innervated by preganglionic neurons with their cell bodies in the *thoracolumbar cord*, they are classified as *postganglionic sympathetic* neurons. Atropine blocks the action of *acetylcholine* on smooth muscle, the heart, and glands. Therefore, it blocks the action of all *postganglionic, cholinergic sympathetic neurons* and all *postganglionic, parasympathetic neurons*.

21

(1:625; 2:3-114; 3:480; 4:715; 5:1047)
NH 2

Statement

_____ If one cuts all the cardiac sympathetic neurons, the central nervous system loses its ability to speed the heart during exercise by a direct neurogenic mechanism because

Reason

during exercise the heart rate is increased by the stimulation of cardiac sympathetic neurons.

Answer and Explanation

D. During running, there is usually (1) an initial speeding of the heart due to the *depression of cardiac parasympathetic tone* and (2) a later speeding due to the *stimulation of cardiac sympathetic neurons*.

22

(1:630; 2:7-246; 3:510; 4:715; 5:1055)
NHCa 2

Statement

_____ In hypovolemic shock, a major problem is an inadequate perfusion of the viscera. It is a condition that is best treated by the intravenous infusion of epinephrine because

Reason

epinephrine increases the cardiac output.

Answer and Explanation

D. Some of the first steps taken in the treatment of *hemorrhagic shock* are to stop any bleeding and restore the blood volume to normal. If an inadequate *perfusion* of the viscera remains, an *epinephrine* infusion is seldom the treatment of choice. Although it does increase *cardiac output*, it also causes *vasoconstriction* in the viscera and may therefore exaggerate the inadequate perfusion of this area. Glucagon and the catecholamine isoproterenol are better than epinephrine in that they both increase cardiac output without constricting the vessels of the viscera. In some cases, glucagon is preferred to a catecholamine because it is less likely to increase the *irritability* of the heart. *Alpha adrenergic blocking agents* have also been used to increase the blood flow to the viscera in shock.

PART THREE
CONTRACTION

5
Skeletal Muscle

1

(1:376; 2:1-77; 3:48; 4:132; 5:91)
Me 2

Skeletal muscle *contraction*

A. lasts approximately as long as the action potential for the muscle
B. lasts approximately as long as the refractory period for the muscle
C. both of the above are correct
D. precedes the refractory period for the muscle
E. none of the above are correct

Answer and Explanation
A. The action potential characteristically lasts about 2 to 4 msec, and the period of contraction ranges between 5 and 50 msec.
B. The muscle is refractory for most of the action potential.
Ⓔ.

2

(1:361; 2:1-59; 3:46; 4:122; 5:83)
Mc 1

Skeletal and cardiac muscle are both striated and at resting length contain in each sarcomere an *A band*. This A band contains

A. essentially all the contractile protein myosin, but no actin
B. essentially all the contractile protein actin, but no myosin
C. essentially all the myosin plus some actin
D. essentially all the actin plus some myosin
E. troponin and tropomyosin, but no actin

Answer and Explanation
C. Much of the actin lies in the *I band*, and all the myosin is in the *A band*.

[handwritten margin note: All A Band – myosin + some actin; I Band – most of Actin]

3

(1:361; 2:1-59; 3:49; 4:125; 5:98)
Mc 1

When skeletal muscle *shortens* in response to stimulation there is

A. a decrease in the width of the I band
B. a decrease in the width of the A band
C. a decrease in the width of the A and I bands
D. an increase in the width of the H zone
E. all of the above occur

(Continued)

Answer and Explanation

(A.) During shortening, there is a greater overlap between actin and myosin and, as a result, a shorter I band.

B. The A band retains approximately the same length.

D. The H zone is formed by the ends of the actin filaments. Since these ends come closer together in shortening, the H zone becomes narrower.

4

(1:368; 2:1-80; 3:51; 4:128; 5:95)
Mc 2

During the resting state, a single skeletal muscle sarcomere can exist at a number of lengths:

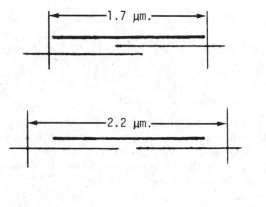

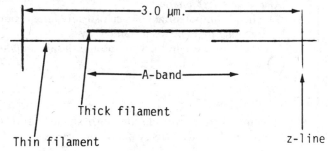

During an isometric contraction, the length at which it can exert its maximum force in response to stimulation is

A. 1.7 μm
B. 2.2 μm
C. 3.0 μm
D. all of the above
E. both 2.2 and 3.0 μm

Answer and Explanation

B. These diagrams are based on the *sliding filament hypothesis*. A sarcomere exerts its maximum contractile force at a length of 2.2 μm because apparently, at this length, there is an optimal relationship between the myosin cross-bridges on the thick filament and the actin receptor sites on the thin filament. At 1.7 μm, some cross-bridges may oppose the action of others, because the thin filament now extends to the contralateral side of the thick filament. At 3.0 μm, there are fewer actin receptor sites available to the myosin cross-bridges.

5

(1:365; 2:1-93;
3:48; 4:129; 5:104)
Mc 2

Check each of the following statements about *skeletal muscle contraction* that is true.

1. the major function of the T system (system of transverse tubules) is to store and release Ca^{++}
2. the intracellular release of Ca^{++} causes the formation of bonds between actin and myosin
3. the bonds between actin and myosin are maintained until the Ca^{++} is sequestered

Which of the following best summarizes your conclusions?

A. all of the above are true
B. none of the above are true
C. statement 2 is true
D. statement 3 is true
E. statements 2 and 3 are true

Answer and Explanation

A. In skeletal muscle, most of the calcium is stored and released by the sarcoplasmic reticulum. The T system is a continuation of the sarcolemma deep into the center of the cell and probably has as its major function the rapid transmission of the action potential from the cell surface.

D. A maintained bond between actin and myosin causes rigor. In contraction, there is a continuous making and breaking of bonds. This is the process that ends when Ca^{++} is sequestered.

6

(1:371; 2:1-66,
3:50; 4:126; 5:104)
Mci 1

The action potential in striated muscle initiates a series of events that leads to a reaction between actin and myosin and the hydrolysis of ATP. This series of events is called the *excitation-contraction coupling* process. This is a process where

A. Ca^{++} is released into the immediate environment of the myosin by the sarcoplasmic reticulum
B. Ca^{++} combines with troponin C
C. both of the above occur
D. Ca^{++} is removed from the environment of myosin by the sarcoplasmic reticulum
E. Ca^{++} is released by troponin C

Answer and Explanation

C. An increase in the concentration of unbound Ca^{++} in the intracellular fluid, a combining of some of that Ca^{++} with troponin C, and a resultant shift of tropomyosin and tropin (i.e., troponin I, C, and T) away from the actin receptor site are important parts of the excitation-contraction coupling process in striated and smooth muscle. In the case of skeletal muscle, most of the Ca^{++} is delivered by the terminal cisterns of the sarcoplasmic reticulum. In smooth muscle and the heart, the cell membrane and external environment are more important sources of Ca^{++} than in striated muscle.

7

(1:367; 2:1-75;
3:51; 4:133)
M 2

If the gastrocnemius *muscle* is removed from the body, it will achieve a *length*

A. greater than it had in the body, because it is more relaxed
B. shorter than it had in the body, because it is less relaxed
C. shorter than it had in the body, because of its elastic characteristics
D. the same as it had in the body

Answer and Explanation

C. The shortening of muscle, in this case, is not associated with an action potential, an

(Continued)

intracellular release of Ca^{++}, a heat of activation, a heat of shortening, or a chemical reaction between actin and myosin and therefore is not a contraction in the sense this term is used in muscle physiology.

8

(1:368; 2:1-79; 3:51; 4:128; 5:95)
M 1

A psoas *muscle is removed from the body* of a 70-kg man, allowed to obtain equilibrium length, stimulated with a maximal stimulus, and then the procedure of stretching and stimulation is repeated many times. If the muscle were stretched 2 additional millimeters each time and kept moist with isotonic saline, which of the following results would be expected? The muscle would

A. not contract in response to stimulation when outside the body
B. not contract in response to the 20th stimulus, because by then it would have been depleted of nutrients
C. exhibit progressively more forceful contractions in response to stimulation as it is stretched the first five times
D. exhibit progressively less forceful contractions in response to stimulation as it is stretched the first five times
E. exhibit no change in force in response to stimulation since it is being stimulated by a maximal stimulus

Answer and Explanation

B. Skeletal muscle maintains its excitability and capacity to contract outside the body as long as it still has nutrient reserves and is not destroyed by careless handling. Although it has a very limited reserve of O_2 in its myoglobin, it does have large reserves of glycogen which can be catabolized to lactic acid and energy in the absence of O_2. In the body, for example, skeletal muscle will remain excitable for up to 2 hours after its blood supply has been cut off. Organs like the cerebral cortex, which have little or no glycogen reserves, stop functioning after 20 seconds of ischemia.
C. Within limits, stretching skeletal muscle increases its force of contraction.

9

(1:377; 2:1-78; 3:457; 4:133; 5:93)
M 1

Skeletal muscle has been characterized physiologically as having (a) a parallel elastic element (PE), (b) a series elastic element (SE), and (c) a contractile element (CE). Label these in part A of the accompanying figure.

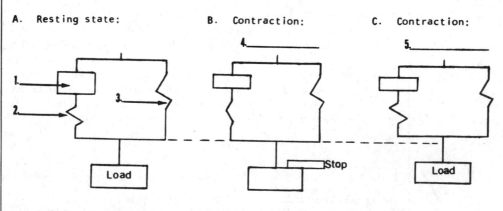

A. Resting state: B. Contraction: C. Contraction:

Answer and Explanation
1. CE
2. SE
3. PE

The *contractile elements* include actin and myosin. The *series elastic elements* lie in an end-to-end relationship with the CE and may include the tendon and cross-bridges. The *parallel elastic element* lies in a side-by-side relationship with the CE and includes the sarcolemma.

10

(1:375; 2:1-79; 3:50; 4:132; 5:93)
M 2

Muscle contractions may be *isometric, isotonic,* or associated with lengthening. Characterize the contractions in parts B and C of the figure in question 9.

Answer and Explanation
4. isometric
5. isotonic
In an isometric contraction, the muscle does not change its length. The contractile element in the model shortens, and the series elastic element is stretched. In an isotonic contraction, the whole muscle and the contractile element shorten.

11

(1:371; 2:1-79; 3:51; 4:128)
M 1

Check each of the following statements that is true.

A. a muscle at resting length exerts its maximum force during an isotonic contraction
B. the maximum velocity of shortening during contraction occurs when there is no afterload
C. the preload is the weight the muscle moves before it starts to relax
D. in most forms of muscle contraction in an intact individual, the preload and afterload are equal

Answer and Explanation
A. A muscle is capable of a greater *contractile force* during an isometric contraction. The more a muscle shortens during contraction, the less force it is capable of exerting.
(B.) The greater the *afterload,* the slower the velocity of shortening. The afterload is equal to the load in question 9, figure C. It is the load that opposes muscle shortening during contraction.
C. The preload is the stretching force to which the muscle is exposed before it starts to contract. It is equal to the load in question 9, figure A.
D. In most skeletal muscle contractions, the preload is close to 0 and the load lifted during contraction (the afterload) exceeds 0.

12

(1:367; 2:1-79; 3:48; 4:132; 5:91)
M 2

Which of the following best defines *contraction*?

A. a series of chemical reactions that cause the muscle to pull
B. a series of chemical reactions that cause the muscle to shorten
C. a series of chemical reactions in which the muscle responds to stimulation
D. shortening
E. production of tension

Answer and Explanation
(A.) Although contraction is frequently associated with shortening (isotonic contraction), it can also be associated with no change in length (isometric contraction) or even with an increase in length. This latter occurs when a muscle is contracting but its antagonists are contracting more forcefully.
E. The elastic elements of muscle are capable of producing tension when they are stretched. This tension is called passive tension to distinguish it from active tension, the tension caused by contraction.

13

*(1:380; 2:1-83;
3:48; 4:131; 5:108)*
MW 3

An *isotonic contraction* differs from an isometric contraction in that in an isotonic contraction.

A. the muscle is less efficient
B. the muscle uses more high-energy phosphate bonds
C. the heat of activation is greater
D. the recovery heat is reduced
E. the heat of activation is less

Answer and Explanation

A. Efficiency is defined as:

(work performed in cal)/(total energy expenditure in cal).

If you define work as:

(force in dynes) (distance in cm) (2.39 cal/dyne-cm),

then the efficiency in an isometric contraction is 0, since there is no distance moved and therefore no work performed. The efficiency of an isotonic contraction may go as high as 40%.

B. A muscle that shortens releases more heat than one that does not. This additional amount of heat is called the heat of shortening and is related to the amount of shortening in centimeters. This heat will come from the hydrolysis of high-energy phosphate bonds.

C. The heats of activation for isometric and isotonic contractions are approximately equal.

D. The recovery heat is elevated.

14

*(1:382; 2:1-113;
3:508; 4:846;
5:1388)*
Mcg W 2

A skeletal muscle participating in a strenuous *exercise* differs from a resting skeletal muscle in that the exercising muscle

A. releases markedly larger quantities of lactic acid into the blood that passes through it
B. transmits markedly increased quantities of heat to the blood passing through it
C. exhibits both of the above characteristics
D. has a venous blood with a markedly reduced concentration of O_2
E. exhibits all of the above characteristics

Answer and Explanation

During exercise, skeletal muscle produces more heat and lactate than during rest. These increases in production are reflected in the venous blood. Although the blood flow to skeletal muscle during exercise increases, this, in itself, is not enough to meet the increased needs of the muscle for O_2. Therefore, in response to exercise, the arteriovenous (a-v) O_2 concentration difference may change from 5 ml/100 ml of blood to 15 ml/100 ml of blood. The heart differs from skeletal muscle in that, in the resting individual, it has an a-v O_2 difference of 13 ml/100 ml (i.e., one approximately equal to that for skeletal muscle during strenuous exercise).

15

*(1:374; 2:1-94;
3:50; 4:131; 5:104)*
Cce 1

removal

The rate at which Ca^{++} is sequestered by the *sarcoplasmic reticulum* of skeletal muscle during a twitch is directly related to

A. the rate of tension development
B. the rate of ATP hydrolysis by myosin
C. both of the above
D. the height of the action potential
E. the rate of relaxation

Answer and Explanation

C. These phenomena are directly related to the concentration of Ca^{++}, not to its removal by the sarcoplasmic reticulum.

E.

16

(1:372; 2:1-93;
3:51; 4:130; 5:104)
Mi 2

In a series of experiments, it is noted that in a skeletal muscle fiber an *intracellular concentration* of Ca^{++} of $10^{-6.5}$ mol/liter is the threshold value needed for inducing contraction. On this basis, one would expect a concentration of $10^{-5.5}$ mol of Ca^{++} per liter to cause

A. a more forceful contraction
B. a less forceful contraction
C. a contraction of equal force
D. relaxation

Answer and Explanation

A. Within limits, increases in the intracellular Ca^{++} levels above threshold will determine the number of troponin molecules combining with Ca^{++} per msec and hence will cause progressively more forceful contractions. Since increases in extracellular Ca^{++} cause increases in intracellular Ca^{++} in the sarcoplasmic reticulum, they can also cause increases in contractile force.

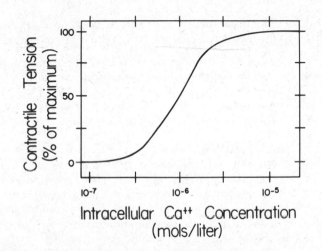

17

(1:375; 2:1-78;
3:36; 4:117; 5:48)
Min 2

The *all-or-none law* stated by Bowditch for the striated muscle cell is that the strength of contraction of a single cell responding to a single stimulus

A. cannot be increased in a healthy subject
B. cannot be changed by changing the environment of the cell
C. cannot be increased by increasing the strength of the stimulus (usually an electrical stimulus) above threshold
D. cannot be increased by increasing the frequency of stimulation
E. is not increased in hypertrophy

Answer and Explanation

C. This law applies to neurons as well as to muscle fibers. In the case of a single cell, you cannot change the characteristics of the action potential or the contraction by increasing the strength of the stimulus beyond threshold. Changes in the environment of the cell, increased frequencies of stimulation, and hypertrophy can all increase the force of contraction and/or the amplitude of the action potential.

18

(1:377; 2:1-80;
3:51; 4:134; 5:976)
Mec 1

Which one of the following statements best defines the *treppe* phenomenon described by Bowditch?

A. a maintained contraction associated with multiple action potentials
B. an increase in the force generated in a twitch due to an increased frequency of stimulation
C. a decrease in the force generated by a muscle due to a prolonged period of stimulation

(Continued)

D. a failure to relax following stimulation

E. a maintained bond between the myosin cross-bridges and actin that is not associated with an action potential

What phenomena do the other statements define?

Answer and Explanation

A. This is a definition of a *complete tetanus*.

B. Albert Szent-Györgyi defined treppe as a condition in which activity creates a situation favorable for activity. This is sometimes referred to as a warm-up phenomenon and is presumably due to a change in the environment.

C. When activity creates an environment unfavorable for activity we have *fatigue*.

D. This is called *contracture*.

E. This is called *rigor*. It may be produced by overwhelming the cell with Ca^{++} or catabolizing all of the cell's ATP.

19

(1:382; 2:1-113; :52; 4:131; 5:1391)
Mgc 2

What evidence is there that, during exercise, human beings create an *oxygen debt*? Which one of the following changes occurs during exercise and contributes to the oxygen debt?

A. a decrease in the arteriovenous oxygen concentration difference

B. an increase in the arterial lactate concentration

C. a decrease in the arterial O_2 concentration

D. a decrease in the concentration of hemoglobin

E. an increased concentration of CO_2 in the venous blood

Answer and Explanation

During and after exercise, there is usually an elevated O_2 consumption. The fact that O_2 consumption immediately after exercise does not return to resting values is the evidence that human beings, during exercise, create an oxygen debt.

A. During exercise, there is an increased removal of O_2 from the blood that causes an increased a-v O_2 concentration difference. This occurs even though the blood flow through active skeletal muscle also increases.

B. During exercise, skeletal muscle depends increasingly on the anaerobic catabolism of glucose to lactate and, as a result, sends large quantities of lactate into the blood. After exercise, the O_2 consumption of the body will remain elevated until the lactate in the blood is returned to normal levels. Part of the lactate will be aerobically catabolized to CO_2, water, and energy. Part of this energy will be used to convert some of the remaining lactate to glucose and glycogen. The healthy heart, unlike skeletal muscle, does not add lactate to the blood.

C. In most exercise, the arterial concentration of O_2 stays fairly constant.

D. Hemoglobin does not decrease in concentration in exercise.

E. This has little or no effect on O_2 consumption.

20

(1:393; 2:7-163; 3:53; 4:136; 5:90)
Mnea 1

Check each of the following statements concerning adult skeletal muscle that is true. Skeletal muscle responds to *decentralization*:

1. with an increase in tone
2. by a progressive atrophy that eventually leads to the muscle's inability to contract in response to stimulation
3. by muscle fiber hyperplasia on reinnervation
4. by fibrillation during the degeneration process
5. with an increased sensitivity to acetylcholine during the first 2 months of decentralization

Answer and Explanation
1. Decentralization of skeletal muscle causes flaccidity. After death, there may also be rigor, but this is due to the disappearance of ATP, not the destruction of nerves.
2. Eventually, all the contractile proteins in the muscle are catabolized.
3. Reinnervation may cause hypertrophy of the remaining viable muscle fibers, but there is no evidence that it will cause hyperplasia (production of more cells).
4. Fibrillation can be shown by recording an electromyogram. On the other hand, the contractions are so weak they cannot be noted through the skin.
5. The stimulation of intact cholinergic autonomic neurons may cause twitches in decentralized skeletal muscle due to its hypersensitivity to acetylcholine. The reason for this may be that decentralized muscle contains less acetylcholine esterase than normal muscle.

21

(1:379; 2:1-79; 3:50; 4:134; 5:48)
Men 2

Which of the following characteristics of skeletal muscle make *tetanic contraction* possible (check the one best answer)?

A. the motor neurons to skeletal muscle have a short refractory period and are therefore capable of delivering a high frequency of stimuli to a muscle fiber
B. the cell membrane of the skeletal muscle fiber recovers its excitability well before the cell ceases its contraction
C. both of the above are correct
D. the prolonged exposure of the muscle end plate to high concentrations of acetylcholine throughout the tetanus
E. the action potential of skeletal muscle outlasts the period of contraction

Answer and Explanation
C. It is through the delivery of a high frequency of stimuli to an excitable cell that we produce a maintained contraction. This requires that the refractory period of both the motor neuron and the muscle fiber be markedly shorter than the period of contraction that occurs in a twitch.
D. The acetylcholine released by a neuron is practically all destroyed or removed before the next action potential causes more to be released. In other words, the accumulation of acetylcholine at the neuromuscular junction is not the mechanism for the production of a maintained contraction.
E. If the action potential outlasted the period of contraction, the refractory period would be too long to permit a fusion of twitch contractions.

22

(1:365; 2:1-86; 3:50; 4:124; 5:104)
Mc 1

Starting with axonal conduction, list the events that cause a single skeletal muscle contraction and relaxation. Also indicate some of the chemical events associated with recovery:

1. axonal conduction
2. neuromuscular transmission
   a. _____
   b. _____
3. activation
   a. _____
   b. _____
4. activation-contraction coupling
   a. _____
   b. _____
   c. _____
5. contraction
   a. _____
   b. _____
6. relaxation
   a. _____
   b. _____
   c. _____
7. recovery
   a. _____
   b. _____

(Continued)

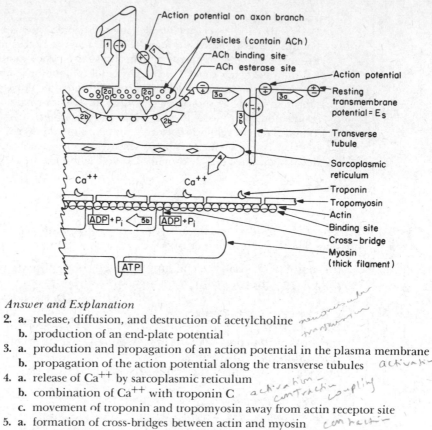

Answer and Explanation

2. **a.** release, diffusion, and destruction of acetylcholine *neuromuscular transmission*
   b. production of an end-plate potential

3. **a.** production and propagation of an action potential in the plasma membrane
   b. propagation of the action potential along the transverse tubules *activation*

4. **a.** release of Ca^{++} by sarcoplasmic reticulum
   b. combination of Ca^{++} with troponin C *activation-contraction coupling*
   c. movement of troponin and tropomyosin away from actin receptor site

5. **a.** formation of cross-bridges between actin and myosin *contraction*
   b. cross-bridges swivel and produce shortening of sarcomere and/or a force

6. **a.** sequestering of Ca^{++} in the sarcoplasmic reticulum
   b. blocking of actin receptor site by troponin and tropomyosin *relaxation*
   c. sarcomere lengthens and/or force of contraction decreases

7. **a.** creatine + phosphate → creatine phosphate
   b. lactic acid + O_2 → CO_2 + H_2O + glycogen *recovery*

23

*(1:370; 2:1-81;
3:459; 4:128; 5:112)*
M 2

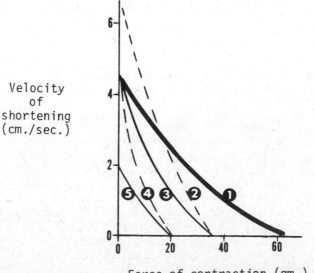

Velocity
of
shortening
(cm./sec.)

Force of contraction (gm.)

The preceding *force-velocity curves* were obtained from a muscle. Assume that supermaximal stimuli were used in each study, and assume that curve 1 was obtained for a muscle at L_{max} (i.e., the muscle length at which the greatest active tension is developed) and that the other curves were obtained at a muscle length less than L_{max}.

What is V_{max} for curve L?

A. greater than 60 g
B. less than 60 g
C. greater than 4 cm/sec
D. less than 4 cm/sec
E. less than 2 cm/sec

Answer and Explanation
C. V_{max} is the maximum velocity of shortening for a muscle. It occurs when the *afterload* equals 0. In this diagram, the afterload is represented by the force of contraction. V_{max} is determined by measuring the *velocity of shortening* for a muscle at different afterloads, plotting the data, and extrapolating the plot to 0 afterload.

24

(1:370; 2:39; 3:459; 4:128; 5:975)
M 3

Which curve in question 23 would be caused by the sarcoplasmic reticulum releasing less Ca^{++} in response to stimulation of the muscle?

A. curve 1
B. curve 2
C. curve 3
D. curve 4
E. curve 5

Answer and Explanation
E. A decrease in available Ca^{++} is an example of a *negative inotropic* condition. Any negative inotropic agent will shift the force-velocity curve down and decrease V_{max}.

25

(1:371; 2:3-92; 3:459; 4:128; 5:975)
M 3

Which curve in question 23 represents the muscle with the smallest preload?

A. curve 1
B. curve 3
C. curve 4
D. curves 2 and 3
E. curves 2 and 5

Answer and Explanation
C. At muscle lengths below L_{max}, the force of contraction during an *isometric contraction* will be directly related to the *preload* and inotropicity. The isometric force for each curve is determined by noting where each curve intersects the 0 velocity of shortening line. This is sometimes called P_0 (pressure, tension, or force at which there is 0 shortening). In other words, curve 1 represents a muscle at an optimal preload, and curves 2, 3, and 4 represent muscles with a lower preload. Curve 5 (low V_{max}) differs from curve 4 (intermediate V_{max}) in that it represents a muscle that was exposed to a negative inotropic condition. Curve 2 (high V_{max}), on the other hand, is from a muscle exposed to a positive inotropic condition. If curve 5 represented the same inotropicity as curve 4 it would touch the 0 velocity line at a force greater than 20 g. Variations in preload have little effect on V_{max}. Variations in inotropicity have a marked effect on V_{max}. (See E Braunwald, J Ross Jr, and EH Sonneblick: *Mechanisms of Contraction of the Normal and Failing Heart*. 2nd ed. Boston: Little, Brown, 1976, p 130.)

26

(1:367; 2:3-89; 3:51; 4:128; 5:92)
M 2

The following *length-tension diagram* was obtained on a muscle. Supermaximal tetanic stimuli were used to initiate a contraction at each muscle length studied.

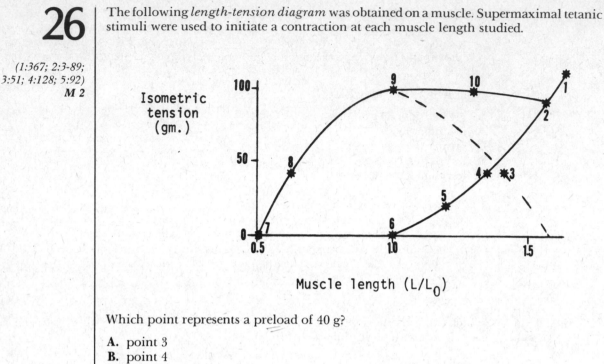

Which point represents a preload of 40 g?

A. point 3
B. point 4
C. point 8
D. points 4 and 8
E. points 3, 4, and 8

Answer and Explanation
   B. Line 6-1 represents the *passive tension*. At point 4, the passive tension is 40 g, and the muscle length is 1.3 times resting length. Passive tension and *preload* are synonymous.

27

(1:367; 2:3-91; 3:51; 4:128; 5:93)
MH 2

Maximal active tension in the diagram in question 26 is developed by skeletal muscle at point(s)

A. 1
B. 2
C. 4
D. 3 and 4
E. 9

Answer and Explanation
   E. Active tension (line 7-9-3) is equal to *total tension* (line 7-9-2-1) minus passive tension (line 6-1).

28

(1:367; 2:3-93; 3:51; 4:128; 5:95)
M 2

Which point(s) in the diagram in question 26 represent(s) no overlap between most of the muscle's *thick and thin filaments*? Point(s)

A. 2
B. 3
C. 6
D. 7
E. 6 and 7

Answer and Explanation

A. As you stretch a muscle from one-half its resting length (L_0) toward twice its resting length, you reach a point (2) where the muscle no longer develops active tension in response to stimulation. This is the point at which the actin of the thin filament can no longer react with the myosin of the thick filament.

29

(1:383; 2:9-87; 3:54; 4:132; 5:674)
Mcn 1

The *slow twitch* muscle fiber differs from the *fast twitch* glycolytic fiber in that the former (check each correct answer):

1. has a smaller number of muscle fibers in each motor unit
2. has a higher concentration of myoglobin and mitochondria
3. has a higher ATPase activity
4. in large limb muscles serves as a reserve which can be recruited if there is a forceful contraction
5. is more readily fatigued
6. is part of a motor unit that consists of only red fibers

Which one of the following best summarizes your conclusions? Statements

A. 1, 2, and 3 are correct
B. 1, 2, and 6 are correct
C. 2, 3, and 4 are correct
D. 3, 4, and 5 are correct
E. 4, 5, and 6 are correct

Answer and Explanation

B. Skeletal muscle fibers in humans can be divided into three categories: (I) slow oxidative, (IIA) fast oxidative, (IIB) fast glycolytic. The slow oxidative fiber will have a lower velocity of contraction, a less rapid relaxation, and a greater *twitch duration* than the type II fiber. Each motor unit will contain only one type of fiber. The *motor unit* with type I fibers (1) innervates a smaller number of muscle fibers and exerts less force. In a weak contraction (4) only type I units are activated. In a strong contraction both type I and II units are activated, with the latter exerting most of the force. The (3) high *ATPase activity* of both type II fibers is, in part, responsible for (5) their rapid fatigue during contraction. The type IIA fiber is the least common in the human. It is similar to the type I fiber and different from the type IIB in that it has a high capacity for oxidative catabolism, (2) a high concentration of myoglobin and mitochondria, and a high capillary density.

Questions 30 through 32 consist of two main parts: a statement and a reason for the statement. Choose the best answer for each of these questions from the options A through E below and mark your choice in the space provided.

A. both the statement and the reason are true and are related as cause and effect (i.e., as indicated)
B. both the statement and the reason are true but are not related as cause and effect
C. the statement is true but the reason is false
D. the statement is false but the reason is an accepted fact or principle
E. both the statement and the reason are false

30

(1:371; 2:1-64; 3:48; 4:124; 5:101)
Mc 1

Statement

_____ During contraction, the myosin cross-bridges combine with actin receptor sites because

Reason

immediately prior to contraction, the regulator proteins in the thin filaments have moved toward the actin-active sites.

(Continued)

Answer and Explanation

C. During the *excitation-contraction coupling* phenomenon that precedes contraction, muscle apparently has the following sequence of events: (1) Ca^{++} is released from intracellular stores, (2) Ca^{++} combines with TnC (troponin C), (3) a conformational change in the troponin weakens the binding of actin by troponin I, and (4) the troponin-tropomyosin complex moves away from the actin-active site. By moving the *regulator proteins* troponin and tropomyosin away from the actin sites, the muscle can now have its *contractile proteins*, actin and myosin, interact.

31

(1:370; 2:1-79; 3:458; 4:128; 5:95)
M 1

Statement

_____ Within limits an increase in a muscle's afterload will decrease the force that the muscle exerts during contraction because

Reason

an increased afterload decreases the velocity of shortening.

Answer and Explanation

D. The *force* the muscle exerts during contraction equals the load it lifts during contraction. The *maximum velocity of shortening* (V_{max}) occurs when the *afterload* equals 0.

32

(1:377; 2:3-89; 3:50; 4:134; 5:91)
Me 2

The following record was obtained from a single muscle fiber.

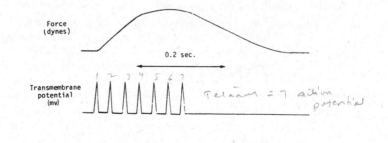

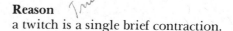

Tetanus = 7 action potential

Statement

_____ This response is called a twitch because

False

Reason

a twitch is a single brief contraction.

True

Answer and Explanation

D. This contractile response (i.e., force pattern) is called a *tetanus*. In this particular example, the tetanus is associated with seven action potentials (i.e., seven spikes in the transmembrane potential). A *twitch* is associated with one action potential.

6
The Heart—Electrical Properties

The following transmembrane potentials were recorded from healthy smooth muscle, cardiac muscle, and skeletal muscle. They each include the resting transmembrane potential followed by an action potential. All were recorded at the same paper speed and sensitivity. Refer to this diagram for questions 1, 2, and 3.

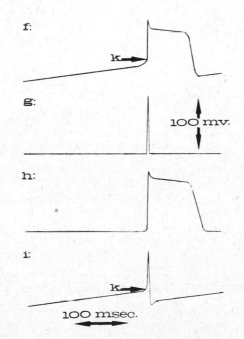

1

Which of the potentials in the diagram were recorded from *potential pacemaker cells*?

A. cells f and h
B. cells g and i
C. cells f and i
D. cells g and h
E. cell f is the only potential pacemaker cell

(1:443; 2:3-58; 3:58; 4:166; 5:965)
MHe 2

(Continued)

Answer and Explanation

A. A pacemaker cell has an unstable resting transmembrane potential. The potential prior to the action potential in pattern h remains constant.

B. Patterns g and h are both from follower cells.

C. Pacemaker cells slowly depolarize on their own in the absence of extrinsic stimuli. Follower cells, on the other hand, characteristically maintain a constant resting potential until an extrinsic stimulus arrives from a pacemaker cell or some other source.

2

(1:375; 2:3-53; 3:435; 4:118; 5:965)
MHe 2

Point k in patterns f and i in the diagram represents

A. the threshold potential for the cell
B. the end of the cell's refractory period
C. the beginning of the period of supernormal irritability
D. the end of the refractory period and the beginning of hyperirritability
E. the cell's rheobase

Answer and Explanation

A. Once the *threshold potential* or firing level has been reached, there is a rapid depolarization of the cell.

D. Point k represents the beginning of the cell's refractory period and the end of its irritability.

E. The rheobase for a cell is the minimal stimulus strength in volts that will produce an action potential when that stimulus is applied for a prolonged period.

3

(1:444; 2:3-52; 3:434; 4:177; 5:965)
MHe 1

Which of the *potentials* in the diagram might have been recorded from the *heart*?

A. cells f and h
B. cells g and i
C. cells g and h
D. cell h
E. cells f, g, h, and i

Answer and Explanation

A. Cardiac cells have a delay in repolarization that results in their having a prolonged period of inexcitability. The plateau type of potential is characteristic of the heart. The spike potential is characteristic of skeletal muscle fibers and axons.

4

(1:452; 2:3-50; 3:3-50; 4:165; 5:967)
MHeW 2

Which of the following is *not characteristic of the heart*?

A. its contraction is initiated by a nerve impulse
B. it conducts impulses from one muscle cell to the next
C. it contains a number of cells with an unstable transmembrane potential
D. it contains a number of cells with a stable transmembrane potential
E. its ventricles are inexcitable for most of their period of contraction

Answer and Explanation

A. The heart's contraction is not initiated by a nerve impulse but by a cardiac cell that depolarizes in the absence of extrinsic stimuli.

D. Characteristic of the heart is that it contains a single pacemaker area that dominates a number of follower cells and potential pacemaker cells. The impulse that the pacemaker originates is transmitted from one cell to the next.

5

(1:453; 2:3-50; 3:434; 4:165; 5:967)
Hen 2

The sinoatrial node is the *pacemaker* for the heart because it

A. is the most richly innervated structure in the heart
B. is the only structure in the heart capable of generating action potentials
C. has the highest rate of automatic discharge
D. has the most stable transmembrane potential
E. is the cardiac cell least sensitive to catecholamines

Answer and Explanation
A. The SA node remains the pacemaker area even after its innervation is destroyed.
B. The atria and ventricles contain numerous areas that will take over the pacing of the heart if the SA node is depressed, destroyed, or cut off from the rest of the heart.
C. The cells of the sinoatrial node become the pacemaker for the heart because they stimulate other potential pacemaker cells before they are able to initiate their own action potentials.

6

(1:450; 2:3-58; 3:436; 4:168; 5:969)
He 1

The most *slowly conducting* part of the heart is the

A. atrium
B. ventricle
C. Purkinje fibers
D. right bundle branch
E. the atrioventricular junctional tissue at the atrioventricular node

Answer and Explanation
A. It conducts at a rate of 0.9 m/sec.
B. It conducts at a rate of from 0.3 to 1 m/sec.
C. It conducts at a rate of 2 to 4 m/sec.
D. It conducts at a rate of from 1 to 2 m/sec. The AV bundle, bundle branches, Purkinje fibers, and ventricular endocardium conduct action potentials rapidly and therefore cause an almost simultaneous activation of the fibers in the right and left ventricles.
E. It conducts at a rate of 0.05 m/sec. This very slow conduction helps delay the transmission of the impulse to the ventricles and therefore prevents the ventricles from contracting at the same time as the atria. All of these rates will vary with the degree of autonomic nervous system tone, the level of epinephrine in the blood, and other changes in the environment.

7

(1:457; 2:3-51; 3:436; 4:168; 5:969)
He 1

The *last part of the ventricles to be activated* after atrial activation is

A. the base of the right ventricle
B. the base of the left ventricle
C. the endocardium of the right ventricle
D. the epicardium of the apex
E. the endocardium of the apex

Answer and Explanation
B. Once the impulse has reached the apex, it travels up the endocardial surface of the right and left ventricles and moves from here to the epicardial surface. Because the left ventricular muscle mass is larger than the right, it takes longer to completely activate the left ventricle than the right. The last parts of the heart to develop action potentials are the base of the left ventricle and the superior interventricular septum. Apparently, the AV bundle and bundle branches in the basal portion of the interventricular septum are sufficiently isolated from their neighboring muscle fibers in the septum that the action potentials they contain pass by the rest of the septal muscle fibers without activating them, and it is not until the impulse comes around a second time that activation of the remaining septum occurs.

8

(1:469; 2:3-86; 3:441; 4:171; 5:967)
Hea 2

The heart has been characterized as a functional atrial syncytium connected by an atrioventricular conducting system to a functional ventricular syncytium. What would happen if a drug such as quinidine or procainamide were given that prevented *intercellular conduction*?

A. there would be an increase in cardiac parasympathetic tone
B. multiple ectopic pacemakers would develop
C. both of the above would occur
D. the force of ventricular contraction would increase
E. all of the above would occur

Answer and Explanation
A. In the absence of intercellular conduction, the heart would stop ejecting blood, the arterial pressure would decline, and there would be a reflex decrease in cardiac parasympathetic tone and an increase in sympathetic tone.
(B.) When a potential pacemaker is isolated from its neighbors, it initiates its own action potentials.
D. The development of multiple ectopic pacemakers leads to a chaotic type of contraction called fibrillation. While some fibers are contracting, others are relaxing. This loss of synchrony leads to such weak contractions that a pulse will not even be detected.

9

(1:465; 2:3-64; 3:436; 4:180; 5:1008)
He 1

In standard *electrocardiogram lead III*, the right leg electrode is grounded through the recorder and the potential difference is recorded between

A. the right arm and left leg
B. the right arm and left arm
C. the left arm and left leg
D. the right arm and a central terminal formed by resistors and the left arm and left leg
E. the left arm and a central terminal formed by resistors and the right arm and left leg

Answer and Explanation
A. This is lead II.
B. This is lead I.
(C.) The left leg lead goes to the positive terminal and the right arm lead to the negative terminal of the recorder. The positive terminal is the part of the recorder which causes a positive deflection when exposed to positivity and a negative deflection when exposed to negativity.
D. This is lead aV_R.
E. This is lead aV_L.

10

(1:466; 2:3-65; 3:437; 4:181; 5:1012)
He 1

Check each of the following electrocardiogram (ECG) leads that is designated *unipolar*:

1. lead V_2
2. lead aV_L
3. lead I

Which one of the following best summarizes your conclusions?

A. statement 1 is correct
B. statement 2 is correct
C. statements 1 and 2 are correct
D. statement 3 is correct
E. statements 2 and 3 are correct

Answer and Explanation
(C.) Lead V_2 is a precordial unipolar lead in which the exploratory electrode is at position 2 on

the chest. Lead a V_L is an augmented unipolar lead in which the equivalent to the exploratory electrode is on the left arm (a: augmented; V: vector; L: left arm).

D. Standard limb lead I is a lead in which you are measuring the potential between the left arm and right arm.

11

(1:466; 2:3-73; 3:437; 4:181)
Hea 2

Note the following ECGs from a normal subject. All were recorded at the same sensitivity and paper speed.

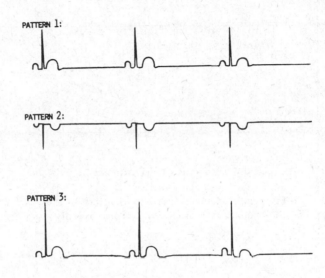

PATTERN 1:

PATTERN 2:

PATTERN 3:

If pattern 1 is from lead II, what conclusions can you draw from these data?

A. pattern 2 is from lead aV_R
B. pattern 2 is from lead aV_F
C. pattern 2 is from lead V_4
D. pattern 3 is from lead aV_R
E. pattern 3 is from lead I

Answer and Explanation

A. Lead a V_R in the normal subject has inverted P, R, and T waves and an R wave of smaller amplitude than that found in lead II.

B. Lead a V_F normally has erect P, R, and T waves.

C. Lead V_4 normally has erect P, R, and T waves.

E. Lead I in the normal subject characteristically has an R wave of lower amplitude than that found in lead II. In other words, this could be lead I if the subject had an electrical axis of 20°. An axis of 60 to 70° is much more likely.

12

(1:457; 2:3-69; 3:439; 4:183; 5:1015)
Hema 3

The depolarization of the atrium consists of numerous waves of depolarization (i.e., waves of negativity) moving away from the sinoatrial node. These waves attract positive charges. It proves useful to characterize all the electrical changes in the heart at any one instant in terms of a single, two-dimensional dipole. This dipole has a positive head and a negative tail. The negativity represents the wave of depolarization. On that basis, one would expect the *atrial dipole* on the frontal plane to be moving

A. cephalad (i.e., –90°)
B. caudad (i.e., +90°)
C. toward the right arm (180°)

D. toward +60°
E. toward −60°

Answer and Explanation
D. Plus 60° is approximately the electrical axis of the atria (i.e., 60° is the direction of its frontal plane dipole). Marked deviations from this may be due to abnormal conduction patterns, the presence of an ectopic pacemaker, or an abnormal anatomical position for the heart.

13

(1:465; 2:3-69; 3:439; 4:183)
Hema 3

If the frontal plane *atrial dipole* had an *axis* of +90°, the highest amplitude P wave on the ECG would be seen on lead

A. I
B. II
C. III
D. aV_L
E. aV_F

Answer and Explanation
A. Lead I represents an electrical axis of 0°.
B. Lead II represents an electrical axis of 60°.
C. Lead III represents an electrical axis of 120°.
D. Lead aV_L represents an electrical axis of −30°.
(E.)

14

(1:465; 2:3-71; 3:439; 4:183)
Hea 2

The *R wave* of the ECG from a healthy subject represents a dipole with an electrical axis of

A. −120°
B. −60°
C. 0°
D. 60°
E. 120°

Answer and Explanation
D. This is a period in the cardiac cycle when activation is spreading somewhat caudally from the endocardium to the epicardium of the right and left ventricles. If the major spread were from apex to base, one would expect an axis of approximately −60°. If the heart were vertical and the right and left ventricles were transmitting signals of equal strength with an axis of +120° (right ventricle) and +60° (left ventricle), respectively, one would expect an axis of +90°. The adult heart, however, is not vertical, and the left ventricle normally transmits a stronger signal.

15

(1:464; 2:3-71; 3:436; 4:182; 5:1011)
Hea 3

A patient was admitted to the hospital and ECG lead III was recorded. It was found to contain no *S wave*. The P, R, and T waves appeared normal. What conclusions can you draw?

A. activation of parts of the base of the heart are abnormal
B. activation of parts of the apex of the heart are abnormal
C. there has been cardiac depression
D. there is left bundle branch block
E. there are no indications of cardiac abnormalities

Answer and Explanation

E. The dipole moment of the Q and S waves is considerably less than that of the P, R, and T waves. Their absence in certain leads is consistent with a healthy heart. It is particularly difficult to record an S wave in lead III, not only because of its low moment but also because of the relationship between the electrodes and dipole. The maximal deflection on the ECG in obtained when the dipole is moving toward one electrode and away from the other. When the electrodes are 90° out of phase with the dipole, they record no deflection. This indicates one advantage of recording more than one ECG lead.

16

(1:467; 2:3-84; 3:441; 4:200; 5:1014)
Hea 2

The following *ECG* was recorded from lead aV_F:

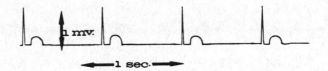

How would you characterize the pattern?

A. a sinus rhythm
B. a ventricular rhythm
C. a nodal rhythm
D. flutter
E. fibrillation

Answer and Explanation

A. In a sinus rhythm, a P wave and an isoelectric line (i.e., the PR segment) will precede each R wave.
B. The R wave would be of longer duration and the T wave would usually be inverted.
C. In a nodal rhythm, the ventricular conduction pattern is normal, but its relationship with the P wave is not.
D. This is a heart rate of about 75/min, not between 250 and 350/min as in flutter.
E. In fibrillation, the waves are of low amplitude (i.e., below 0.2 mv).

17

(1:468; 2:3-78; 3:443; 4:198; 5:1010)
HNe 2

A patient had an ECG with a prolonged *PR segment*. What might have caused this?

A. increased sympathetic tone to the atrioventricular node
B. increased parasympathetic tone to the atrioventricular node
C. increased sympathetic tone to the sinoatrial node
D. increased parasympathetic tone to the sinoatrial node
E. decreased carotid sinus pressure

Answer and Explanation

A. This would decrease the PR segment.
B. Parasympathetic neurons are capable of producing first-, second-, and third-degree block in response to increases in arterial pressure.
C. This speeds the heart under most conditions.
D. This slows the heart under most conditions.
E. This would reflexly decrease parasympathetic tone and increase sympathetic tone.

18

(1:463; 2:3-62; 3:436; 4:176; 5:1010)
Hev 2

The *T wave* of the ECG occurs

A. at the beginning of the heart's refractory period
B. during the depolarization of the heart
C. during atrial systole
D. during the repolarization of the ventricle
E. during the first heart sound

Answer and Explanation
A. The QRS complex occurs at the beginning of the heart's refractory period.
B. The QRS complex occurs during the depolarization of the ventricles.
C. The PR interval occurs during atrial systole.
(D.)
E. The first heart sound begins during the QRS complex and ends prior to the beginning of the T wave.

19

(1:568; 2:3-78; 3:176; 4:197)
Hea 2

Check each of the following events that may either cause or be a sign of *nodal rhythm*:

1. increased sympathetic tone to the SA node
2. increased parasympathetic tone to the SA node
3. a prolonged QRS complex
4. a prolonged PR segment

Which one of the following best summarizes your conclusions?

A. statements 1 and 3 are correct
B. statements 1, 3, and 4 are correct
C. statements 1 and 4 are correct
D. statements 2 and 3 are correct
E. statement 2 is correct

Answer and Explanation
A. In nodal rhythm there is a QRS complex of normal duration.
C. In nodal rhythm there may be retrograde atrial conduction or no apparent atrial conduction. In the case of the former, the P wave would be inverted and the PR segment shortened. In the case of the latter, there would be no PR segment.
(E.) Parasympathetic tone to the SA node can be increased to such an extent that the SA node stops firing. When this happens, another pacemaker (i.e., the AV node) takes over.

20

(1:467; 2:3-78; 3:443; 4:198; 5:1010)
Hea 2

The following *ECG* was recorded from lead II of a patient:

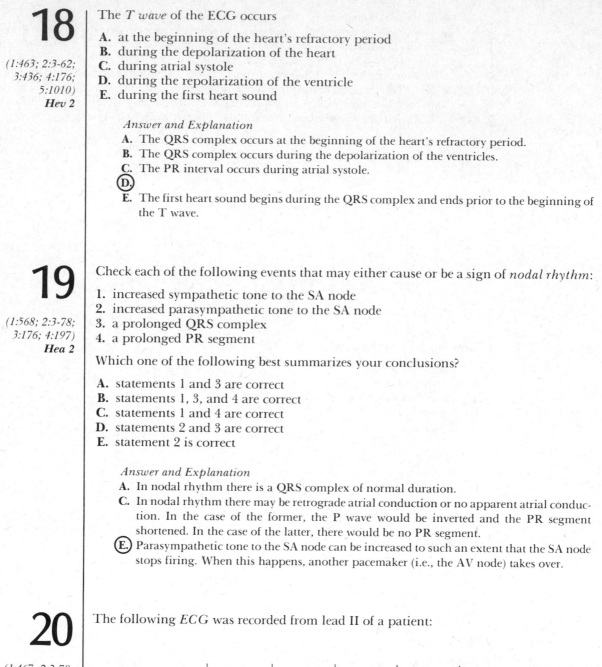

1 sec.

How would you characterize this pattern?

A. complete heart block
B. second-degree block
C. sinus arrhythmia
D. ventricular tachycardia
E. ventricular bradycardia

Answer and Explanation

A. In most complete blocks the R-R′ interval exceeds 1.4 seconds, but one may obtain a pattern such as seen here when the block is associated with a marked cardiac sympathetic tone.

B. In a second-degree heart block every second (2:1 block), or every third (3:1 block), etc., P wave is followed by a normal PR segment, QRS complex, and T wave. All other P waves are not followed by QRS and T waves.

C. In sinus arrhythmia the P-P′ interval equals the R-R′ interval in each cardiac cycle, but they vary from one cycle to the next.

E. In ventricular rhythm the duration of the QRS complex is more prolonged than above.

21

(1:467; 2:3-78; 3:443; 4:198)
Hea 2

Check each of the following events which the pattern in question 20 suggests:

1. the heart has two separate pacemakers
2. ventricular conduction is abnormally slow
3. there is atrial asystole

Which one of the following best summarizes your conclusions?

A. statement 1 is correct
B. statement 2 is correct
C. statements 1 and 2 are correct
D. statement 3 is correct
E. statements 1, 2, and 3 are correct

Answer and Explanation

A. One *pacemaker* is probably in or near the SA node and the second is in or near the AV node.

B. The duration of the *QRS complex* is normal and, therefore, the velocity of conduction in the ventricles is also probably normal.

D. The presence of well-defined *P waves* usually indicates the presence of atrial systole.

22

(1:566; 2:3-114; 3:176; 4:176; 5:1047)
HeN 2

Check each of the following that is produced by an increased *sympathetic tone* to the heart:

1. a decrease in the duration of R-R′ interval
2. a decrease in the duration of the ST segment
3. a decrease in the duration of the PR segment
4. a decrease in the rate at which the ventricles develop pressure
5. a decrease in the duration of the refractory period of the AV node

Which one of the following best summarizes your conclusions?

A. all of the above are correct
B. none of the above is correct
C. statements 1, 2, 3, and 5 are correct
D. statements 1, 2, and 3 are correct
E. statement 4 is correct

Answer and Explanation

A. The stimulation of cardiac sympathetic nerves causes a more forceful contraction of the ventricles. One sign of this is an increased rate of pressure development in the ventricles.

C. Sympathetic neurons (1) act on the SA node to increase the heart rate (i.e., decrease the R-R′ interval), (2 and 3) act on the atria and ventricles to increase their speed of conduction and decrease their refractory periods (decrease the duration of the ST segment and the PR

(Continued)

segment), and (5) decrease the refractory periods of the AV node. By decreasing the refractory period of the AV node it is possible to prevent a second-degree heart block when the atrial rate is rapid.

23

(1:467; 2:3-78; 3:441; 4:198; 5:1010)
Hea 2

An independence of the *P wave* and the *QRS complex* of the ECG indicates

A. an early repolarization of the ventricular fibers
B. a failure of the AV node to conduct
C. depression of the sinoatrial node
D. slowing of conduction at the atrioventricular node
E. a conduction block in the left bundle branch

What type of ECG would be caused by each of the other abnormalities listed above?

Answer and Explanation
A. This would cause a decrease in the ST segment.
(B.)
C. This would cause either an increase in the R-R′ interval or an ectopic rhythm.
D. This would produce a prolongation of the PR segment.
E. This would produce a prolongation of the QRS complex.

24

(1:464; 2:3-62; 3:436; 4:182; 5:1010)
Hea 2

The following ECG was obtained from lead II in a patient with an *implanted, battery-operated pacemaker:*

Identify each of the following:

1. isoelectric line
2. T_p
3. R wave
4. T wave
5. pacemaker artifact

Which one of the following best summarizes your conclusions?

A. 1: g, 2: i, 3: f, 4: h, 5: j
B. 1: j, 2: g, 3: f, 4: i, 5: h
C. 1: g, 2: j, 3: h, 4: i, 5: f
D. 1: g, 2: i, 3: h, 4: f, 5: j
E. 1: g, 2: j, 3: f, 4: i, 5: h

Answer and Explanation
C. 1. The isoelectric line is the pattern (g) that results from no potential between the two terminals of the recorder. When there is no injury potential, it lies between the P and QRS waves and between the QRS and T waves, and between the T and P waves.
2. The T_p (j) is caused by the wave of atrial repolarization and therefore follows the P wave. In the above pattern, there is a complete heart block, and therefore the P wave bears no consistent relationship with the R wave.
3. The R wave (h) follows the pacemaker artifact (f). It is of abnormally long duration in the above pattern because the stimulating electrodes have been placed distal to the

rapidly conducting fibers of the heart (i.e., the AV bundle, its branches, the Purkinje fibers, and the endocardium.)

4. The T wave (i) follows the R wave and, in this case, is partially incorporated into the R wave. In the above record it is a negative deflection. This is characteristic of ventricular rhythm. The inverted T wave also occurs in cardiac ischemia.

5. The pacemaker artifact (f) is a short-duration, high-amplitude deflection. When the pacemaker electrodes are in the ventricle, it precedes the R wave. When they are in the atrium, it precedes the P wave.

25

(1:469; 2:3-85; 3:441; 4:200; 5:1010)
Hepa 2

The following *ECG* was recorded from lead II of a patient:

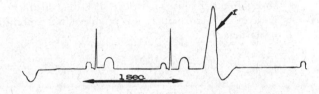

Check each of the following statements concerning this pattern that is true:

1. it is an example of bigeminy
2. the wave at point f is an R wave and will cause a stronger arterial pulse than the other R waves shown
3. the deflection at point f will be associated with a premature ventricular systole

Which one of the following best summarizes your conclusions?

A. statements 1, 2, and 3 are correct
B. statements 1 and 2 are correct
C. statement 1 is correct
D. statement 2 is correct
E. statement 3 is correct

Answer and Explanation

C. It is a trigeminy (one premature ventricular beat in every three cardiac cycles).
D. The R wave at point f will cause a weaker than normal pulse. This is due in part to the reduced filling time for the heart and the less rapid activation for the ventricles. When one takes a pulse and finds every third wave weak, this pattern is also characterized as a trigeminy.

Ⓔ

26

(2:3-77; 3:449; 4:161; 5:981)
Heia 2

The following *ECGs* were recorded from lead II of a patient before and after he developed anuria:

(Continued)

Which of the following statements best characterizes the second pattern?

A. ventricular fibrillation
B. ventricular flutter
C. ventricular tachycardia
D. a slowing of cardiac conduction and an elevation of the T wave due to hyperkalemia
E. a slowing of cardiac conduction and an elevation of the T wave due to hypokalemia

Answer and Explanation
A. In ventricular fibrillation, the deflections of the ECG are more frequent and of lower amplitude than normal.
C. In ventricular flutter and tachycardia, the cardiac cycle is of shorter duration.
D. Normally, the potassium concentration of the blood plasma is 5 mEq/liter. An increase to 7 mEq/liter will cause a prolonged P-R interval and QRS complex and an elevation of the T wave. The above pattern was produced by a concentration of potassium of 8 mEq/liter of plasma.

27

(1:944; 2:7-84; 3:449; 4:161; 5:1657)
Heia 3

Check each of the following conditions that might produce the hyperkalemia and is responsible for the abnormal ECG in question 26:

1. hemolysis
2. diabetes mellitus
3. hyperaldosteronism

Which of the following best summarizes your conclusions?

A. statement 1 is correct
B. statement 2 is correct
C. statements 1 and 2 are correct
D. statement 3 is correct
E. statements 1, 2, and 3 are correct

Answer and Explanation
C. The K^+ is in higher concentration in the intracellular fluid than in the extracellular fluid. Degenerative conditions which result in the breakdown of the cell membrane or conditions which increase the permeability of the membrane can cause a marked and sometimes fatal movement of K^+ into the serum. This may result from hemolysis or the systemic acidosis produced in diabetes mellitus. It is most dangerous when the kidneys have either a lost or an impaired ability to excrete potassium.
D. Aldosterone facilitates the loss of potassium from the body in the urine and sweat. It does not, either in the presence or absence of kidney malfunction, produce hyperkalemia.

28

(1:469; 2:3-173; 3:445; 4:200; 5:1014)
Hepa 3

The following records were made before and during a patient's response to a drug:

Check each of the following statements that is true:

1. before therapy there was a sinus arrhythmia

2. before therapy there was a periodic ventricular rhythm
3. during therapy there was a conversion to a nodal rhythm
4. during therapy there was a conversion to ventricular tachycardia
5. during therapy there was a conversion to a ventricular rhythm

Which of the following best summarizes your conclusions?

A. statements 1 and 3 are correct
B. statements 1 and 4 are correct
C. statements 1 and 5 are correct
D. statements 2 and 3 are correct
E. statements 2 and 4 are correct

Answer and Explanation
    C. **1.** The presence of a P wave and a PR segment prior to each R wave indicates a sinus rhythm. The variability of the duration of the R-R′ interval is a form of arrhythmia, which is also sometimes called inspiratory tachycardia, since the R-R′ intervals shorten during each inspiration. *Sinus arrhythmia* (i.e., inspiratory tachycardia) occurs in healthy individuals.
    2. Sinus arrhythmia is not due to a ventricular pacemaker.
    3. A nodal rhythm would not be associated with an inversion of the R wave and an increase in its duration.
    4. The therapy resulted in a heart rate of about 50/min.
    5. The long duration of the R wave is indicative of a ventricular pacemaker.

29

(1:497; 2:3-188; 3:469; 4:220; 5:1047)
HCpn 2

In the *records* seen in the study in question 28, the drug produced (check each answer that is correct).

1. a decreased pulse pressure
2. an increased diastolic pressure
3. a decreased heart rate, possibly due to a reflex stimulation of cardiac parasympathetic neurons
4. a decreased heart rate, possibly due to a reflex inhibition of cardiac parasympathetic neurons

Which of the following best summarizes your conclusions?

A. statements 1, 2, and 3 are correct
B. statements 1, 2, and 4 are correct
C. statements 1 and 4 are correct
D. statements 2 and 3 are correct
E. statements 2 and 4 are correct

Answer and Explanation
    D. The drug apparently caused an increase in diastolic, systolic, and pulsatile *pressure* in the systemic arteries. This, acting through the baroreceptors, caused a reflex stimulation of vagal parasympathetic neurons to the heart.

30

(1:568; 2:3-173; 3:568; 4:170; 5:1047)
Hena 3

The mechanism responsible for a stimulation of cardiac parasympathetic neurons causing a *ventricular* rhythm in the preceding study is that parasympathetic neurons

A. act directly on the ventricles to decrease their rate of contraction
B. act directly on the ventricles to increase their irritability
C. act directly on the sinoatrial node to decrease its rate of firing
D. prevent atrial activation
E. block the atrioventricular node

(Continued)

Answer and Explanation

A. The parasympathetic neurons exert little direct action on the ventricles.

B. Acetylcholine would act directly on the ventricles to decrease irritability.

(D.) The absence of P waves during the ventricular rhythm probably indicates that the parasympathetic neurons have prevented atrial activation which releases a potential pacemaker in the ventricle from sinoatrial dominance.

E. In heart block there might also be ventricular rhythm, but one would expect to see P waves if the block were restricted to the AV node.

31

(1:478; 2:3-160; 3:469; 4:220; 5:1018)
HCpn 3

Which of the following would you suggest represents the major *action of the drug* given in the study in question 28?

A. vasoconstriction of the systemic arterioles
B. a decrease in total systemic peripheral resistance
C. vasoconstriction and a decrease in resistance
D. an increased sensitivity of the heart to acetylcholine
E. a decreased sensitivity of the heart to acetylcholine

Answer and Explanation

(A.) Arteriolar *vasoconstriction*, by reducing arterial drainage, produces a distention of the arteries that causes an increase in the arterial diastolic pressure.

C. Vasoconstriction causes an increase in peripheral resistance.

D. This would explain the cardiac slowing but not the increase in diastolic pressure. Cardiac slowing causes a decrease in diastolic pressure. What probably occurred during therapy is that a vasoconstrictor such as *norepinephrine* was injected and produced an increase in arterial pressure and a reflex slowing of the heart.

32

(1:497; 2:3-104; 3:453; 4:154; 5:987)
Hepv 1

The following recording was obtained from a patient:

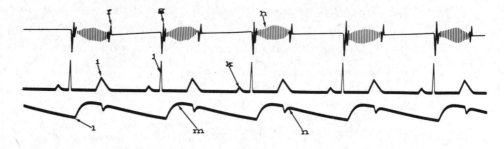

Identify each of the following on this record.

A. _____ P wave of the electrocardiogram
B. _____ R wave of the electrocardiogram
C. _____ T wave of the electrocardiogram
D. _____ first heart sound
E. _____ second heart sound
F. _____ a murmur (systolic: yes/no? diastolic: yes/no?)
G. _____ diastolic pressure
H. _____ systolic pressure
I. _____ dicrotic wave

Answer and Explanation

A. k. It normally precedes the R wave and is a shorter duration than the T wave.

B. j. It is the highest amplitude wave (see question 24 for an exception to this rule).

C. i. It follows the R wave.

D. g. It is heard immediately after the closure of the AV valves and between the R and T wave.

E. f. It is heard after the closure of the semilunar valves, near the end of the T wave.

F. h. It is a systolic murmur. Note that it is heard between the R and T wave (i.e., during ventricular systole).

G. l. The pressure in the aorta that results from ventricular diastole (i.e., the lowest pressure during any cardiac cycle) is called the diastolic pressure. It occurs during early ventricular ejection.

H. m. The highest pressure in the aorta or ventricle during a single cardiac cycle is called systolic pressure.

I. n. The dicrotic wave follows the dicrotic notch or incisura and occurs early in the catacrotic limb of the pressure curve. It should not be confused with a weak ventricular contraction.

33

(1:443; 2:3-54; 3:57; 4:177; 5:965)
M 1

Draw the resting *transmembrane potential* and the action potential for a cardiac Purkinje follower cell. Label the abscissa and ordinate for this drawing and the various phases (phases 0, 1, 2, 3, and 4) of the transmembrane potential.

Answer and Explanation

Note that each phase of the transmembrane potential is associated with specific conductances (g) for Na^+, Ca^{++}, and K^+. Note, too, the changes in excitability that occur during the action potential.

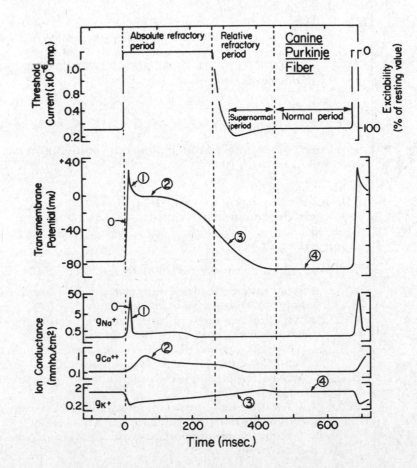

34

(1:444; 2:3-56;
3:55; 4:161; 5:966)
Mdei 2

One can cause a cardiac Purkinje fiber to depolarize during phase 4 of the transmembrane potential by

A. decreasing its Na^+ conductance
B. decreasing its Ca^{++} conductance
C. decreasing either its g_{Na^+} or its $g_{Ca^{++}}$
D. decreasing its K^+ conductance
E. decreasing its g_{Na^+}, its $g_{Ca^{++}}$, or its g_{K^+}

Answer and Explanation

D. *Depolarization* is caused by an increase in Na^+ or Ca^{++} *conductance* or a decrease in K^+ conductance. (1) In skeletal muscle, *acetylcholine* causes depolarization by increasing g_{Na^+}. (2) In the sinus node of the heart the *pacemaker cell* shows a phase 4 depolarization because of its low g_{K^+}. (3) *Norepinephrine* increases the heart rate because it increases $g_{Ca^{++}}$ in the pacemaker cell during phase 4.

35

(1:454; 2:3-58;
3:434; 4:166; 5:966)
Mdei 2

A pacemaker cell in the sinoatrial node differs from a follower cell in the ventricle in that the pacemaker cell has

A. a smaller K^+ conductance during its diastolic potential
B. a smaller diastolic potential
C. both of the above characteristics
D. a phase 0 potential that is due primarily to Na^+ influx
E. a steeper slope for its phase 0 potential

Answer and Explanation

C. The (A) smaller K^+ *conductance*, by moving the SA node cell's transmembrane potential away from the equilibrium potential for K^+, (B) decreases the *diastolic potential*, (D) eliminates the *fast Na^+ current* during phase 0, and (E) produces a phase 0 that is caused by a *slow Ca^{++} current*. This slow current causes the lower slope during phase 0 (i.e., a lower dV/dt: rate of change of voltage).

36

(1:446; 2:3-56;
3:55; 4:165; 5:966)
Hdei 1

Select from statements 1 through 5 the one most appropriate answer for each of the questions, A through H:

1. Ca^{++} gates open
2. K^+ gates open
3. K^+ gates close (or are partly closed)
4. Na^+ gates open
5. Inner gates for Na^+ close (or are closed)

Each statement can be used once, more than once, or not at all.

A. _____ What initiates the fast current (in some cardiac muscle cells at a transmembrane (TM) potential of about –70 mv)?
B. _____ What initiates the slow current (in some cardiac muscle cells at a transmembrane potential of about –50 mv)?
C. _____ Why will a cardiac cell NOT develop a fast current in response to an adequate stimulus, if the cell has a diastolic potential of –60 mv?
D. _____ What stops the fast current, but not the slow current, during most cardiac action potentials?
E. _____ What is the most important cause of the low transmembrane potential in the cells of the sinoatrial and atrioventricular nodes?
F. _____ What is the most important cause of the repolarization (phase 3) of a cardiac muscle cell?

G. _____ What is the action of acetylcholine on the cells of the SA node during the middle of ventricular diastole?

H. _____ What is the action of norepinephrine on the cells of the SA node during the middle of ventricular diastole?

Answer and Explanation

A. 4.

B. 1.

C. 5.

D. 5.

E. 3. This moves the TM potential away from the *equilibrium potential for K^+*.

F. 2.

G. 2. This causes *hyperpolarization*.

H. 1. This causes *depolarization*.

Questions 37 through 41 consist of two main parts: a statement and a reason for the statement. Choose the best answer for each of these questions from the options A through E below and mark your choice in the space provided.

A. both the statement and the reason are true and are related as cause and effect (i.e., as indicated)

B. both the statement and the reason are true but are not related as cause and effect

C. the statement is true but the reason is false

D. the statement is false but the reason is an accepted fact or principle

E. both the statement and the reason are false

37

(1:450; 2:3-57; 4:161; 5:967)
Hdei 3

Statement
_____ If you remove the Na^+ from the external environment of a cardiac Purkinje fiber but keep the osmotic pressures approximately constant, the cell will respond to stimulation by producing an action potential with a low dV/dt during phase 0 because

Reason
phase 0 is now produced by an influx of Ca^{++} into the cell.

Answer and Explanation

A. Many cardiac cells respond to stimulation by exhibiting (1) a *fast inward current*, which is due to an opening of Na^+ *gates* in the cell membrane, and (2) a *slow inward current*, which is at least partly due to the opening of Ca^{++} *gates*. The fast current causes a rapid change in the voltage across the membrane (high dV/dt). The slow current, in the absence of the fast current, causes a low dV/dt. It has been suggested that at *transmembrane potentials* below 60 mv, inner Na^+ gates are "locked closed." On the basis of this hypothesis, we can explain why the pacemaker cell in the SA node has a low phase 0 dV/dt (its phase 4 potential is below 60 mv) and why hyperkalemia produces a depolarization associated with a reduced phase 0 dV/dt and slowed conduction.

38

(1:944; 2:3-77; 3:449; 4:161; 5:981)
Heia 2

Statement
_____ Hyperkalemia causes slow conduction throughout most of the heart because

Reason
hyperkalemia causes a decrease in the diastolic potential of cardiac cells.

Answer and Explanation

A. If plasma K^+ increases from a value of 5 to one of 8 mEq/liter most of the cardiac muscle cells during their diastolic potential depolarize to the point where the inner Na^+ gates are

(Continued)

closed (therefore the *fast current is lost* from phase 0), but the *Ca^{++} gates can still be opened.*

39

(1:442; 2:3-57; 3:55; 4:177; 5:966)
Hdei 2

Statement
_____ A heart cell during its action potential will remain depolarized more than eight times longer than a skeletal muscle cell because

Reason
during phase 2 of the cardiac action potential, there is an elevated K^+ conductance (i.e., a higher g_{K+} than during phase 4).

Answer and Explanation
C. The *K^+ conductance* is reduced during phase 2. This reduction, along with an increase in *Ca^{++} conductance,* is responsible for phase 2 (prolonged depolarization).

40

(1:442; 2:3-115; 3:55; 4:170; 5:966)
Hdei 2

Statement
_____ Catecholamines have a direct negative chronotropic action on the heart because

Reason
they increase calcium conductance during phase 4 of the transmembrane potential.

Answer and Explanation
D. Catecholamines act directly on the beta$_1$ receptors of the SA node to increase heart rate. This positive chronotropic action is caused by an increase in dV/dt during phase 4 of the pacemaker cell.

41

(1:442; 2:3-116; 3:55; 4:170; 5:966)
Hdei 2

Statement
_____ Acetylcholine increases the diastolic transmembrane potential of cells in the sinoatrial node because

Reason
it increases their g_{K+} during phase 4 of the potential.

Answer and Explanation
A. The diastolic potential is the phase 4 potential.

7
The Heart—Mechanical Properties

(1:497; 2:3-104;
3:453; 4:154; 5:987)
Hefp 1

1

The following *pressure pattern* was recorded through a catheter from a patient with normal cardiovascular function. Questions A through J refer to this pattern:

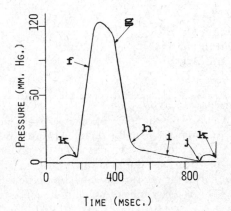

A. The catheter was in
1. the right ventricle
2. the left ventricle
3. a systemic artery
4. either the systemic artery or the left ventricle
5. either the right or left ventricle
B. ___ The P wave of the electrocardiogram begins immediately prior to what point on the pressure pattern?
C. ___ The T wave ends at approximately what point?
D. ___ The aortic valve opens at what point?
E. ___ The period of most rapid ventricular filling begins at what point?
F. ___ The second heart sound begins at what point?
G. ___ The third heart sound, if present, begins at what point?
H. ___ The c wave of the right atrium begins at what point?
I. ___ The middle of the period of ventricular diastasis occurs at what point?
J. ___ The period of isovolumic ventricular relaxation begins at approximately what point?

(Continued)

Answer and Explanation

A. 1. The peak pressure in the right ventricle of the resting, healthy subject averages about 27 mm Hg, not 120 mm Hg as seen above.

2.

3. The lowest pressure in a systemic artery averages 80 mm Hg, not 5 mm Hg as seen above.

B. j. Interval j-k represents atrial systole.

C. g. Point g also represents the end of ventricular systole.

D. f. Point f also normally represents the aortic diastolic pressure.

E. h. Point h represents the opening of the *mitral valve*.

F. g. Point g represents the closing of the *aortic valve*.

G. h. This is caused by the vibrations set up in the ventricles during rapid ventricular filling.

H. k. This pressure wave is caused by the right ventricle pushing the tricuspid valve up toward the atrium. The letter, c, derives from the mistaken notion that this wave is due to the carotid artery pulse.

I. j. Diastasis is a period of little electrical or mechanical change in the heart.

J. g. It begins with the closing of the aortic valve (g) and ends with the opening of the mitral valve (h).

2

(1:673; 2:3-104; 3:453; 4:290; 5:987)
Hp 2

In a resting, healthy individual, the pulmonary valve opens when the pressure in the right ventricle is approximately

A. 10 mm Hg
B. 30 mm Hg
C. 50 mm Hg
D. 80 mm Hg
E. 120 mm Hg

Answer and Explanation

A. This is the diastolic pressure of the pulmonary artery. In other words, this is the minimum pressure in the pulmonary artery during a single cardiac cycle.

B. This is the systolic pressure of the pulmonary artery. In other words, it is the maximum pressure during a single cycle.

D. This is the pressure at which the aortic valve opens.

E. This is the systolic pressure in the aorta.

3

(1:498; 2:3-109; 3:454; 4:157; 5:989)
Hvp 1

The *second component of the first heart sound* is most closely associated with what event during the cardiac cycle?

A. opening of the mitral valve
B. closure of the mitral valve
C. closure of the aortic valve
D. closure of the pulmonary valve
E. closure of the tricuspid valve

Answer and Explanation

A. The opening of the mitral valve is followed by the third heart sound.

B. The closure of the mitral valve is associated with the first component of the first heart sound.

C. The closure of the aortic valve is associated with the first component of the second heart sound.

D. The closure of the pulmonary valve is associated with the second component of the second heart sound.

E. Since right ventricular activation lags behind left ventricular activation by a few milliseconds, the increase in pressure that helps close the tricuspid valve and sets up the vibrations that produce the tricuspid component of the first sound lags behind the mitral component of the first sound.

4

(1:498; 2:3-19; 3:454; 4:157; 5:990)
Hv 2

The *second component of the second heart sound* is usually most clearly heard on the ventral surface of the chest at the

A. second intercostal space to the right (patient's right) of the sternum
B. second intercostal space to the left of the sternum
C. fifth intercostal space to the left of the sternum
D. fifth intercostal space at the sternum
E. ventral tip of the xiphoid process

Answer and Explanation
A. This is where the first component of the second sound is best heard.
B. The second heart sound, for the most part, is due to vibrations of the walls of the ascending aorta and pulmonary artery. The pulmonary artery, as it leaves the base of the right ventricle, crosses to the *left* of the ascending aorta. The second heart sound, which is due to vibrations of the pulmonary artery (second component), is best heard to the *left* of the point where the first component of the second heart sound is best heard.

5

(1:498; 2:3-107; 3:454; 4:157; 5:989)
Hv 2

The *first component of the first heart sound* is usually most clearly heard on the ventral surface of the chest at the

A. ventral tip of the xiphoid process
B. fifth intercostal space over the sternum
C. fifth intercostal space to the left (patient's left) of the sternum
D. second intercostal space to the left of the sternum
E. second intercostal space to the right of the sternum

Answer and Explanation
A. The first heart sound is due to vibrations from the ventricles. The two ventricles lie cephalad to the tip of the xiphoid process.
B. The right ventricle lies beneath the sternum at this point. It is here that the second component of the first heart sound is best heard.
C. It is at the apex of the heart where the first component of the first heart sound is usually most distinctly heard. It is the vibrations from the left ventricle that are responsible for this sound.

6

(1:498; 2:3-19; 3:455; 4:324; 5:1022)
Hfbva 2

A *systolic murmur* was heard over the manubrium of the sternum. The most likely diagnosis is

A. an increased hematocrit
B. aortic stenosis
C. aortic insufficiency
D. mitral stenosis
E. patent ductus arteriosus

Answer and Explanation
B. If, during ventricular ejection, the blood from the left ventricle must be pumped through a narrow lumen, a high-velocity jet of blood will be pushed into the aorta, and turbulence will result. When the aortic valve closes and ejection ceases, the turbulence will also cease.

(Continued)

The turbulence may be either pansystolic or midsystolic but will not occur during diastole unless there is some other disturbance in cardiac function.

7

(1:498; 2:3-19; 3:455; 4:324; 5:1022)
Hvfba 2

Systolic and diastolic murmurs are heard over the body of the sternum. The most likely diagnosis is

A. aortic and pulmonary stenosis
B. aortic and pulmonary insufficiency
C. aortic insufficiency and stenosis
D. an increased hematocrit
E. a decreased hematocrit

Answer and Explanation
A. This causes only a systolic murmur.
B. This causes only a diastolic murmur.
C. The aortic insufficiency causes a murmur as the blood rushes into the relaxing left ventricle from the aorta. The stenosis causes a murmur when the blood is being ejected into the aorta from the left ventricle.
D, E. A decreased hematocrit is more likely to produce a continuous murmur. It never causes systolic and diastolic murmurs.

8

(1:498; 2:3-27; 3:455; 4:327; 5:1966)
Hvfba 2

A *continuous murmur* is heard over the manubrium of the sternum. The most likely diagnosis is

A. an increased hematocrit
B. aortic stenosis
C. aortic insufficiency
D. mitral stenosis
E. patent ductus arteriosus

Answer and Explanation
A. A murmur (i.e., turbulent blood flow) is caused by a decreased hematocrit.
B. This causes a systolic murmur.
C, D. Aortic insufficiency and mitral stenosis cause a diastolic murmur.
E. A patent ductus arteriosus in the adult is a vessel through which blood usually moves from the aorta to the pulmonary artery (i.e., a left-to-right *shunt*). Since the pressure in the aorta is considerably higher than in the pulmonary artery, there is a high-velocity flow. A sign of a high-velocity flow is turbulence (i.e., a murmur).

9

(1:480; 2:3-17; 3:466; 4:210; 5:1022)
Cvfba 2

In this question the following symbols are used:

- I: flow in cm^3/sec
- $\bar{v}$: average velocity of flow in cm/sec
- r: radius of vessel in cm
- ρ: density of fluid in g/cm^3
- η: viscosity of fluid in poises

The tendency of a fluid to become *turbulent* is directly proportional to

A. $(v\,\rho r)/\eta$
B. $(I\rho)/(\pi r\eta)$
C. both of the above
D. $(\bar{v}\,\eta r)/\rho$
E. $(I\rho r)/\eta$

Answer and Explanation
C. One can derive formula B from formula A by substituting for v̄:

$$\bar{v} = I/(\text{cross-sectional area}) = I/(\pi r^2).$$

10

*(1:480; 2:3-17;
3:466; 4:210;
5:1022)*
Cvfba 2

Turbulent blood flow in human subjects results from a decreased

A. hematocrit
B. critical Reynold's number (this occurs in a saccular aneurysm)
C. hematocrit or critical Reynold's number
D. cardiac output
E. cardiac output or hematocrit

Answer and Explanation
C. A decrease in the hematocrit causes a decreased blood viscosity. Turbulence is frequently heard in the area of a saccular aneurysm.
D. Increases in volume flow cause turbulence if the increases are sufficiently great.

11

*(1:514; 2:3-191;
3:466; 4:207;
5:1004)*
Hvfba 2

A physician notes that when taking a patient's arterial *blood pressure* using a *sphygmomanometer cuff*, although the sound over the artery becomes muffled at a cuff pressure of 70 mm Hg, it does not disappear as the cuff pressure returns to 0 mm Hg. What conclusion is most likely drawn from these data? The patient

A. has a diastolic pressure at or near 0 mm Hg (i.e., may have aortic insufficiency)
B. probably has a weakened heart (i.e., left ventricular congestive heart failure)
C. has a patent ductus arteriosus
D. has an aortic stenosis
E. has a low hematocrit

Answer and Explanation
A, B. When taking an arterial pressure by this method, you have two indices of true diastolic pressure: (1) the pressure in the cuff at which a muffling of the sound is heard through the stethoscope, and (2) the pressure at which the disappearance of the sound is noted. Whenever this method yields widely divergent values for diastolic pressure, the muffling is the best index of arterial diastolic pressure. In other words, this patient does not have a diastolic pressure that would indicate either aortic insufficiency or left ventricular failure. One would need more evidence to come to either of these conclusions.
C, D. A patent ductus arteriosus and an aortic stenosis produce a murmur that can be heard on or near the surface of the sternum but will not be heard in the forearm, where pressure is usually obtained.
E. As the hematocrit decreases, the viscosity of the blood decreases and murmurs are more likely. The failure of the sound to disappear when one takes a pressure reading by the cuff method may be due to this continuous background murmur.

12

*(1:597; 2:3-135;
3:498; 4:299;
5:1099)*
Hpf 2

At a heart rate of 80/min, *coronary artery blood flow*

A. is greatest shortly after the second heart sound is heard
B. is zero in the subendocardial portion of the left ventricle during ventricular systole
C. is well characterized by both A and B
D. is greatest during the period of peak left ventricular ejection
E. is well characterized by both A and D

(Continued)

Answer and Explanation

C. Prior to the second heart sound, there is ventricular systole. During this period, ventricular contraction is occurring and causing the occlusion of the subendocardial coronary vessels. During ventricular diastole the aortic pressure is decreasing and the coronary vessels are becoming more patent. The net effect is an increase in coronary blood flow during ventricular relaxation.

B, C, D, E. The events noted in statements B through E all occur during ventricular systole, when many of the coronary vessels are occluded by the contracting cardiac muscle fibers.

13

(1:604; 2:7-123; 3:499; 4:307)
HCbgfa 2

Which one of the following findings is indicative of *coronary ischemia*?

A. a higher lactate concentration in the coronary sinus than in the aorta
B. an increased coronary a-v$_{O_2}$ concentration difference
C. both of the above
D. an increased coronary blood flow
E. a decreased coronary blood flow

Answer and Explanation

C. Normally, the heart catabolizes more lactic acid than it produces. During hypoxia the heart becomes more dependent on the anaerobic production of pyruvate and lactate from glucose as a source of energy, and the coronary sinus lactate concentration goes up, as does the a-v$_{O_2}$ difference.

D, E. Changes in coronary blood flow are usually a sign of a dynamic system of vessels capable of changing its resistance in response to the changing demands of the organ being served. Coronary ischemia occurs when the vessels of the heart lose their ability to maintain a level of O_2 delivery that permits the heart to catabolize more lactate than it produces.

14

(1:600; 2:3-140; 3:499; 4:299; 5:1057)
HCbgf 2

Which one of the following is the most common cause of an increased coronary blood flow?

A. a decreased coronary perfusion pressure
B. a decreased stimulation of beta$_2$ adrenergic receptors in the heart
C. an increased stimulation of alpha adrenergic receptors in the heart
D. an increased stimulation of beta$_1$ adrenergic receptors in the heart
E. an increased O_2 concentration in the coronary arteries

Answer and Explanation

A. An increased *perfusion pressure* will cause an increased flow if there is not an associated increase in *resistance*. In the heart, when the perfusion pressure is maximum, many of the coronary vessels are being constricted by the contracting ventricular fibers.

D. The stimulation of *beta$_1$* cardiac receptors causes (1) a positive chronotropic and inotropic cardiac response, (2) an increased cardiac metabolism, (3) a decreased Po$_2$ and the accumulation of *metabolites*, (4) coronary arteriolar dilation in response to this change in environment, and (5) an increased coronary flow. Item (3) is the usual cause of a *coronary vasodilation*. Other apparently less important causes are stimulation of *beta$_2$* receptors in the coronary arterioles and a decreased stimulation of alpha receptors in the coronary arterioles.

15

(1:604; 2:3-39; 3:245; 4:302; 5:1647)
Hc 1

The principal *energy source* for the heart of a healthy fasting subject is

A. free fatty acids
B. glucose
C. lactate and pyruvate
D. amino acids
E. polypeptides

Answer and Explanation
A. The catabolism of free fatty acids in the fasting subject is responsible for about 65% of the heart's O_2 consumption; the catabolism of glucose, 18%; the catabolism of lactate, 16%. These figures will change when the concentration of nutrients in the blood changes. The catabolism of lactate, for example, will increase in strenuous exercise. The catabolism of glucose will decrease in diabetes mellitus and increase in response to insulin. Skeletal muscle and the brain, on the other hand, have a much greater preference for glucose catabolism as a source of energy than does the heart.

16

(1:619; 2:3-232; 3:503; 4:1040; 5:1966)
GCfg 1

Which one of the following statements best characterizes the *ductus venosus* of the fetus?

A. it receives most of its blood from the umbilical vein
B. it sends its blood directly into the inferior vena cava
C. both of the above statements are correct
D. it contains blood with a higher O_2 content than the blood in the abdominal aorta of the fetus
E. all of the above statements are correct

Answer and Explanation
E. The ductus venosus carries O_2 and other nutrients from the umbilical vein to the inferior vena cava, where it is mixed with less well-oxygenated blood. The blood in the ductus venosus is about 80% saturated with O_2 and that in the fetal aorta is about 60% saturated.

17

(1:621; 2:3-233; 3:504; 4:1039; 5:1978)
GHCfS 1

Check each of the following conditions that is prevalent within a few minutes after the *birth* of a healthy child:

1. permanent fusion of the septum primum and secundum is complete
2. the foramen ovale is closed
3. the ductus arteriosus is functionally closed
4. the umbilical arteries are functionally closed by contraction of the smooth muscles in their walls

Which one of the following best summarizes your conclusions?

A. statements 1, 2, 3, and 4 are all correct
B. statement 1, 2, and 3 are correct
C. statements 1 and 4 are correct
D. statements 2, 3, and 4 are correct
E. statements 2 and 4 are correct

Where are each of the structures mentioned above located?

Answer and Explanation
A. The *septum primum and secundum* separate the right and left atria and form the foramen ovale. Their permanent fusion will not be complete until two to eight weeks after birth.
B, C, (D.) The *foramen ovale* lies in the interatrial septum and the *ductus arteriosus* connects the pulmonary artery and aorta. Prior to birth, they provide a right-to-left *shunting* of the blood past the lungs. Immediately after birth, there is a fivefold increase in lung volume

(Continued)

after the first breath is taken and a five- to tenfold increase in pulmonary blood flow due in part to the closure of the foramen ovale and ductus arteriosus. The *umbilical arteries* carry blood from the common iliac artery of the fetus toward the intervillous space of the placenta.

18

(1:621; 2:2-233; 3:504; 4:1039; 5:1978)
GCfo 1

List the cardiovascular changes that occur in the transition from the *fetal* to the *newborn state*. Check each of the following changes that is characteristic of this transition.

A. left and right ventricles change from a series to a parallel circulation
B. the pressure in the aorta rises until it exceeds the pressure in the pulmonary artery
C. right atrial pressure becomes greater than left atrial pressure
D. circulatory resistance through the lungs increases

Answer and Explanation
A. The closure of the *ductus arteriosus* and the *foramen ovale* converts a parallel circulation to a series circulation.
(B.) When the *placental circulation* is cut off, the systemic resistance rises.
C. It is the reversal of the *pressure head in the atria* (increase in left atrial pressure and decrease in right atrial pressure) during birth that is responsible for the closure of the foramen ovale.
D. The asphyxia associated with birth facilitates gasping movements in the fetus which will usually produce an intrapleural pressure between -30 and -50 mm Hg. At the same time, the *pulmonary vascular resistance* through the lungs becomes less than 20% of what it was prior to labor. This is, in part, responsible for a marked increase in *blood flow through the pulmonary arteries*, capillaries, and veins.

19

(1:673; 2:482; 3:533; 4:329; 5:1711)
Hgvpa 2

A patient is diagnosed as having an opening in the *interatrial septum* uncomplicated by other cardiac abnormalities. Which of the following would best confirm this diagnosis?

A. a systolic murmur
B. an elevated P_{O_2} in the pulmonary artery
C. a decreased pressure in the right atrium
D. an elevated pressure in the left atrium
E. cyanosis

Answer and Explanation
A. The pressure differences between the right and left atria are small and are therefore unlikely to cause a sufficiently high-velocity flow to produce a murmur.
(B.) The right atrial P_{O_2} is normally markedly lower than the left atrial P_{O_2}, and therefore the left-to-right shunt of blood through the interatrial septum would markedly increase the right atrial P_{O_2}.
E. Cyanosis is neither a specific indication nor likely in this condition.

20

(1:497; 2:3-104; 3:453; 4:221; 5:987)
Hpag 3

During a cardiac catheterization the following data are collected from two 50-year-old men:

	Patient 1	Patient 2
Aorta:		
Pressure (mm Hg)	140/50	95/65
O_2 saturation (%)	95	96
Left Ventricle:		
Pressure (mm Hg)	140/15	95/9
O_2 saturation (%)	95	96

Left Atrium:
Pressure (mm Hg)   12   31
O$_2$ saturation (%)   95   96

Pulmonary Artery:
Pressure (mm Hg)   26/11   68/21
O$_2$ saturation (%)   75   74

Right Ventricle:
Pressure (mm Hg)   26/5   68/7
O$_2$ saturation (%)   75   74

Right Atrium:
Pressure (mm Hg)   5   6
O$_2$ saturation (%)   75   74

For patient 1, circle all the data in the list that are grossly abnormal and select the one best diagnosis below for this individual.

A. aortic insufficiency associated with left ventricular hypertrophy
B. aortic stenosis associated with left ventricular hypertrophy
C. tricuspid insufficiency
D. patent ductus arteriosus
E. systemic hypertension due to arteriolar constriction

Answer and Explanation

(A.) The aortic diastolic pressure is about 30 mm Hg lower than normal, and the left ventricular end-diastolic pressure is 5 to 10 mm Hg higher than normal. This is characteristic of aortic insufficiency, a condition in which there is a leakage of blood from the aorta back into the ventricle during ventricular diastole. This results in the decrease in aortic diastolic pressure (i.e., increased diastolic runoff of pressure and blood) and the increase in the ventricular end-diastolic pressure (i.e., ventricular distention). The aortic pulse pressure in this patient (140 − 50 = 90 mm Hg) is also quite high. This is probably in part due to the increased end-diastolic volume of the ventricle, which causes an increased stroke volume through a Starling mechanism and through ventricular hypertrophy.

B. Aortic stenosis is a condition in which there is an increased resistance to left ventricular ejection. This results in a markedly lower aortic systolic pressure than is found in the left ventricle. Since the systolic pressures in the left ventricle and the aorta are the same, it is unlikely that the patient has aortic stenosis.

C. In tricuspid insufficiency, one would expect an elevated right atrial pressure.

D. In a patent ductus arteriosus, one would expect an elevated pressure in the pulmonary artery.

E. Arteriolar constriction produces an increase in the arterial diastolic pressure as well as in the mean arterial pressure.

21

(1:497; 2:3-104; 3:453; 4:326; 5:987)
Hpga 3

For patient 2 in question 20, circle all the data in the list that are grossly abnormal and select the one best *diagnosis* below for this individual.

A. aortic insufficiency
B. aortic stenosis
C. tricuspid insufficiency
D. tricuspid stenosis
E. mitral stenosis

Answer and Explanation

E. The pressures in the chambers carrying blood to the left ventricle (left atrium, pulmonary artery, and right ventricle) all are elevated. The pressures in the left ventricle and aorta, on the other hand, are reduced. Normally, during ventricular diastole, left atrial and ventricular pressures are similar (7 *vs.* 7 mm Hg, for example). In this case, they are not (31 *vs.* 9 mm

(Continued)

Hg). Therefore, there is an area of high resistance during ventricular diastole between the left atrium and left ventricle (i.e., mitral stenosis). In other words, an area of high resistance is an area where there is a large pressure drop:

[resistance = (pressure #1 − pressure #2)/flow].

22

(1:497; 2:3-104; 3:453; 4:154; 5:987)
Haop 2

The following records (A through E) were obtained from five patients. Each record contains a left atrial, left ventricular, and aortic *pressure* tracing. The paper speed, 0 baseline, and amplification for all parameters in each study are the same. Questions 1 through 5 refer to these records.

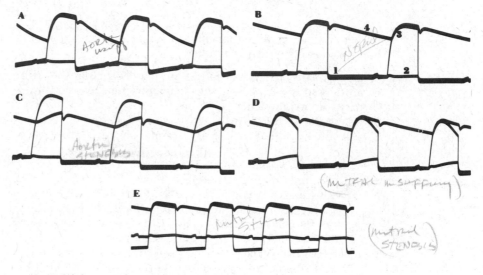

_____ 1. Which record was obtained from a patient with aortic stenosis?
_____ 2. Which record was obtained from a patient with aortic insufficiency?
_____ 3. Which record was obtained from a patient with mitral insufficiency?
_____ 4. Which record was obtained from a patient with mitral stenosis?
_____ 5. Which record was obtained from a patient with normal heart function?

Answer and Explanation

1. **C.** The lower tracing on each record is from the left atrium. The upper tracing is from the aorta. The connecting tracing is from the left ventricle. A *stenosis* is an area of abnormally *high resistance*. Since resistance is directly related to perfusion pressure ($P_1 − P_2$), there results an abnormally high pressure difference across the stenosis. In aortic stenosis, the abnormally high perfusion pressure exists between the left ventricle and aorta during *ventricular systole*.

2. **A.** An *insufficiency* is an area of abnormally *low resistance* and therefore a reduced $P_1 − P_2$. In aortic insufficiency, the aortic valve does not close completely and therefore there is a regurgitation of blood back into the ventricle and a reduced $P_1 − P_2$ between the ventricle and aorta *during diastole*. Note that the reduced pressure head is most marked at the end of diastole.

3. **D.** In MI the reduced *pressure head* is between the atrium and ventricle during ventricular systole. Note that this reduction is most marked at the end of ventricular systole.

4. **E.** The increased pressure head is between the atrium and ventricle during ventricular diastole. The low aortic pulse pressure is also seen in aortic stenosis and when ventricular contractions are abnormally weak. If aortic stenosis is associated with arteriosclerosis, the aortic pulse pressure could be normal.

5. **B.** In the normal heart there is (1) practically no *resistance* between the atrium and ventricle during most of ventricular diastole (period of rapid filling through period of atrial systole), (2) an infinite resistance between the atrium and ventricle during

ventricular systole, (3) practically no resistance between the ventricle and aorta during ventricular ejection, and (4) an infinite resistance between the ventricle and aorta during all of ventricular diastole. In the resting subject, the aortic pulse pressure is usually about one-third the systolic pressure, not over one-half as seen in record A (aortic insufficiency).

23

(1:574; 2:3-153; 3:455; 4:286; 5:999)
HCmfg 3

The following data are collected from a patient:

Respiratory tidal volume:	230 ml
O_2 consumption:	110 ml/min
Femoral artery O_2:	20 volumes % (P_{O_2} = 100 mm Hg)
Femoral vein O_2:	13 volumes % (P_{O_2} = 35 mm Hg)
Pulmonary artery O_2:	14 volumes % (P_{O_2} = 40 mm Hg)

What would the *cardiac output* of this patient be?

A. less than 1600 ml/min
B. between 1600 and 1900 ml/min
C. between 1900 and 2200 ml/min
D. between 2200 and 2500 ml/min
E. more than 2500 ml/min

Answer and Explanation

B. Cardiac output $= \dfrac{O_2 \text{ consumption}}{\text{arteriovenous } O_2 \text{ difference}}$

$$= \dfrac{110 \text{ ml } O_2/\text{min}}{(20-14) \text{ (ml } O_2/100 \text{ ml blood)}}$$

$$= 1833 \text{ ml/min.}$$

In this calculation, one must use the O_2 concentration in the pulmonary artery, because this value is more representative of the average for all the systemic veins than is the femoral vein value. Since O_2 is not removed from the blood between the left heart and the systemic capillaries, the O_2 concentration in all systemic arteries should be the same. The partial pressures of O_2 in mm Hg cannot be used in this calculation. They are representative of the plasma concentration of O_2, not the total blood content. Volume % = ml/100 ml. The calculation of cardiac output by a *dye-dilution method* is discussed on *page 119, question 24*.

24

(1:673; 2:3-104; 3:455; 4:158; 5:987)
Hpa 3

The following data were collected in the *right heart* (ventricle and pulmonary artery) and the *left heart* (ventricle and aorta):

	Right	**Left**
Ventricular pressure (mm Hg):	70/8	100/8
Arterial pressure (mm Hg):	70/20	100/70

On the basis of these data, what conclusions can you draw (check each correct conclusion)?

1. right ventricular systolic pressure is higher than normal
2. pulmonary arterial pressure is lower than normal
3. the abnormal data may be due to mitral stenosis
4. the abnormal data may be due to tricuspid stenosis

Which of the following best summarizes your conclusions?

A. statements 1 and 2 are correct
B. statements 1 and 3 are correct
C. statements 2 and 3 are correct
D. statements 2 and 4 are correct
E. statement 4 is correct

(Continued)

Answer and Explanation
B. Tricuspid stenosis would not cause an elevated right ventricular systolic pressure.

25

(1:486; 2:3-91; 3:458; 4:158; 5:991)
Hpfq 2

Within limits, an increase in the *end-diastolic volume* of the healthy right ventricle will usually (check each correct statement):

1. decrease the stroke volume of the ventricle
2. increase the mean ejection pressure of the ventricle
3. increase the stroke work index of the ventricle

Which one of the following best summarizes your conclusions?

A. statement 1 is correct
B. statement 2 is correct
C. statements 2 and 3 are correct
D. statements 1 and 3 are correct
E. statements 1, 2, and 3 are correct

Answer and Explanation
C. Starling stated: "The energy of contraction is a function of the length of the muscle fiber." We can demonstrate in a heart-lung preparation that, within limits, an increase in the end-diastolic volume of a chamber of the heart causes an increase in the product of stroke volume and mean ejection pressure for that chamber. This product is called the stroke work. The stroke work index is the stroke work divided by the surface area of the subject. The surface area of the average adult man is approximately 1.7 m². The stroke work performed by the right ventricle of a man at rest is approximately 0.2×10^7 ergs. The stroke work index would therefore be about 0.12×10^7 ergs/m².

26

(1:625; 2:3-102; 3:457; 4:155; 5:1078)
Hfq 1

When an average, healthy, 20-year-old individual changes from a relaxed standing position to *running* with a maximal effort, he is capable of

A. increasing his stroke volume twofold
B. increasing his stroke volume fourfold
C. increasing his stroke volume sixfold
D. increasing his stroke volume ninefold
E. increasing his stroke volume twelvefold

Answer and Explanation
A. A more than twofold increase in *stroke volume* either seldom or never develops. This is in marked contrast to another pumping system, the respiratory system. In this system a ninefold increase in tidal volume (i.e., respiratory stroke volume) can occur.

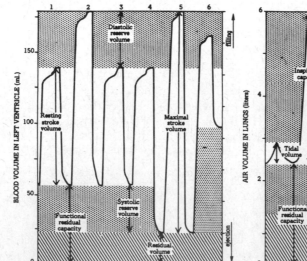
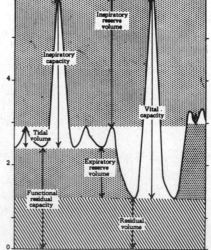

27

(1:485; 2:3-94;
3:467; 4:214;
5:1023)
Hpqm 2

The following data are collected from a resting patient before and after surgery:

	Before	After
Diameter of the lumen of the left ventricle (cm):	8	10
Heart rate (cycles/min):	70	90
Peak pressure in left ventricle (mm Hg):	110	100

Assuming that the ventricle is a sphere, calculate the tension the left ventricle has to develop before and after surgery in order to produce the pressures indicated.

Which one of the following best summarizes your conclusions? The surgery causes

A. a 10% to 20% reduction in the tension exerted by the left ventricle in the resting patient

B. a 10% to 20% increase in the tension exerted by the left ventricle in the resting patient

C. little or no change in the tension exerted by the left ventricle in the resting patient

D. more than a 20% reduction in the tension exerted by the left ventricle in the resting patient

E. more than a 20% increase in the tension exerted by the left ventricle in the resting patient

Answer and Explanation

B. Tension produced by a contractile sphere (dynes/cm): $T = P \times R \times \frac{1}{2}$

= [Pressure produced (dynes/cm^2)] × radius (cm) × 1/2

Before	After
(110 × 1330) × (4) × (1/2)	(100 × 1330) × (5) × (1/2)
= 2.9 × 10^5	= 3.3 × 10^5

% change = $\dfrac{3.3 - 2.9}{2.9} \times 100 = +14\%$.

On the basis of these calculations, we can see that after surgery, although the left ventricle is producing less pressure, it is exerting more tension. This demonstrates that a distended ventricle is at a mechanical disadvantage. In other words, distention results in the ventricle having to develop more tension in order to produce the same pressure. Note that in the above calculations, mm Hg is converted to dynes/cm^2 by incorporating the constant 1330 dynes/cm^2/mm Hg in the calculation.

28

(1:490; 2:3-89;
3:457; 4:155
5:1023)
Hpq 2

Which of the following will cause an increase in left ventricular preload?

A. an increased end-diastolic volume

B. an increased end-diastolic pressure

C. both of the above

D. an increased end-systolic radius

E. an increased stroke volume

Answer and Explanation

C. The tension exerted on a muscle prior to contraction is called its *preload*. In the case of the left ventricle, it is directly related to the radius or volume of the ventricle and to the pressure in the ventricle.

29

(1:490; 2:3-122; 3:459; 4:160; 5:993)
Hnq 2

Which of the following is a response of the heart to the stimulation of *adrenergic sympathetic neurons*?

1. it shifts the Starling curve of the heart to the left ↑ contraction of Heart
2. it increases conduction time
3. it gives a positive chronotropic response
4. it decreases the refractory period

Which of the following best summarizes your conclusions?

A. statements 1, 2, 3, and 4 are correct
B. statements 1, 2, 3, and 4 are incorrect
C. statements 1, 3, and 4 are correct
D. statement 2 is correct
E. statements 3 and 4 are correct

Answer and Explanation

A. The stimulation of cardiac sympathetic neurons causes a decrease in conduction time.
B, C. Sympathetic stimulation has a positive inotropic, dromotropic, and chronotropic action on the heart. In other words, it increases (1) cardiac force at any given end-diastolic volume (a shift of the Starling curve to the left = a form of homeometric regulation), (2) the velocity of conduction (decreases conduction time), and (3) the frequency of stimulation (increases heart rate).

30

(1:318; 2:3-122; 4:715; 5:896)
Hnq 2

Check each of the following factors that is a response of the healthy heart to the stimulation of *cardiac parasympathetic neurons*:

1. a shift of the Starling curve for the atria to the right
2. a shift of the Starling curve for the ventricles to the left
3. a slowing of conduction at the AV node
4. a positive chronotropic action on the ventricles

Which one of the following best summarizes your conclusions?

A. statements 1, 3, and 4 are correct
B. statements 1 and 3 are correct
C. statements 2, 3, and 4 are correct
D. statements 2 and 3 are correct
E. statements 2 and 4 are correct

Answer and Explanation

A. Parasympathetic neurons exert little or no direct influence in controlling ventricular conduction or force.
B. The acetylcholine liberated by parasympathetic neurons going to the atria causes the atria to exert less force at each end-diastolic volume (shifts the Starling curve to the right).

31

(1:488; 2:3-117; 3:459; 4:162; 5:980)
Hi 2

Which one of the following statements is most accurate?

A. Ca^{++} and K^+ have synergistic actions on the heart
B. Ca^{++} has a positive inotropic action on the heart
C. K^+ has a positive inotropic action on the heart
D. Ca^{++} and K^+ have a positive inotropic action on the heart and act synergistically
E. The intravenous injection of KCl causes hyperpolarization of cardiac cells

Answer and Explanation

A. For the most part, these two ions act antagonistically.
B. Within limits, as one increases the Ca^{++} the heart's contractions strengthen at any given end-diastolic volume. In other words Ca^{++} shifts the Starling curve to the left.
C, D. K^+ has a negative inotropic action on the heart.
E. KCl injections cause hypopolarization.

32

(1:673; 2:3-118; 3:459; 4:290; 5:987)
Hpf 2

Check each of the statements concerning the *left ventricle* of a resting adult that is true:

1. it has a stroke volume equal to that of the right ventricle
2. it produces a pressure equal to that of the right ventricle
3. its stroke work is four to five times greater than that of the right ventricle
4. its stroke work index is equal to that of the right ventricle

Which one of the following best summarizes your conclusions?

A. statements 1, 2, and 4 are true
B. statements 1 and 3 are true
C. statements 1 and 4 are true
D. statements 1, 3, and 4 are true
E. statements 1 is false

Answer and Explanation

B. In the healthy adult, the mean ejection pressure produced by the left ventricle is four to five times that produced by the right ventricle. Since the stroke volume produced by both ventricles in the resting subject is equal, the stroke work and stroke work index of the left ventricle will be four to five times that of the right ventricle.

33

(1:492; 2:3-90; 3:459; 4:163; 5:991)
Hpgf 2

Check each of the following that results from an increase in the inotropicity of the left ventricle.

A. an increase in its ejection fraction
B. a decrease in its ejection time/preejection period ratio
C. an increase in the rate at which the ventricle develops pressure
D. an increase in the end-systolic volume of the ventricle
E. a decrease in the ventricle's V_{max}

Answer and Explanation

A.
B. It causes an increase in this ratio.
C. This is called dP/dt.
D. It usually causes a decrease in end-systolic volume.
E. It will increase V_{max}. V_{max} is the velocity of shortening of a muscle when that muscle has no afterload (see question 23 in Chapter 5).

34

(1:490; 2:3-99; 3:459; 4:163; 5:991)
Hpfoma 3

A patient has the following changes in response to *therapy*:

	Before	After
Mean arterial pressure (mm Hg):	100	70
Cardiac output (liters/min):	5	4
Heart rate (#/min):	70	90
End-diastolic volume of left ventricle (ml):	140	240

Check each of the following statements about these data that is true. The therapy produced:

1. an increased left ventricular stroke work
2. a negative inotropic response
3. a decreased total systemic peripheral resistance

Which one of the following best summarizes your conclusions?

A. statement 1 is true
B. statement 2 is true
C. statements 1 and 2 are true
D. statements 1 and 3 are true
E. statements 2 and 3 are true

Answer and Explanation

E. Let $\bar{P}$ = mean arterial pressure, I = cardiac output, R = heart rate:

$$\text{Stroke work (dyne-cm)} = \frac{[\bar{P} \text{ (mm Hg)}] \times [1330 \text{ (dynes/cm}^2/\text{mm Hg})] \times [I \text{ (cm}^3/\text{min)}]}{[R \text{ (\#/min)}]}$$

Before

$$\frac{(100) \times (1330) \times (5000)}{(70)}$$

$$= 0.95 \times 10^7$$

After

$$\frac{(70) \times (1330) \times (4000)}{(90)}$$

$$= 0.41 \times 10^7$$

$$\text{Total systemic peripheral resistance (PRU)} = \frac{[\bar{P} \text{ (mm Hg)}] \times (60 \text{ sec/min})}{[I \text{ (ml/min)}]}$$

Before

$$\frac{(100) \times (60)}{5000}$$

$$= 1.2$$

After

$$\frac{(70) \times (60)}{4000}$$

$$= 1.0$$

When an increase in the end-diastolic volume of the left ventricle is associated with a decrease in left ventricular stroke work, the Frank-Starling curve has usually shifted to the right. This shift to the right is called a negative inotropic response.

35

The following record represents the changes in a patient's left ventricular volume during several cardiac cycles. Answer questions A, B, and C on this recording.

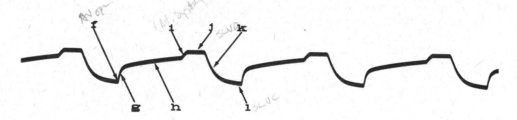

A. At what point does atrial systole occur?
B. At what point does the mitral valve open?
C. At what point does isovolumic systole occur?

Answer and Explanation

A. i. It causes a small increase in *ventricular volume* and is preceded by (h) a period of diastasis.
B. f. It marks the beginning of (g) the period of rapid ventricular filling (i.e., rapid increase in ventricular volume).
C. j. It is followed by (k) ventricular ejection and (l) isovolumic systole.

36

The following *pressure-volume loop* was recorded from the left ventricle:

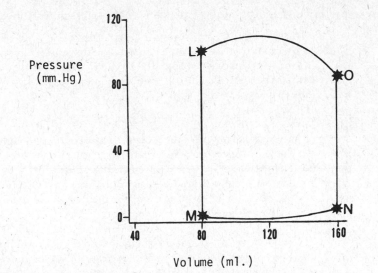

At what point in the loop (L, M, N, or O) do the following events occur?

_____ closure of the mitral valve
_____ opening of the mitral valve
_____ closure of the aortic valve
_____ opening of the aortic valve

> *Answer and Explanation*
> **N.** The mitral valve closes after the ventricle has filled with blood (period M-N is the period of *ventricular filling*) and before it has markedly increased its pressure.
> **M.** It opens prior to ventricular filling.
> **L.** The aortic valve closes after ventricular ejection (period O-L).
> **O.** It opens prior to ventricular ejection.

37

(1:501; 2:3-108; 3:452; 4:154; 5:998)
Hpq 1

Name period L-M and period N-O in the diagram in question 36.

> *Answer and Explanation*
> L-M is *isovolumic ventricular relaxation.*
> N-O is *isovolumic ventricular contraction.*

38

(1:501; 2:3-108; 3:452; 4:208; 5:998)
Hpq 3

The following pressure-volume loops were recorded from the left ventricle before and during medication:

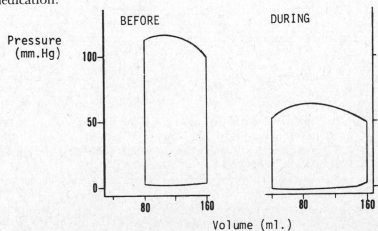

(Continued)

Check each of the following statements that is a possible effect of the medication:

1. a decrease in the total systemic resistance
2. an increased stroke volume caused by a decreased ventricular afterload
3. an increased stroke volume due to the Frank-Starling mechanism
4. an increased stroke volume due to the positive inotropic action of the medication
5. a decreased preload

Answer and Explanation

(1.) The average height of the loop (the *mean ejection pressure*) decreased during medication. The medication probably lowered the mean ejection pressure by producing vasodilation.

(2.) The decreased resistance to ventricular outflow represents a decreased afterload, which permits increased muscle fiber shortening and therefore an increased *stroke volume*. Although the medication decreased the *end-systolic volume* from 80 to 40 ml, it did not change the *end-diastolic volume* from 160 ml.

3. In the *Frank-Starling mechanism*, an increased end-diastolic volume causes an increased stroke work. There was no change in end-diastolic volume.

4. A positive *inotropic* agent would increase the stroke work of the left ventricle when there is no change in the ventricle's end-diastolic volume. The *stroke work* is represented by the area enclosed by the pressure-volume loop. That area has not been increased by the medication.

5. Since neither the end-diastolic volume (160 ml) nor the end-diastolic pressure (5 mm Hg) changed, the *preload* did not change.

39

(1:588; 2:3-108; 3:452; 4:309; 5:998)
Hpq 3

The following pressure-volume loops were obtained from the left ventricle of a subject in 1976 and 1979:

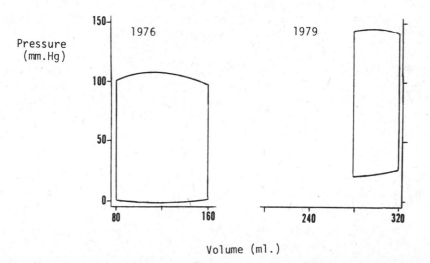

These data are consistent with the diagnosis that the subject between 1976 and 1979 has

A. improved the function of his heart through a demanding exercise regime
B. developed a Frank-Starling curve that has shifted to the left
C. increased his stroke work by a Frank-Starling mechanism
D. statements B and C are both correct
E. developed a left ventricular failure in response to an aortic stenosis

Answer and Explanation
A, B, C, D. The change that occurred between 1976 and 1979 is that the subject has devel-

oped an increase in his end-diastolic volume associated with a decrease in his stroke work and *ejection fraction*. In other words, there is a shift of the Frank-Starling curve to the right. This is indicative of a deteriorating cardiac function. Catecholamines such as norepinephrine shift the curve to the left and are said to have a positive inotropic action. *Heart failure* shifts it to the right and therefore has a negative inotropic action.

INOTROPY

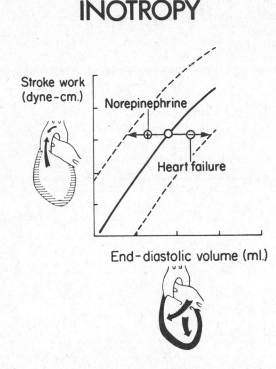

(E.) The *elevated ejection pressure* can be caused by an increased afterload as a result of aortic stenosis, coarctation of the aorta, or arteriolar constriction. An excessive afterload, preload, metabolic load, or contractile impairment are all possible causes of ventricular failure. Characteristic of ventricular failure is an end-diastolic blood volume greater than 200 ml and an end-diastolic pressure greater than 10 mm Hg.

40

(1:402; 4:330)
Hpq 2

In the study in question 39, the end-diastolic left ventricular sarcomere length was 2.1 μm in 1976 and 1979. What mechanism permits the left ventricle to have the same sarcomere length at two widely different end-diastolic volumes?

A. atrophy
B. hypertrophy: a laying down of sarcomeres in series with preexisting sarcomeres
C. hypertrophy: a laying down of sarcomeres in parallel with preexisting sarcomeres
D. cardiomegaly: an increase in the width of the z line
E. cardiomegaly: an increase in extracellular elements (see Braunwald E, Ross J Jr, Sonneblick EH: *Mechanisms of Contraction of the Normal and Failing Heart.* 2 ed., Boston: Little, Brown, 1976, p 87).

Answer and Explanation
   B. When the heart is exposed to an increased load, it will get larger (*cardiomegaly*). This is caused by the laying down of additional intercellular and intracellular elements. The resultant increase in the size of cells is called *hypertrophy*. It may be associated with an increase in the number of sarcomeres in series or in parallel or a combination of the two.

41

(1:607; 2:3-118; 3:221; 4:157; 5:997)
HWmp 3

The following data were obtained from a patient:

Cardiac output:	6000 ml/min
Mean aortic pressure:	100 mm Hg
Mean pulmonary artery pressure:	15 mm Hg
O_2 consumption of the heart:	24 ml/min
Energy value of O_2:	4.8 kcal/liter

On the basis of these data, calculate the following (*Hint*: Assume the atria perform a negligible amount of work, 1 mm Hg = 1330 dynes/cm², and 1 kcal = 4.19×10^{10} dyne-cm).

Energy requirement of the heart:	_____ kcal/min
Left ventricular work:	_____ kcal/min
Right ventricular work:	_____ kcal/min
Cardiac work:	_____ kcal/min
Cardiac efficiency:	_____ %

Answer and Explanation

Energy requirement of the heart =

$$\frac{24 \text{ ml/min}}{1000 \text{ ml/liter}} (4.8 \text{ kcal/liter}) = 0.12 \text{ kcal/min}.$$

Left ventricular work =

$$\frac{(6000 \text{ cm}^3/\text{min}) (100 \text{ mm Hg}) (1330 \text{ dynes/cm}^2/\text{mm Hg})}{(4.19 \times 10^{10} \text{ dyne-cm/kcal})} = 0.019 \text{ kcal/min}.$$

Right ventricular work =

$$\frac{(6000 \text{ cm}^3/\text{min}) (15 \text{ mm Hg}) (1330 \text{ dynes/cm}^2/\text{mm Hg})}{(4.19 \times 10^{10} \text{ dyne-cm/kcal})} = 0.003 \text{ kcal/min}.$$

Cardiac work = (0.019 + 0.003) (kcal/min) = *0.022 kcal/min*.

$$\text{Cardiac efficiency} = \frac{\text{Cardiac work}}{\text{Energy required}} = \frac{0.022 \text{ kcal/min}}{0.12 \text{ kcal/min}} = 0.18 = 18\%.$$

42

(1:603; 2:3-118; 3:460; 4:157; 5:997)
HWpf 2

Select the one best statement:

A. an acute increase in the ventricular end-diastolic volume causes an increased ventricular efficiency
B. the three major determinants of myocardial oxygen consumption are (1) cardiac afterload, (2) inotropic state of the heart, and (3) heart rate
C. both of the above are true
D. an aortic stenosis associated with a normal mean arterial pressure and cardiac output can markedly increase myocardial O_2 consumption
E. all of the above are true

Answer and Explanation
A. True: The *stroke work* increases more than the O_2 *consumption* under these circumstances.

D. True: The aortic stenosis increases the pressure head between the left ventricle and the aorta during ventricular ejection. Therefore, the mean ejection pressure in the ventricle is elevated.

Ⓔ.

43

(1:612; 2:6-83; 3:446; 4:347; 5:1838)
HCRfga 3

An 8-year-old boy walking on the ice of a freshwater lake falls through. He is totally submerged for 10 minutes. After an additional minute, he is brought to land and found to have no carotid pulse, no respiratory movements, and markedly dilated pupils. Check each of the following that apply to this case:

1. there is practically no chance of reviving the subject, because complete brain ischemia has occurred for well over 4 to 6 minutes.
2. dilated pupils occur in response to cerebral ischemia
3. if resuscitation is attempted, the first step is to free the lungs of water
4. if resuscitation is attempted, the first step is to give external cardiac massage
5. if resuscitation is attempted, cardiac massage should not be used because open-chest massage is not practical under these conditions and closed-chest massage is usually ineffective and may produce a ruptured spleen, cracked ribs, or internal bleeding

Which one of the following best summarizes your conclusions?

A. statement 1 is correct
Ⓑ. statement 2 is correct
C. statements 1, 2, and 3 are correct
D. statement 4 is correct
E. statements 3 and 5 are correct

Answer and Explanation

A. There are a number of reasons why *revival* is quite likely. They include: (1) Cooling: during submersion the body temperature lowers, and therefore the body's requirement for O_2 decreases. (2) Heart function: the heart probably continues to pump blood for the first 10 minutes or more of asphyxia. Therefore, there probably are not even 4 minutes of complete brain ischemia. (3) Oxygen reserves: the blood (contains about 400 ml of O_2 in the adult), lungs (contain about 100 ml of O_2 in the adult), myoglobin (contains about 200 ml of O_2 in the adult), and body water (contains about 50 ml of O_2 in the adult) all contain O_2 that can be utilized during apnea. (4) Circulatory control: there is an autonomic nervous system-induced vasoconstriction in skeletal muscle, the viscera, and the skin that serves to shunt most of the available blood and, hence, most of the available O_2 to the brain and heart, where this O_2 is most needed.

Ⓑ. **C.** In human beings, drowning apparently can cause a laryngospasm, which prevents the inspiration of large quantities of water. The fresh water that is inspired rapidly diffuses into the blood, so in this drowning victim, there is no water in the lungs to expel.

D. Cardiac massage will be ineffective in restoring body function unless there is also aeration of the blood through the ventilation of the lungs. The first step is diagnosis (apnea, no detectable heartbeat, dilated pupils). The next step is either simultaneous *ventilation* and external *massage* if two people are present, or two or three artificial ventilations followed by 10 seconds of massage (10 compressions) given repeatedly if one person is present.

E. Closed-chest massage is an effective means of restoring the circulation. Since the life of the subject is at stake, the risk intrinsic to the method is worth taking.

Comment: Victims have been resuscitated after periods of more than 22 minutes of submersion in cold water. Under these conditions, there may be some nerve damage, but eventual, complete recovery is possible.

44

*(1:625; 2:3-104;
3:457; 4:154)*
Hfe 2

Check each of the following changes that would be caused by an increase in *heart rate* from 75 to 200/min:

1. a loss of the period of diastasis Rest period
2. a decrease in dependence of the ventricles on atrial systole
3. a decrease in the refractory period of the ventricle
4. an increase in the ventricular conduction time

Which of the following best summarizes your conclusions?

A. statements 1, 2, 3, and 4 are correct
B. statements 1, 3, and 4 are correct
C. statements 1 and 3 are correct
D. statements 2 and 4 are correct
E. statement 4 is correct

Answer and Explanation

A. When there is an increased heart rate the period of ventricular filling is shortened, and it therefore becomes progressively more important to accelerate ventricular filling. Atrial systole is one of the important means for doing this. An increased rate of ventricular relaxation is a second important means. At a heart rate of 75/min, about 15% of ventricular filling (10 ml) occurs during atrial systole. At twice this heart rate, the period of diastole is reduced by more than 60%, and well over half of ventricular filling (40 to 90 ml) may occur during atrial systole.

B. When the heart rate increases in the absence of changes in hormones or the nervous system (i.e., when a patient turns a variable-rate artificial pacemaker to a higher rate), there is a decrease in atrial and ventricular conduction time. The conduction time will be further decreased by increased sympathetic tone and epinephrine.

C. The period of diastasis is a period of little electrical or mechanical activity in the heart (essentially a rest period for the entire heart). Its loss means that atrial systole will occur soon after the AV valves open (i.e., during the period of rapid ventricular filling). The decreased refractory period of the ventricles permits more impulses to be conducted per minute. It occurs in response to an increased heart rate as well as in response to epinephrine and an increased sympathetic tone to the heart.

45

*(1:625; 2:3-120;
3:457)*
Hfpa 2

Which one of the following statements best describes a 30-year-old patient's response to a *heart rate* change from 220 to 280 per minute? The increase in heart rate causes

A. a decrease in stroke volume
B. a decrease in stroke volume and cardiac output
C. an increase in arterial pulse pressure
D. an increase in cardiac output
E. an increase in arterial pulse pressure and cardiac output

Answer and Explanation

B. Within limits, an increase in heart rate will cause an increase in cardiac output. As one progressively increases the heart rate, however, a point is reached at which the increase in rate is so great that a serious decrease in filling time occurs. At heart rates of 250 per minute, the filling time is so reduced that there usually is a lower stroke volume and cardiac output than would occur at rates of 190 per minute. The point at which a further increase in heart rate causes a decrease in cardiac output will vary from one individual to another and will vary depending on the degree of cardiac sympathetic tone. It will usually lie somewhere between 100 and 200 beats per minute.

C. Decreases in stroke volume cause decreases in pulse pressure.

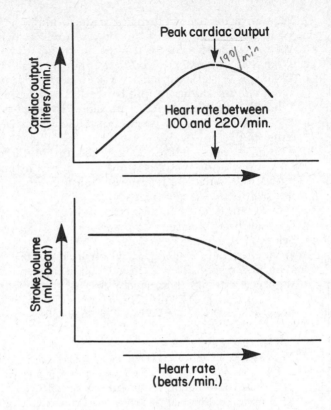

46

(1:490; 2:3-1;
3:459; 4:163; 5:999)
Hpbma 3

The following data apply to questions A through N. Answer as many of these questions as you can before checking your answers. A patient arrived at the clinic complaining of a *numbness* in the right thigh during running. She was told to take a certain medication daily for this problem, but returned 3 weeks later stating that although the medication had cured the problem with her thigh, she now experienced *chest pain* when she exerted herself. The following data were collected from this patient before and 3 weeks after the beginning of the medication (i.e., during medication):

	Before	During
O_2 concentration (ml of O_2/ml of blood)		
Right femoral artery:	0.190	0.190
Right femoral vein:	0.120	0.140
Pulmonary artery:	0.130	0.135
Coronary sinus:	0.06	0.04
O_2 consumption of the patient (ml O_2/min):	250	260
End-diastolic volume of left ventricle (ml):	150	160
Mean ejection pressure of left ventricle (mm Hg):	90	100
Aortic pressure (mm Hg):	120/80	115/85
Heart rate (cycles/min):	75	100

A. Do you think the therapy had any effect on the blood flow to the right thigh?

B. What evidence do you have to support this?

C. What mechanisms might be responsible for a change in blood flow to the right thigh?

D. What evidence is there that the mechanism is not solely the result of a change in cardiac output?

(Continued)

E. What probably caused the chest pain?

F. What evidence is there that there was cardiac ischemia?

G. What was the effect of the therapy on the heart's inotropicity?

H. What direct evidence do you have to support this conclusion?

I. What was the left ventricular stroke work before and during therapy? What was the cardiac output before and during therapy? What was the left ventricular minute work before and during therapy?

J. What was the ejection fraction of the left ventricle before and during therapy? Is either of these normal?

K. The data support the thesis that the therapy produced an increase in the minute work for the heart and that this stress was not met by an adequate increase in coronary blood flow. What might be responsible for the heart's failure to increase its nutritive flow in the presence of such a stress?

L. Why is it unlikely that the failure of the heart to increase its nutritive blood flow is due to either a blockage or a destruction of all or part of the nerves innervating the heart?

M. What effect did the therapy have on arterial pulse pressure? What was responsible for this change in pulse pressure?

N. What effect did the therapy have on systemic peripheral resistance? Assume a mean arterial pressure of 89 mm Hg before therapy and 95 mm Hg during therapy. The right atrial pressure was 5 mm Hg both before and during therapy.

Answer and Explanation

A. Yes, it probably increased *blood flow* to the right thigh.

B. The femoral a-v_{O_2} difference decreased during therapy [The Fick principle states that: flow = (O_2 consumption)/(a-v_{O_2} difference). Therefore, if the metabolism of the thigh is constant, a decreased a-v_{O_2} difference in the thigh is due to an increased flow to the thigh.]

C. A decreased resistance to flow in the thigh (an atherosclerotic plaque or a blood clot might have been lysed or there may have been vasodilation).

D. The decreased systemic a-v_{O_2} difference (0.06 *vs.* 0.055) is much less marked than the decreased right thigh a-v_{O_2} difference (0.07 *vs.* 0.05).

E. Cardiac *ischemia* probably caused the chest pain.

F. An increased coronary a-v_{O_2} difference (0.13 *vs.* 0.15).

G. The therapy had a negative *inotropic action* on the heart.

H. The therapy increased the end-diastolic volume of the left ventricle. The therapy decreased the stroke volume and stroke work of the left ventricle. There was therefore a *shifting of the Frank-Starling curve* to the right.

I. Stroke work = stroke volume × mean ejection pressure × constant.

Constant = 1330 dynes/cm² /mm Hg.

$$\text{Stroke volume} = \frac{\text{cardiac output}}{\text{heart rate}}.$$

$$\text{Cardiac output} = \frac{O_2 \text{ consumption}}{\text{a-}v_{O_2} \text{ difference across lung}}.$$

	Before	During
Cardiac output =	$\dfrac{250}{0.19-0.13} =$	$\dfrac{260}{0.19-0.135} =$
	4167 ml	4727 ml
Stroke volume =	55.6 ml	47.3 ml
Stroke work =	55.6 × 90 × 1330 =	47.3 × 100 × 1330 =
	6.66 × 10⁶ dyne-cm or ergs	6.29 × 10⁶ dyne-cm
Minute work =	Heart rate × stroke work	
	500 × 10⁶ dyne-cm	629 × 10⁶ dyne-cm

J. Ejection fraction $= \dfrac{\text{stroke volume}}{\text{end-diastolic volume}}$

Before	After
$= \dfrac{55.6 \text{ ml}}{150 \text{ ml}}$	$= \dfrac{47.3 \text{ ml}}{160 \text{ ml}}$
$= 37\%.$	$= 30\%.$

Ejection fractions below 60% are abnormal. They occur in heart failure. In severe cases of failure, they may be as low as 10%.

K. Coronary arteriosclerosis.

L. Hypoxia and various metabolites (CO_2, H^+, etc.) do not require nerves to produce coronary dilation.

M. The pulse pressure changed from 40 to 30 mm Hg (probably as a result of decreased stroke volume).

N. The therapy decreased the systemic resistance:

Systemic resistance $= \dfrac{\text{mean arterial pressure} - \text{venous pressure}}{\text{cardiac output}}$

Before	After
$= \dfrac{84 \text{ mm Hg} \times 60 \text{ sec/min}}{4167 \text{ ml/min}}$	$= \dfrac{90 \text{ mm Hg} \times 60 \text{ sec/min}}{4727 \text{ ml/min}}$
$= 1.21$ PRU.	$= 1.14$ PRU.

Summary statement: The therapy increased the heart's left ventricular end-diastolic volume (an index of preload), mean ejection pressure (an index of afterload), and heart rate. In these ways, it placed an increased *metabolic load* on an already failing left ventricle. The therapy should be discontinued immediately.

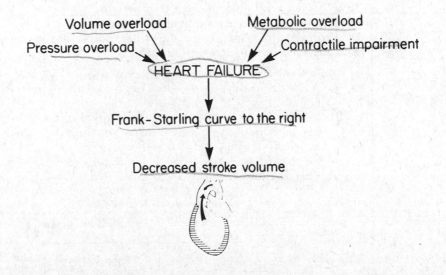

47

(1:588; 2:3-130; 3:515; 4:309; 5:998)
Ha 2

In many cases of left ventricular failure, although the Frank-Starling curve is shifted to the right, the stroke volume, tissue perfusion, and aortic pressure are normal. Apparently, by mobilizing its *reserves*, the cardiovascular system is sometimes able to compensate for the failing heart. In the four blank spaces of the following diagram explain how the system brings about this *compensation*. Item (1) is a physical change in the heart that can develop in one cardiac cycle, whereas item (3) is a physical change in the heart that requires a number of days to develop. Items (2) and (4) result from the production and release of a neurotransmitter substance and hormones.

(Continued)

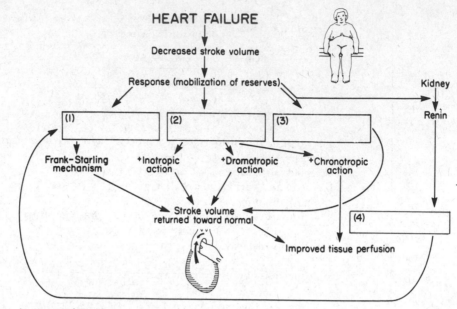

HEART FAILURE

Decreased stroke volume

Response (mobilization of reserves)

Kidney

Renin

(1)   (2)   (3)

Frank–Starling   +Inotropic   +Dromotropic   +Chronotropic
mechanism   action   action   action

Stroke volume
returned toward normal

(4)

Improved tissue perfusion

Answer and Explanation

1. Increased end-diastolic volume
2. Increased catecholamines (released by postganglionic adrenergic sympathetic neurons and adrenal medulla)
3. Cardiac hypertrophy
4. Increased blood volume

48

(1:588; 2:3-242; 3:515; 4:309; 5:998)
Ha 2

Which of the following is characteristic of left ventricular failure?

A. inadequate tissue perfusion
B. a low left ventricular end-diastolic pressure
C. a high left ventricular mean ejection pressure
D. a reduced ability to increase the cardiac output during stress testing
E. a high left ventricular ejection fraction

Answer and Explanation

C. A high ventricular pressure load can cause failure but is not found in many forms of failure.

D. In failure, many of the *cardiovascular reserves* have been used in order to produce a fairly normal resting state and are therefore not available when an additional stress such as exercise is superimposed on the system. For example, an individual in ventricular failure may be unable to walk up two flights of stairs without stopping.

E. Characteristic of failure is a ventricular ejection fraction below 60%.

Heart	Tissue perfusion	Ventricular end-diastolic pressure	Reserve*
Normal	Good	<12 mm. Hg	High
Compensated failure	Adequate	High normal	Low
Decompensated failure	Inadequate	High	Low

*A resting subject should be able to increase his cardiac output 3 to 4 fold.

Questions 49 through 52 consist of two main parts: a statement and a reason for the statement. Choose the best answer for each of these questions from the options A through E below and mark your choice in the space provided.

A. both the statement and the reason are true and are related as cause and effect (i.e., as indicated)
B. both the statement and the reason are true but are not related as cause and effect
C. the statement is true but the reason is false
D. the statement is false but the reason is an accepted fact or principle
E. both the statement and the reason are false

49

(1:619; 2:3-233; 3:503; 4:1040; 5:1978)
HCGf 2

Statement
_____ During birth, the pressure in the aorta increases and the pressure in the pulmonary artery decreases because

Reason
the resistance in the systemic system decreases and the resistance in the pulmonary system increases.

Answer and Explanation
C. During birth, the constriction of the umbilical vessels increases systemic resistance, and the expansion of the lungs decreases pulmonary resistance.

50

(1:478; 2:3-190; 3:464; 4:325; 5:987)
Hfa 2

Statement
_____ In aortic stenosis, during ventricular ejection the pressure in the left ventricle is much higher than the pressure in the aorta because

Reason
in aortic stenosis the resistance to flow at the aortic orifice is abnormally high during ventricular ejection.

Answer and Explanation
A.

51

(1:577; 2:3-89; 3:457; 4:325; 5:1018)
Hfa 2

Statement
_____ Aortic stenosis causes an increased left ventricular afterload because

Reason
in aortic stenosis the pressure in the thoracic aorta increases.

Answer and Explanation
C. It is the increased pressure in the ventricle that is responsible for the increased ventricular load. The pressure in the aorta is either reduced or near normal.

52

(1:510; 2:3-188; 3:469; 4:221; 5:898)
HCF 2

Statement
_____ In an areflexic patient, epinephrine causes an increased aortic pulse pressure because

Reason
epinephrine causes an increased left ventricular stroke volume.

Answer and Explanation
A. Epinephrine has a direct positive inotropic action on the atria and ventricles.

8
Smooth Muscle

1

(1:397; 2:9-54; 3:58; 4:141; 5:141)
Me 2

Which of the following characteristics apply to both skeletal muscle and all smooth muscle?

A. depolarize in response to an increased Na^+ conductance
B. produce action potentials that contain an overshoot
C. are depolarized by acetylcholine
D. respond to stimulation by producing a spike potential
E. all of the above

Answer and Explanation

Ⓐ.
B. Some smooth muscles have action potentials that begin at −50 mv and go to only −10 mv. The overshoot is that part of an action potential that goes to a transmembrane potential of greater than +1 mv.
C. Systemic arterioles are hyperpolarized by acetylcholine.
D. Some produce a plateau potential.

2

(1:397; 2:9-54; 3:58; 4:141; 5:123)
Men 2

Which of the following characteristics are shared by all smooth muscle?

A. it is controlled by pacemaker cells outside the central nervous system
B. it exhibits intercellular conduction
C. it is controlled by sympathetic neurons
D. it contracts in response to acetylcholine
E. none of the above

Answer and Explanation
A. The intrinsic muscles of the eye (*ciliary and iris muscles*), as well as many other smooth muscles, are controlled by impulses originating from the central nervous system.
B. The multiunit smooth muscles (intrinsic muscles of the eye, for example) do not exhibit intercellular conduction.
C. The circular muscle of the iris does not receive a sympathetic innervation.
D. The smooth muscle of arterioles relaxes in response to acetylcholine.
Ⓔ.

3

*(1:397; 2:9-54;
3:58; 4:141; 5:137)*
Me 2

Smooth muscle

A. is not found in the deltoid muscle
B. does not exhibit a resting transmembrane potential
C. has a shorter twitch duration than skeletal muscle
D. has no well-defined motor end plate
E. lacks pacemaker cells

*[handwritten: twitch — one spike
tetanus — 7 spikes]*

Answer and Explanation
A. The arteries and veins of all skeletal muscle contain smooth muscle.
B. The resting transmembrane potential for smooth muscle averages about −50 mv.
C. Smooth muscle has a twitch duration about ten times that of skeletal muscle. This is due in part to the poorly developed sarcoplasmic reticulum in smooth muscle and therefore to an inability to sequester Ca^{++} rapidly.
(D.) The transmitter substance released by a branch of a postganglionic neuron is not restricted to a single cell.
E. Some smooth muscle contains pacemaker cells.

4

*(1:400; 2:8-63;
3:59; 4:141; 5:135)*
Mes 1

Check each of the following statements that applies to multiunit smooth muscle:

1. the muscle fibers of the uterus are an example
2. the muscle fibers of the iris are an example
3. it generally has intercellular conduction
4. it is characteristically not controlled by intrinsic pacemakers

Which one of the following best characterizes your conclusions?

A. statements 1 and 3 are correct
B. statements 1, 3, and 4 are correct
C. statements 2, 3, and 4 are correct
D. statements 2 and 4 are correct
E. statements 1, 2, 3, and 4 are correct

Answer and Explanation
A, B, C. The uterus is an example of visceral (single unit) smooth muscle. Multiunit smooth muscle, unlike the smooth muscle of the uterus, characteristically does not have intercellular conduction, hence the name "multiunit."
(D.)

5

*(1:318; 2:899;
3:176; 4:715; 5:899)*
MD 2

Check each of the following statements that applies to the small intestine:

1. stimulation of parasympathetic neurons increases the frequency of peristaltic waves in the small intestine
2. decentralization of the small intestine (i.e., cutting its sympathetic, parasympathetic, and sensory neurons) causes a disappearance of segmentation
3. decentralization of the small intestine causes a disappearance of peristalsis

Which one of the following best summarizes your conclusions?

A. statement 1 is correct
B. statements 1 and 2 are correct
C. statement 2 is correct
D. statements 1, 2, and 3 are correct
E. statements 2 and 3 are correct

(Continued)

Answer and Explanation

(A.) Parasympathetic stimulation has a positive chronotropic action on the intestine and a negative chronotropic action on the heart.

B. Segmentation does not require neurons.

C. Peristalsis does not require nervous connections with the central nervous system. It does, unlike segmentation, require the intrinsic nerve plexi of the small intestine. These plexi continue to function after decentralization and differ from the neurons innervating skeletal muscle fibers in this regard.

6

(1:891; 2:5-98; 3:589; 4:473; 5:912)
MKNa 2

Transection of the thoracic *spinal cord* causes (check each true statement):

1. a loss of volitional control over the urinary bladder
2. a bladder that is always full and constantly dribbling urine out through the urethra
3. a bladder that responds to distention by a reflex emptying that is as complete as that found under normal conditions
4. a bladder that responds to distention by a reflex emptying that is not as complete as that found under normal conditions

Which one of the following best summarizes your conclusions?

A. statements 1 and 2 are true
B. statements 1, 2 (an early response), and 3 (a later response) are true
C. statements 1, 2 (an early response), and 4 (a later response) are true
D. statement 3 is true
E. statement 4 is true

Answer and Explanation

A. This answer is incomplete.

B. Under normal conditions, there is from 0.09 to 2.3 ml of urine left in the bladder at the end of *micturition*. Unfortunately, this residual volume is much greater in the paraplegic. This creates an important clinical problem. Since urine supports the growth of microorganisms, bladder infection is common among *paraplegic patients*.

(C.) Some paraplegic patients are able to initiate contraction of the urinary bladder by pinching their thighs or scratching the skin in their genital area, but volitional control over micturition, in the usual sense of the term, is lost.

7

(1:318; 2:9-60; 3:176; 4:715; 5:899)
MNRDs 2

Curare

Atropine is a drug that prevents *acetylcholine* from acting on smooth muscle, glands, and the heart. Check each of the following effects that might be produced by atropine:

1. bronchial constriction
2. diarrhea
3. difficulty in reading a book
4. inability to rotate the eyeball in its socket
5. a dry mouth
6. dilation of the pupils
7. apnea
8. arteriolar vasodilation

Answer and Explanation

1. Atropine, by blocking postganglionic parasympathetic neurons, causes bronchial dilation.
2. There is constipation.
3. Postganglionic parasympathetic neurons stimulate the ciliary muscle of the eye to contract. This contraction causes an increase in the convexity of the lens and, in this way, brings an image from a near object into focus on the retina.

4. Movement of the eyeball in its socket is due to the contraction of the extrinsic muscles of the eye. These skeletal muscles remain sensitive to the acetylcholine released by their motor neurons after atropine administration.

⑤ ⑥ 7. The skeletal muscle fibers of the diaphragm remain sensitive to the acetylcholine released by their motor neurons.

8. Acetylcholine causes arteriolar vasodilation.

8

(1:377; 2:1-78; 3:60; 4:133; 5:1022)
MHCq 2

Smooth muscle, cardiac muscle, and skeletal muscle are said to be viscoelastic. Which one of the following definitions best characterizes the properties of an *elastic material*?

A. exerts tension in response to distention
B. exerts tension in response to distention that is, within limits, proportional to the speed of distention
C. exerts tension in response to distention that is, within limits, proportional to the distance distended
D. exerts tension in response to distention that is, within limits, proportional to the speed of and the degree of distention
E. exerts a large quantity of tension in response to a small degree of distention.

Answer and Explanation
A. Both viscous and elastic materials exert tension during distention.
B. The tension exerted by a perfectly elastic substance is independent of the speed of distention.
C, D, E. The quantity of tension (dynes/cm) that occurs in response to distention (cm) is expressed by the *modulus of elasticity*, which is related to the slope of the tension-distention curve.

9

(1:506; 2:1-78; 3:60; 4:133; 5:1022)
MHCq 2

Which one of the definitions listed in question 8 best characterizes a *viscous or plastic material*?

Answer and Explanation
B. Viscous materials, like chewing gum, exert a tension when distended that is independent of the length distended.

10

(1:506; 2:1-78; 3:60; 4:133; 5:1022)
MHCq 2

Which of the statements in question 8 explains why muscle is characterized as *viscoelastic*?

Answer and Explanation
D. It has been well established by electron-microscopic studies that when striated muscle is distended the thick and thin filaments slide over one another. It is this internal resistance (i.e., friction) in striated muscle, and possibly a similar event in smooth muscle, that is responsible for muscle's viscous characteristics. Viscosity (i.e., internal resistance) is also an important factor in fluid dynamics (air flow and blood flow, for example). Here, however, we are concerned with a relationship between pressure and flow rather than tension and linear displacement. This relationship for water and other Newtonian fluids can be expressed as follows:

$$\text{Viscosity} = \frac{\text{pressure head} \times (\text{radius of the tube})^4 \times \text{constant}}{\text{flow} \times \text{length of the tube}}.$$

These principles are important for understanding such smooth muscle-containing structures as bladders, blood vessels, ducts, tubes, and tracts. Some of the elements that

(Continued)

contribute to the viscoelastic characteristics of muscle include the sarcolemma, cell membrane, collagen and elastin fibers, and actomyosin. These are further classified as being either *series* viscoelastic components or *parallel* viscoelastic components. The sarcolemma of skeletal muscle, for example, serves an important parallel viscoelastic function, as do the pericardium of the heart and the elastic and collagen fibers of arteries. Actomyosin, on the other hand, is not only an important contractile component of muscle but also an important contributor to the series viscoelastic characteristics of muscle. During rigor, it either prevents distention or ruptures. During relaxation and contraction it resists distention.

11

(1:373; 3:27; 4:145; 5:129)
Mci 1

Calmodulin is an acidic peptide which

A. activates an enzyme that causes the contraction of smooth muscle
B. activates an enzyme that causes the contraction of skeletal muscle
C. activates an enzyme that causes the relaxation of smooth muscle
D. activates an enzyme that causes the relaxation of skeletal muscle
E. causes the relaxation of either smooth or skeletal muscle

Answer and Explanation

A. Ca^{++} binding to troponin C in skeletal and cardiac muscle is part of the activation-contraction coupling process. Calmodulin is structurally similar to *troponin C*, and in smooth muscle serves an *activation-contraction coupling* function (see question 22 in Chapter 5). Other questions on smooth muscle include question 10 in Chapter 3, questions 2 through 22 in Chapter 4, and question 12 in Chapter 5.

9
Arteries, Arteriovenous Anastomoses, and Veins

1

(1:823; 2:5-3; 3:13; 4:391; 5:1118)
Cmd 2

Eleven hundred millimicrocuries of tritiated water are injected intravenously into a 50-kg man. If you assume that there is complete mixing and distribution throughout the body and that 200 millimicrocuries of the tritiated water were excreted prior to sampling, what approximate plasma *concentration* of tritiated water would you expect to find?

A. less than 15 millimicrocuries/liter
B. 30 millimicrocuries/liter
C. 45 millimicrocuries/liter
D. 60 millimicrocuries/liter
E. more than 80 millimicrocuries/liter

Answer and Explanation

B. The average amount of water in the human body is 63% of the body weight. For the approximations required in this problem, you can use values anywhere from 55% to 70% and still select the correct answer:

$$\text{Concentration in body liquids} = \frac{\text{quantity injected} - \text{quantity lost}}{\text{quantity of body water}}$$

$$= \frac{(1100 - 200) \text{ millimicrocuries}}{(0.63 \times 50) \text{ liters}}$$

$$= 29 \text{ millimicrocuries/liter.}$$

2

(1:823; 2:5-3; 3:13; 4:391; 5:1118)
Cbmd 2

Nine hundred millimicrocuries of albumin labeled with radioactive iodine are injected intravenously into a lean, 70-kg man. If there is complete mixing in the plasma and no excretion of the albumin, what approximate *plasma concentration* of the labeled albumin do you expect to find 10 minutes after injection?

A. less than 29 millimicrocuries/liter
B. 60 millimicrocuries/liter
C. 130 millimicrocuries/liter
D. 260 millimicrocuries/liter
E. 500 millimicrocuries/liter

(Continued)

Answer and Explanation

D. For the approximations needed in this problem, you can assume that the plasma represents 5% of the body weight and has a density of 1 g/ml. You can also assume that, at the time of sampling (i.e., 10 minutes after injection), all of the labeled albumin is still in the blood, since the capillaries are relatively impermeable to it. It has been estimated that less than 5% of the radioactive albumin will be lost from the blood after the first hour, less than 60% after the first day.

Plasma volume = 0.05 × 70 kg × 1 liter/kg = 3.5 liters.

$$\text{Plasma concentration} = \frac{\text{quantity injected}}{\text{plasma volume}}$$

$$= \frac{900 \text{ millimicrocuries}}{3.5 \text{ liters}}$$

$$= 260 \text{ millimicrocuries/liter.}$$

3

(1:825; 2:5-5; 3:14; 4:395; 5:1119)
Cbm 2

A 60-kg patient has a *hematocrit* reading of 40 and a plasma volume of 3 liters. What is his *total blood volume*?

A. 4.0 liters
B. 5.0 liters
C. 6.0 liters
D. 7.0 liters
E. greater than 7.5 liters

Answer and Explanation

B. The hematocrit is the concentration of red blood cells in the plasma. In this example, the blood is stated to be 40% erythrocytes by volume, and if one assumes that this is an accurate count, then the total blood volume would be:

$$\text{Blood volume} = \frac{100}{100 - \text{hematocrit}} \times \text{plasma volume}$$

$$= 1.67 \times 3 \text{ liters} = 5.0 \text{ liters.}$$

In the above calculations the assumption is made that the hematocrit is precisely determined. In most hospitals the stated hematocrit is approximately 5% too high. This is because they make no allowance for the plasma trapped between the erythrocytes. In addition, it has been suggested that the hematocrit in large vessels is higher than that in small vessels. This may result in an additional sampling error that brings the total possible error to 13%. This would mean that the estimated true body hematocrit is 34.8% and the blood volume is 4.6 liters.

4

(1:478; 2:3-14; 3:466; 4:2:13; 5:1018)
Cfpo 1

The French physician Jean L. M. *Poiseuille* (1799–1869) defined the *factors that regulate the flow of water* through a single, rigid cylindrical tube. List these factors, indicate the units used to measure each, and indicate which are directly (+) and which are inversely (–) related to flow.

Answer and Explanation
Let the flow in a tube from point 1 to point 2 be equal to I, which will be measured in units of *ml/sec*. Then,

$$I = (P_1 - P_2) \times \frac{(\pi r^4)}{(8L\eta)},$$

where:
- $P_1 - P_2$ is the *pressure head* in *dynes/cm²* (1 mm Hg = 1330 dynes/cm²)
- r is the *radius* of the cylinder in *cm*
- L is the *length* of the cylinder in *cm*
- η is the viscosity of the liquid in poises (water has a viscosity of about 10^{-2} poises)
- π (= 22/7) and 8 are constants

5

(1:478; 2:3-14; 3:466; 4:208; 5:1018)
Cfpo 2

Check each of the following that is a proper derivation of the *Poiseuille formula* presented in question 4:

1. hindrance = $(8L\eta)/(\pi r^4)$
2. $P_1 - P_2$ = hindrance $\times$ I
3. resistance = $(P_1 - P_2)/I$

Which one of the following best summarizes your conclusions?

A. statement 1 is correct
B. statement 2 is correct
C. statement 3 is correct
D. statements 2 and 3 are correct
E. all of the above statements are correct

Answer and Explanation
A. This is the formula for *resistance*. *Hindrance* = $(8L)/(\pi r^4)$.
B. $P_1 - P_2$ = resistance $\times$ I.
(C.) Resistance = $(P_1 - P_2)/I = (8L\eta)/(\pi r^4)$.

6

(1:475; 2:3-16; 3:466; 4:208; 5:1018)
Copf 2

Check each of the following that is true:

1. the formula, resistance = $(P_1 - P_2)/I$, is used to determine accurately resistance in the cardiovascular system
2. in the above formula, $(P_1 - P_2)$ = pulse pressure
3. the formula, resistance = $(8L\eta)/(\pi r^4)$, is used to determine accurately resistance in the cardiovascular system
4. blood is a Newtonian fluid

Which one of the following best summarizes your conclusions?

A. statement 1 is true
B. statements 1 and 2 are true
C. statements 1 and 3 are true
D. statements 1, 3, and 4 are true
E. statement 1 is false

Answer and Explanation
(A.)
B. Pulse pressure is systolic minus diastolic arterial pressure. $P_1 - P_2$ represents the difference in mean pressures between two points.
D. The formula for *resistance* in statement 3 is based on a number of assumptions that do not apply to the cardiovascular system. Some of these assumptions are that the system under study consists of (1) nondistensible vessels and (2) cylindrical vessels and contains (3) a Newtonian fluid. Blood vessels are usually distensible and are conical in shape. Blood, unlike water, has different *viscous characteristics* in different vessels. In other words, it does not obey the laws Sir Isaac Newton established for water.

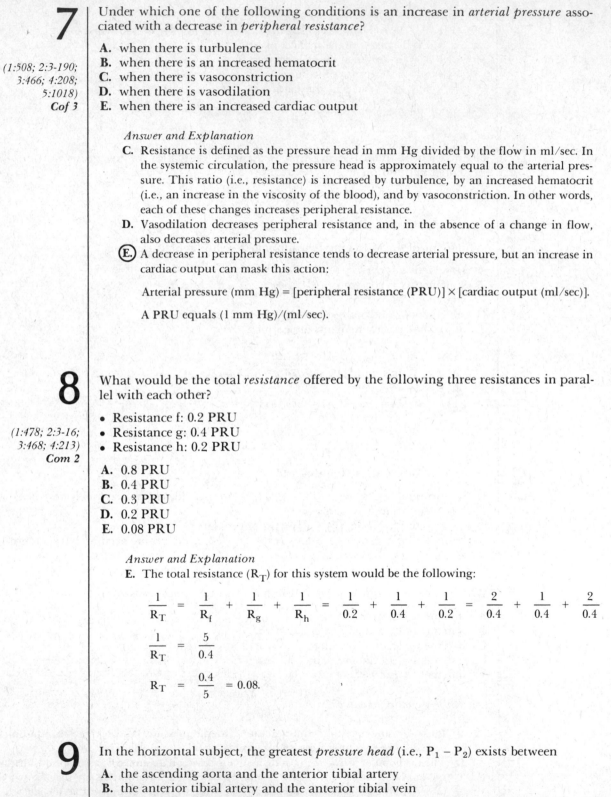

7

(1:508; 2:3-190;
3:466; 4:208;
5:1018)
Cof 3

Under which one of the following conditions is an increase in *arterial pressure* associated with a decrease in *peripheral resistance*?

A. when there is turbulence
B. when there is an increased hematocrit
C. when there is vasoconstriction
D. when there is vasodilation
E. when there is an increased cardiac output

Answer and Explanation

C. Resistance is defined as the pressure head in mm Hg divided by the flow in ml/sec. In the systemic circulation, the pressure head is approximately equal to the arterial pressure. This ratio (i.e., resistance) is increased by turbulence, by an increased hematocrit (i.e., an increase in the viscosity of the blood), and by vasoconstriction. In other words, each of these changes increases peripheral resistance.

D. Vasodilation decreases peripheral resistance and, in the absence of a change in flow, also decreases arterial pressure.

E. A decrease in peripheral resistance tends to decrease arterial pressure, but an increase in cardiac output can mask this action:

Arterial pressure (mm Hg) = [peripheral resistance (PRU)] × [cardiac output (ml/sec)].

A PRU equals (1 mm Hg)/(ml/sec).

8

(1:478; 2:3-16;
3:468; 4:213)
Com 2

What would be the total *resistance* offered by the following three resistances in parallel with each other?

• Resistance f: 0.2 PRU
• Resistance g: 0.4 PRU
• Resistance h: 0.2 PRU

A. 0.8 PRU
B. 0.4 PRU
C. 0.3 PRU
D. 0.2 PRU
E. 0.08 PRU

Answer and Explanation

E. The total resistance (R_T) for this system would be the following:

$$\frac{1}{R_T} = \frac{1}{R_f} + \frac{1}{R_g} + \frac{1}{R_h} = \frac{1}{0.2} + \frac{1}{0.4} + \frac{1}{0.2} = \frac{2}{0.4} + \frac{1}{0.4} + \frac{2}{0.4}$$

$$\frac{1}{R_T} = \frac{5}{0.4}$$

$$R_T = \frac{0.4}{5} = 0.08.$$

9

(1:497; 2:3-161;
3:469; 4:220; 5:954)
Cpo 2

In the horizontal subject, the greatest *pressure head* (i.e., $P_1 - P_2$) exists between

A. the ascending aorta and the anterior tibial artery
B. the anterior tibial artery and the anterior tibial vein
C. the anterior tibial vein and the right atrium
D. the pulmonary artery and the pulmonary vein
E. the efferent renal arteriole and the renal vein

Answer and Explanation

A. The vessels in this channel, though long, have a sufficiently great radius to keep this a low-resistance channel. As a result, the pressure lost in transit is low (i.e., about 5 mm Hg).

B. The small-diameter lumina of this channel (arterioles and capillaries: 7.5 μm) make it a high-resistance system with a pressure loss of about 80 mm Hg.

C. $P_1 - P_2 = 5$ to 10 mm Hg.

D. $P_1 - P_2 = 15$ mm Hg.

E. $P_1 - P_2 = 10$ mm Hg.

10

(1:590; 2:3-200; 3:470; 4:227; 5:1039)
Cpa 2

A 70-kg, 6-ft, normal, healthy subject standing quietly erect for 30 seconds has a mean arterial pressure of 100 mm Hg in the ascending aorta and a venous pressure of 2 mm Hg in the superior portion of the inferior vena cava. What is the *pressure* in the veins of the dorsum of the foot?

A. less than 0 mm Hg (i.e., below atmospheric pressure)

B. 2 mm Hg

C. 4 mm Hg

D. about 20 mm Hg

E. above 40 mm Hg

Answer and Explanation

E. Under these circumstances, the pressure in the veins of the foot (P_F) is about 91 mm Hg. It will equal the pressure in the vena cava at heart level (P_H) plus the change in pressure due to the resistance between the veins of the foot and the veins leading into the heart (P_R) plus the pressure due to the weight of the blood (P_W). If the distance between the veins of the foot and the heart is 1200 mm, P_W equals approximately 88 mm Hg [= (1200 mm H_2O)/(13.6 mm H_2O/mm Hg)]. Therefore $P_F = P_H + P_R + P_W = 1 + 2 + 88 = 91$. The effect of *gravity* on blood pressure is one of the reasons that edema and overly distended veins (*varicose veins*) are more common in the lower extremities than the upper extremities.

11

(1:593; 2:3-199; 3:464; 4:227; 5:1039)
Cpof 2

A soldier *stands* at attention for 30 seconds. How can his femoral vein pressure best be decreased?

A. by decreasing the heart rate

B. by dilating the systemic arterioles

C. by constricting the systemic arterioles

D. by having him hold his breath

E. by having him take one step forward

Answer and Explanation

A. Most changes in heart rate have little or no effect on central venous pressure.

C. Changes in the caliber of the arterioles will affect how much blood leaves the arterial system and therefore how much enters the venous system. Arteriolar constriction, for example, by increasing arterial blood volume, produces important increases in arterial blood pressure but produces little or no change in venous blood pressure. This apparent discrepancy occurs because the veins are much more *compliant* than the arteries.

E. When a subject contracts his skeletal muscles, they tend to compress veins and therefore push blood toward the heart. The *valves in the veins of the extremities* prevent the blood from flowing away from the heart. In other words, skeletal muscle contraction and relaxation tend to increase the *venous return* of blood to the heart and decrease the venous pressure. Taking one step forward can decrease the pressure in the veins of the dorsum of the foot by 40 mm Hg.

12

(1:585; 2:3-195;
3:475; 4:225;
5:1039)
CMfp 1

Venous return of blood to the right heart is normally increased by

A. increased minute ventilation
B. increased venous tone
C. increased cardiac sympathetic tone
D. all of the above
E. none of the above

Answer and Explanation

D. Venous return is directly related to the amount of push and pull between the great veins, the right atrium, and the right ventricle (i.e., (1) the pumping action of the muscles, (2) the compliance of the veins, (3) the pumping action of the heart). When an increase in venous return precedes an increase in cardiac output, the end-diastolic volume of the heart and/or the pulmonary blood volume will increase. Since the circulation is a closed, compliant system, venous return (ml of blood flowing into the heart per minute) can exceed or be less than cardiac output (blood flowing out of the heart per minute) for only short periods (usually less than 1 minute).

13

(1:387; 2:3-191;
3:314; 4:221; 5:954)
Cq 2

Check each true statement concerning the collagen fibers in a healthy, 25-year-old man. *Collagen fibers* are

A. the intercellular structures most important in preventing the rupture of an artery
B. the intercellular structures most important in determining the distensibility of an artery
C. not intercellular structures
D. not found in veins

Answer and Explanation

(A.)

B. In a young, healthy subject the distensibility of an artery or vein is determined by the *elastic fibers and smooth muscles* of these vessels. The collagen, under most circumstances, remains slack and does not contribute to distensibility. In cases where an artery or vein is markedly distended, the slack in the collagen is taken up, and it prevents further distention. It performs this function by virtue of its low distensibility and great strength. In this respect, it is similar to the pericardium of the heart. It, of course, differs from the pericardium in that it is a part of the organ it protects from overdistention (i.e., the artery or vein), rather than the part surrounding the organ (i.e., the heart). Examples of conditions where the collagen fibers have failed to protect from overdistention include saccular and fusiform aneurysms.

14

(1:519; 2:3-94;
3:466; 4:210;
5:1023)
Cpfba 2

A *saccular aneurysm* in the abdominal aorta is more likely to *rupture* than that part of the aorta that carries blood to it because

A. more tension is exerted on the wall of aneurysm
B. there is less pressure in the aneurysm
C. there is less turbulence in the aneurysm
D. there is more tension exerted on and less turbulence in the aneurysm
E. there is more pressure and less turbulence in the aneurysm

Answer and Explanation

(A.) If there were no flow in the system, the pressures would be the same in the lumen of the aneurysm and the afferent vessel. Under these circumstances, according to the *law of Laplace*, the tension will be greater on the wall of the aneurysm because the radius of the aneurysm is greater than the radius of the afferent vessel:

Tension (dynes/cm) = [pressure (dynes/cm^2)] × [radius (cm)].

Since there is flow, the pressure will be greater in the aneurysm than in the afferent vessel (see B below).

B. In a closed system, when the velocity of flow decreases (i.e., in the aneurysm), the lateral pressure increases. This inverse relationship between velocity of flow and lateral pressure is called *Bernoulli's principle*.

C. There is more turbulence in the aneurysm. This can cause a further weakening of the wall and a murmur that can frequently be heard through a stethoscope.

15

(1:511; 2:6-38; 3:468; 4:221; 5:1687)
Cqpm 2

A 10-cm strip of artery with no branches was isolated and clamped centrally and peripherally. The pressure in the strip was 90 mm Hg before and 120 mm Hg after the injection of 3 ml of blood. What was the *compliance* of this strip of artery?

Answer and Explanation

$$\text{Compliance} = \frac{\text{change in volume}}{\text{change in pressure}} = \frac{3 \text{ ml}}{30 \text{ mm Hg}} = 0.1 \text{ ml/mm Hg.}$$

16

(1:511; 2:3-188; 3:454; 4:221; 5:1036)
CGqp 2

Check each of the following processes that usually occurs in *aging* (i.e., between the ages of 45 and 95):

1. an increase in distensibility of the arteries
2. replacement of collagen fibers by elastic fibers
3. an increase in concentration of calcium in the arteries
4. the valves of the veins become incompetent
5. an increase in arterial pulse pressure

Answer and Explanation

1. They become less distensible.
2. The opposite is true.
3. At age 20, about 0.4% of the aorta is calcium, and by age 80 the calcium is in excess of 6%.
4. By age 70, about 70% of the valves of the veins have atrophied.
5. In aging there is a loss of distensibility in the arteries that causes an increase in pulse pressure. In men at age 40, the average arterial pressure is 126/84 (pulse pressure: 42), and at age 80, the arterial pressure is about 147/84 (pulse pressure: 63).

17

(1:510; 2:3-16; 3:469; 4:222; 5:1022)
HCpa 3

The following records were obtained from a patient with a *decentralized heart* (i.e., an individual who has received a heart transplant). In each of these records the aortic pressure was recorded at a constant paper speed and amplification.

(Continued)

Identify in questions 1 through 5 in which record (A through E) each of the indicated changes occurred. The arrow points to where the change began.

1. _____ an increased left ventricular stroke volume
2. _____ an increased heart rate
3. _____ an increased total systemic resistance
4. _____ aortic stenosis
5. _____ a decreased arterial compliance

Answer and Explanation

1. **D.** The *mean aortic pressure* is directly related to the volume of blood in the systemic arteries. The aortic *pulse pressure* is directly related to the change in aortic volume per cardiac cycle. The aortic *diastolic pressure* is directly related to the volume of blood left in the arteries at the end of diastole. A rise in the *stroke volume* increases all three parameters.

2. **C.** Increases in *heart rate* and stroke volume by increasing *cardiac output* increase *mean* aortic pressure and aortic *diastolic pressure*. Increases in heart rate by increasing the heart's *afterload* and by decreasing the *duration of ventricular filling* during each cardiac cycle decrease the stroke volume and aortic pulse pressure.

3. **A.** *Arteriolar constriction* will impede flow out of the arteries and therefore increase *arterial* blood volume and aortic diastolic and mean pressures. The decreased pulse pressure is due to the increase in the heart's afterload.

4. **B.** Records A and B both represent increases in *resistance*. In record B this increase causes a decreased aortic diastolic, pulse, and mean pressure, because the *stenosis* is an increased resistance into, rather than out of, the arterial system.

5. **E.** A decreased arterial *compliance* in an areflexic individual represents a decrease in the *dampening action* of the artery on the ventricular pressure signal, hence the increased systolic pressure. The arteries also lose their ability to *store pressure* during diastole, hence the decrease in diastolic pressure. If the arteries had 0 compliance the diastolic pressure would be 0. In other words, the distensibility of the arteries is responsible for much of the energy output of the ventricle being converted to *potential energy*, which is then changed back to *kinetic energy* during ventricular diastole. In *arteriosclerosis*, you also have a decrease in arterial compliance and an increase in aortic pulse pressure, but since other factors are operating you usually do not see the decreased diastolic pressure.

18

(1:510; 2:3-16;
3:469; 4:222;
5:1022)
HCpa 2

Aortic insufficiency

A. increases aortic pulse and diastolic pressure
B. increases pulse pressure and decreases diastolic pressure
C. decreases aortic pulse and diastolic pressure
D. is similar to vasoconstriction in that it decreases pulse pressure and increases diastolic pressure
E. is similar to vasodilation in that it decreases pulse pressure and increases diastolic pressure

Answer and Explanation

B. In *aortic insufficiency*, the loss of blood from the arterial system during each diastole is abnormally high and therefore the aortic diastolic pressure is low and the pulse pressure (= systolic pressure − diastolic pressure) is high. This increased *diastolic run-off* of pressure during each cycle is also seen when the heart rate, total systemic resistance, or arterial compliance is decreased.

19

(1:510; 2:3-16;
3:469; 4:222;
5:1022)
Hpa 2

An increase in the rate of blood flow from the arterial system causes a more negative slope of the *catacrotic limb* of the aortic pressure curve. This more negative slope may be caused by

A. a decreased heart rate
B. a decreased stroke volume
C. both of the above
D. aortic insufficiency
E. aortic stenosis

Answer and Explanation
A. A decreased heart rate causes an increased volume loss per cycle, but not an increased rate of loss.
B. A decreased stroke volume decreases the perfusion pressure across the capillaries and therefore decreases the rate of diastolic run-off of blood and pressure.
(D.) In AI the rate of diastolic run-off is increased because in AI the arteries are losing blood to both the capillaries and the relaxing ventricle during diastole.

20

(1:497; 2:3-193;
3:454; 4:223;
5:1035)
CpHe 3

The following pressure pattern was recorded from the femoral artery of a patient:

How would you characterize the above pattern?

A. a bigeminy containing an extra systole
B. a bigeminy containing a premature systole
C. a trigeminy containing an extra systole
D. a trigeminy containing a premature systole
E. a normal sinus rhythm

Answer and Explanation
E. On the basis of the ECG alone (bottom tracing), one can conclude that this is a sinus rhythm. The pressure pattern is different from what one usually observes in the thoracic aorta but is typical of that found further along the arterial path. As the pressure wave passes down the arterial system, its character is changed by the elastic characteristic of the arterial system. In the horizontal subject, the mean pressure will decrease minutely, the systolic pressure will rise, the diastolic pressure will fall, and the dicrotic wave will separate from the rest of the catacrotic limb of the pressure curve.

21

(1:473; 2:3-158;
3:466; 4:219; 5:953)
Cqf 2

Check each of the following statements about the *systemic circulation* that is true:

1. the velocity of flow in the large arteries is from 100 to 500 times faster than in the capillaries
2. the blood volume in the arteries is approximately equal to that in the veins
3. the velocity of flow in the veins is greater than in the arteries

Which one of the following statements best summarizes your conclusions?

A. none of the above statements is true
B. statement 1 is true

(Continued)

C. statement 2 is true
D. statements 1 and 2 are true
E. statement 3 is true

Answer and Explanation
B. Under resting conditions, the *velocity of flow* in the capillaries is about one four hundredth that in the arteries. This is due to the greater total cross-sectional area of the capillaries (1800 cm² *vs.* 4 cm²). It means that in the healthy subject, the blood will remain in the capillary long enough to come into equilibrium with the perivascular fluid.
C. The *blood volume* in the systemic arteries is about 1000 ml, and that in the systemic veins is 3400 ml.
E. The velocity of flow in the large arteries is about 20 cm/sec and in the large veins is about 14 cm/sec.

22

(1:577; 2:3-156; 3:456; 4:287; 5:1000)
CHf 2

The following *dye-dilution curve* is obtained on a patient. The dye, indocyanine green, is injected rapidly into the cephalic vein. At the same time, blood samples are withdrawn from the subclavian artery at a constant rate and moved through a cuvette-densitometer.

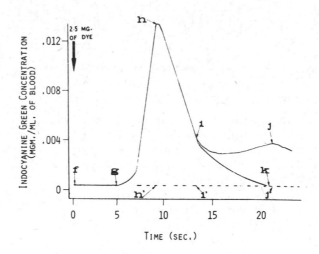

What part of this curve is not an actual recording of dye concentration but instead represents an estimate (i.e., an extrapolation) of the dye-dilution pattern in the absence of *recirculation*?

A. line g-h′
B. curve g-h-i
C. curve h-i-j
D. curve i-j
E. curve i-k

Answer and Explanation
A. Line g-h′ is the baseline of dye concentration (i.e., optical density) prior to injection.
B. Curve g-h-i is the recorded change in dye concentration.
D. Curve i-j is that part of the recorded curve which is distorted by the recirculation of the dye.
E. Curve i-k is the extrapolated part of the curve.

23

*(1:575; 2:3-156;
3:456; 4:287;
5:1000)*
CHf 2

Which interval in the dye-dilution curve in question 22 represents the *appearance time* and which one the *mean recirculation time*?

A. the appearance time is period f-g and the mean recirculation time is period g-h′
B. the appearance time is period f-g and the mean recirculation time is period h-j′
C. the appearance time is period f-g and the mean recirculation time is period i′-j′
D. the appearance time is period g-h′ and the mean recirculation time is period g-i′
E. the appearance time is period g-h′ and the mean recirculation time is period h′-j′

Answer and Explanation

B. The appearance time is the time it takes the tracer (i.e., indocyanine green) to move from the point of injection to the sampling densitometer. The mean recirculation time is the interval between the peak of the first dye curve and the peak of the second (the curve due to recirculation of dye past the sampling point).

24

*(1:575; 2:3-156;
3:456; 4:287;
5:1000)*
CHfm 2

An investigator can use the dye-dilution curve in question 26 to estimate the cardiac output of the patient. In doing this, she determines the average concentration under the dye-diluation curve ($\overline{C}$), the duration of the curve (T), and quantity of dye injected in mg (M). Which one of the following formulae will best approximate the cardiac output in ml of blood per second? In these formulae the subscripts represent the curves being studied (i.e., $\overline{C}_{ghik}$ = average concentration under curve g-h-i-k).

A. Cardiac output = $\dfrac{(M) \times (\overline{C}_{ghijj'})}{T}$.

B. Cardiac output = $\dfrac{(M) \times (\overline{C}_{ghik})}{T}$.

C. Cardiac output = $\dfrac{(M) \times (\overline{C}_{ghii'})}{T}$.

D. Cardiac output = $\dfrac{M}{(\overline{C}_{ghijj'}) \times (T)}$.

E. Cardiac output = $\dfrac{M}{(\overline{C}_{ghik}) \times (T)}$.

Answer and Explanation

E. One uses the average concentration under the extrapolated curve to determine cardiac output. In the example shown, the cardiac output would be:

$$\frac{2.5 \text{ mg}}{(0.0037 \text{ mg/ml of blood}) \times (15.6 \text{ sec})} = 43.3 \text{ ml of blood/sec} = 2600 \text{ ml of blood/min.}$$

25

*(1:623; 2:3-156;
3:456; 4:287;
5:1000)*
CHf 3

How would the *dye-dilution curve* change in *exercise*? There would be

A. an increased mean recirculation time
B. an increased appearance time
C. an increased mean recirculation time and appearance time
D. an increased duration of the total extrapolated curve (i.e., period g-k)
E. none of the above changes

Answer and Explanation

A, B, C. In exercise, the velocity of flow increases. Therefore, the mean recirculation time and appearance times will decrease (i.e., the dye will move more rapidly through the circulation).

(Continued)

D. The duration of the curve will decrease. In other words, in exercise, there is an increased *cardiac output*, and therefore the dye will move more rapidly from its injection site through the systemic arteries (i.e., its sampling site). Note that, in the formula for cardiac output (see question 24E), a decrease in either C (average concentration under the extrapolated curve) or T (duration of the extrapolated curve) can yield a higher value for cardiac output.

Ⓔ

26

(1:473; 2:3-158; 3:466; 4:327; 5:1000)
CHfa 3

How would the dye-dilution curve change in a patient with a femoral *arteriovenous fistula*? There would be

A. a reduced mean recirculation time
B. a reduced appearance time
C. an increased appearance time
D. no need to extrapolate the recorded curve
E. a reduced mean recirculation time and an increased appearance time

Answer and Explanation

Ⓐ The systemic circulation can be characterized as having short (the coronary circulation), intermediate (the circulation through the arm), and long (the circulation through the leg) circuits back to the heart. The femoral a-v fistula represents a *left-to-right shunt* in one of the longer circuits. Some of the dye, by taking this shortcut back to the sampling site, causes a shorter *mean recirculation* time.

C. The appearance time will not change. In this condition, there is little or no alteration in the circulation between the cephalic vein and the sampling site.

D. In this condition, the first and second dye curves will be more fused than ever. Therefore, the need to separate them by extrapolation becomes even more important than in the normal subject.

27

(4:329; 5:100)
CHfa 3

A patient is suspected of having a tetralogy of Fallot that is resulting in a *right-to-left shunt* of the blood. How would the dye-dilution curve change?

Answer and Explanation
There would be a decreased appearance time. In addition, the initial densitometer wave would usually be of smaller amplitude than the second wave.

28

(2:3-171; 3:509; 5:1037)
Cf 1

Which one of the following structures in the resting subject receives the greatest *blood flow* per gram of tissue?

A. brain
B. heart
C. liver
D. gastrocnemius muscle
E. kidney

Answer and Explanation
A. It receives 60 ml of blood per gram of tissue.
B. It receives 70 ml of blood per gram of tissue.
C. It receives 100 ml of blood per gram of tissue.
D. It receives 5 ml of blood per gram of tissue.
Ⓔ It receives 730 ml of blood per gram of tissue.

29

(1:473; 2:3-4;
3:466; 4:221; 5:953)
Cfm 3

The following data are collected from a 22-year-old patient:

Cardiac output:	6 liters/min
Total blood volume:	5.5 liters
Average velocity of flow in systemic arteries:	20 cm/sec
Average velocity of flow in systemic capillaries:	0.05 cm/sec

What is the total *cross-sectional area* for the lumina of the patient's systemic capillaries?

Answer and Explanation

$$\text{Cross-sectional area (cm}^2) = \frac{\text{flow (cm}^3/\text{sec})}{\text{velocity (cm/sec)}} \cdot$$

Flow: If you assume that all the cardiac output passes through the systemic capillaries, then they receive a flow of 6 liters/min (100 cm^3/sec). This is a valid assumption if the arteriovenous anastomoses are closed and there are no abnormal shunts open such as a patent ductus arteriosus. Since these are usually valid assumptions, then the *total cross-sectional area of the lumina of patient's systemic capillaries that are in parallel with one another is*:

$$\frac{100 \; cm^3/sec}{0.05 \; cm/sec} = 2000 \; cm^2.$$

30

(1:605; 2:3-167;
3:501; 4:352;
5:1101)
Cfh 2

Check each statement that is characteristic of *arteriovenous anastomoses* in the skin:

1. they dilate in response to cutaneous cooling
2. when dilated, they cause an appreciable reddening of the skin
3. when dilated, they cause an appreciable increase in the venous O_2 concentration of associated veins
4. they are a means of heat loss

Answer and Explanation

1. They dilate in response to cutaneous warming.
2. These connecting links between arteries and veins, unlike capillaries, are sufficiently thick-walled vessels that one cannot see the red blood in them. One mechanism for warming the skin is an increased nutritive blood flow (i.e., an increased flow in the cutaneous blood capillaries). A second mechanism is an increased nonnutritive blood flow (i.e., an increased flow through the a-v anastomoses).
3. The cutaneous a-v shunts do not exchange nutrients with the perivascular space. They function, as far as we know, solely for heat exchange. They also, when open, tend to arterialize the venous blood, but this is not considered one of their functions.
4. Under most circumstances, the body eliminates heat by shunting additional blood to the skin, which warms the skin. If the skin is warmer than the external environment, it will lose heat to that environment. If the external environment is warmer than the skin, the only mechanism for heat loss is through evaporation.

31

(1:605; 2:3-211;
3:501; 4:352;
5:1101)
Chap 2

Check each of the following statements about cutaneous *arteriovenous anastomoses* that is true:

1. they are usually closed
2. when open, they lower the peripheral resistance
3. they open in response to an increased pressure in the carotid sinus
4. they open in response to a decreased pressure in the carotid sinus
5. they contain smooth muscle

(Continued)

Answer and Explanation

①. They open in response to a warming of the skin or an increased core body temperature.

②.

4. Although this is true of many arterioles, there is no evidence that it is true for a-v anastomoses.

⑤.

32

(1:602; 2:3-218; 5:102)

CfG 2

Ligation of the femoral artery in most subjects causes

A. a cessation of blood flow to the leg
B. a marked, permanent decrease in blood flow through the leg
C. a marked, temporary decrease in blood flow through the leg
D. a decreased P_{CO_2} in the femoral vein
E. a marked, permanent decrease in femoral vein P_{CO_2} and blood flow

·*Answer and Explanation*

A. There are arteries parallel to the femoral artery that carry blood to the leg.
Ⓒ. After ligation, preexisting channels expand, so that after an hour, flow is approximately 70% of normal. After several weeks, flow is probably back to normal. These data are based on experiments on dogs but are probably also true of human beings. The speed with which flow returns after ligation will vary from one vascular bed to another. It will depend upon (1) the number and characteristics of the *collateral channels* and (2) the ability of the body to form new vessels.
D. Ligation causes an increase in femoral vein P_{CO_2}.

33

(1:575; 2:3-156; 3:456; 4:287; 5:1000)

Hpbm 3

Two adult patients had their *cardiac output* determined by the dye-dilution technique. In each patient, a catheter was positioned in the thoracic aorta and attached to a cuvette-densitometer with an automatic withdrawal system for sampling blood from the aorta. The densitometer had been equipped with the appropriate optical filter system and calibrated for use with the dye known as cardio-green. Five milligrams of this dye were rapidly injected into the left cephalic vein of each patient, and blood was sampled from the thoracic aorta via the automatic withdrawal system. The resultant curves relating aortic dye concentration to time are shown in the accompanying diagrams. The solid lines represent the actual output of the densitometer in terms of dye concentration. The dashed lines represent the best estimate of what the curve would have looked like if there had not been any *recirculation effect*. This estimate is based on the assumption that the decrease in dye concentration (curve BC) is exponential. In patient 1 the average dye concentration under the extrapolated curve (ABC) was 0.0037 mg/ml of blood. In addition to the dye concentration data, catheters for collecting data on intravascular and intracardiac *pressures*, as well as *oxygen saturation*, were positioned within patient 2, and the following values were observed:

Pressures (mm Hg):	
Left ventricle	90/5
Pulmonary artery	45/15
Ascending aorta	91/65
Percent saturation of *hemoglobin* with oxygen:	
Left ventricular blood	92%
Pulmonary artery blood	84%
Radiographic analysis provided the following values:	
Right ventricular *end-diastolic volume:*	200 ml
Right ventricular *end-systolic volume:*	80 ml

In addition, there was evidence of right ventricular *hypertrophy*.

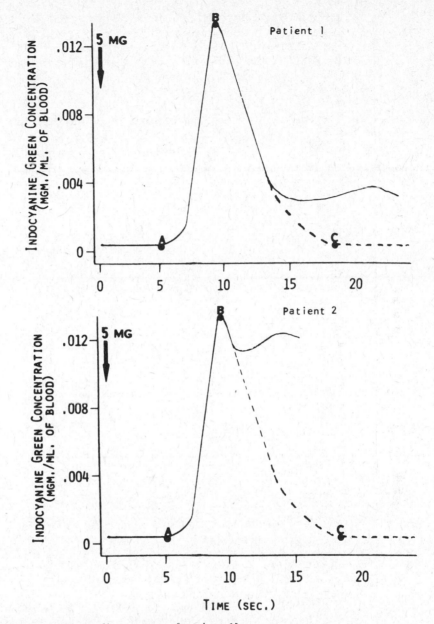

A. What was the cardiac output of patient 1?

B. What was the apparent cardiac output of patient 2?

C. On the basis of previous experience, the dye curve for patient 1 was judged to be normal. How do the curves from the two patients differ?

D. Patient 1 appeared to be *cyanotic*. Is it likely that his cyanosis was due to a right-to-left shunt through an *interatrial septal defect*? What evidence is there for your reply? If your answer was "no," what other problem could explain the cyanosis?

E. What might account for the difference between the curves for patients 2 and 1?

F. Is it likely that patient 2 has an aortic *stenosis* or a pulmonary valve *insufficiency*?

G. Is there any evidence to indicate either overt or impending right ventricular *failure* in patient 2?

H. Assuming you could measure any additional cardiovascular parameters you felt were relevant to rendering a diagnosis of patient 2's problem, what values would support a diagnosis of severe pulmonary valve stenosis?

(Continued)

I. Assuming that patient 2 has a single cardiovascular defect, what types of disorders could explain the data presented thus far?

J. One rather simple piece of physical evidence that would have shed much light on the question of whether this patient's condition is associated with stenosis, regurgitation, or septal defect has been omitted from the data. What is it?

K. How do you distinguish a pulmonary stenosis from a pulmonary insufficiency on the basis of *auscultation*?

L. The percent saturation of hemoglobin with oxygen in the right atrium was 73% and in the right ventricle, 84%. This final piece of information should allow you to make a definitive decision as to the specific cardiovascular defect present in patient 2.

Answer and Explanation

A. The cardiac output is calculated from curve ABC:

$$\text{Cardiac output} = \frac{\text{amount of dye injected}}{\text{average concentration under ABC}} \times \frac{60}{\text{duration of ABC}}$$

$$= \frac{5.0 \text{ mg}}{0.0037 \text{ mg/ml}} \times \frac{60 \text{ sec/min}}{13.5 \text{ sec}}$$

$$= 6006 \text{ ml/min}.$$

B. Patients 1 and 2 have the same extrapolated curves and therefore the same cardiac output (6006 ml/min).

C. Patient 2 has a decreased *mean recirculation time*.

D. No, he has a normal *appearance time*. A reduced appearance time would be caused by a right-to-left shunt. The cyanosis could be caused by an alveolar *diffusion block* due to pulmonary edema, fibrosis, etc.

E. The short recirculation time for patient 2 suggests a left-to-right shunt.

F. The absence of a large *systolic pressure gradient* between the left ventricle and aorta does not support a diagnosis of aortic stenosis. The high pulmonary artery *pulse pressure* (45 − 15 = 30 mm Hg) is consistent with a diagnosis of insufficiency, but the high O_2 saturation of the blood in the pulmonary artery and other data are suggestive of another possibility.

G. The elevated right ventricular end-diastolic volume and right ventricular hypertrophy are consistent with a diagnosis of compensated right ventricular failure. Note, however, that there is a normal *ejection* fraction:

$$\frac{200 - 80 \text{ ml}}{200 \text{ ml}} = 0.60.$$

H. Evidence of a large systolic pressure gradient between the right ventricle and the pulmonary artery would support a diagnosis of severe pulmonary stenosis.

I. Tentative diagnosis: a left-to-right shunt (interventicular or interatrial *septal defect* or *patent ductus* arteriosus). Evidence for diagnosis: (1) a reduced mean recirculation time, (2) a high O_2 saturation of pulmonary artery blood, (3) a high right ventricular stroke volume, (4) a high right ventricular end-diastolic volume, (5) a high pulmonary artery pressure, and (6) right ventricular hypertrophy.

J. No mention was made as to whether there was a systolic or diastolic *murmur*.

K. Pulmonary stenosis is associated with a systolic murmur, whereas pulmonary insufficiency is characterized by a diastolic murmur.

L. The higher O_2 saturation in the right ventricle than in the right atrium is consistent with a diagnosis of an interventricular septal defect. The net result of this defect has been an *effective right ventricular stroke volume* that exceeds the effective left ventricular stroke volume. The *preload* (a volume load) and *afterload* (a pressure load) for the right ventricle have been increased and, if sufficiently marked, can send the ventricle into a decompensated failure.

34

(1:510; 2:3-190;
3:469; 4:221;
5:1018)
CHop 3

In the table that follows, indicate the changes (increase: ↑; decrease: ↓) that occur in aortic *pressure* in response to arteriolar *constriction* (↑ resistance), an increase in *stroke volume*, an increase in *heart rate,* and *arteriosclerosis* (↓ arterial *compliance*). Assume that the reflexes have been depressed and that no autoregulatory mechanism will mask these *hemodynamic relationships.* In addition, write a formula that defines or approximates each of these hemodynamic relationships.

	Mean aortic pressure ($\overline{P}$)	Aortic pulse pressure (dP)	Formula (R, Q, f, C, $\overline{P}$, dP)
↑ Resistance (R):	_____	↓	$\overline{P} \cong$
↑ Stroke volume (Q):	_____	_____	$\overline{P} \cong$
			dP∝
↑ Heart rate (f):	_____	↓	$\overline{P} \cong$
↓ Compliance (C):		_____	dP∝

Answer and Explanation

	$\overline{P}$	dP	Formula
↑ R	↑	↓	$\overline{P} \cong R \times$ cardiac output
↑ Q	↑	↑	$\overline{P} \cong R \times Q \times f$
			dP ∝ Q/C
↑ f	↑	↓	$\overline{P} \cong R \times Q \times f$
↓ C		↑	dP ∝ Q/C

35

(1:510; 2:3-16;
3:469; 4:221;
5:1018)
CHop 2

Why does the formula, $\overline{P} \cong R \times$ cardiac output, represent only an approximate relationship between these factors?

Answer and Explanation
The formula is derived from: $P_1 - P_2 = R \times$ cardiac output, where P_1 is mean arterial pressure, and P_2 is central venous pressure. The omission of P_2 from the formula assumes a *central venous pressure* of 0 mm Hg.

36

(1:510; 2:6-38;
3:469; 4:221;
5:1022)
CHqp 2

In question 34, why can we not say dP = Q/C?

Answer and Explanation
Compliance (C) equals (change in volume)/(change in pressure). Arterial *compliance* would equal (stroke volume)/(pulse pressure) if there were no outflow from the arterial system during ventricular ejection. In other words, the change in volume in the compliance equation is directly related to stroke volume but not equal to it.

Questions 37 through 39 consist of two main parts: a statement and a reason for the statement. Choose the best answer for each of these questions from the options A through E below and mark your choice in the space provided:

A. both the statement and the reason are true and are related as cause and effect (i.e., as indicated)
B. both the statement and the reason are true but are not related as cause and effect
C. the statement is true but the reason is false
D. the statement is false but the reason is an accepted fact or principle
E. both the statement and the reason are false

37

(1:564; 2:3-91;
3:458; 4:458; 5:991)
HCp 2

Statement

_____ A premature ventricular systole causes a weak aortic pulse because

Reason

it causes a decreased left ventricular end-diastolic fiber length.

Answer and Explanation

A. A decrease in the period of ventricular filling is a common cause of a decreased end-diastolic volume.

38

(1:493; 2:6-38;
3:468; 4:325;
5:1022)
CHqp 2

Statement

_____ In aortic stenosis, dP/dt for the anacrotic limb of the aortic pulse increases because

Reason

a decrease in aortic compliance tends to increase dP/dt for the anacrotic limb of the aortic pressure wave.

Answer and Explanation

D. In aortic *stenosis*, aortic dP/dt decreases because of the high resistance to flow from the ventricle at the aortic orifice.

39

(1:493; 2:6-38;
3:468; 4:221;
5:1022)
Cqp 2

Statement

_____ A marked decrease in total systemic resistance causes a marked decrease in arterial pressure and a very small or negligible increase in venous pressure because

Reason

the systemic arteries are more compliant than the systemic veins.

Answer and Explanation

C. The less marked change in venous pressure occurs because of the greater *compliance* of the venous system.

10
Control of the Circulation

1

(1:596; 2:3-209; 3:495; 4:345; 5:1094)
Cnf 1

Write in Column I each of the appropriate matches from Column II (f, g, h, etc.). Each item in Column II can be used once or more than once. The numbers in parentheses represent the number of appropriate matches for the items in Column I.

Column I (Mechanisms)

1. _____ warming the blood to the hypothalamus of the brain causes a reflex vasodilation in this circulation (1)
2. _fg_ a 30% decrease in the total blood volume does NOT cause a reflex vasoconstriction in this part of the systemic circulation (2)
3. _k_ resistance to flow increases in response to a local decrease in P_{O_2} (1)
4. _g, k_ resistance to flow decreases during running even when body temperature remains constant (3)
5. _j_ contains high and low pressure capillaries that are in series with each other (1)
6. _k_ contains only low pressure capillaries (1)
7. _g_ its veins have the lowest concentration of oxygen found in the body (1)
8. _j_ its veins have the highest concentration of oxygen found in the systemic circulation (1)

Column II (Systems)

f. cerebral circulation
g. coronary circulation
h. cutaneous circulation
i. gastrocnemius muscle circulation
j. Renal circulation
k. Pulmonary circulation

Answer and Explanation

1. **h.** The heat regulatory center in the *hypothalamus* increases heat loss from the body by decreasing *adrenergic sympathetic tone* to the blood vessels of the skin.
2. **f, g.** *Autoregulation* is a major factor in the control of both cerebral and coronary blood flow. For example, a local elevation of P_{CO_2} acts as an important cerebral vessel dilator. Vasomotor reflexes play little or no role in the regulation of either cerebral or coronary blood flow, but increases in metabolism or decreases in blood flow do cause an accumulation of *metabolites* and a resulting vasodilation.
3. **k.** In the *pulmonary circulation*, a low alveolar P_{O_2} or pH causes a local vasoconstriction. This serves to shunt blood away from poorly ventilated alveoli.
4. **g, i, k.** During running there is (g) an increased cardiac rate and inotropicity, which leads to an accumulation of metabolites in the myocardium and a coronary dilation. In skeletal muscle (i), both the accumulation of metabolites and the autonomic neu-

(Continued)

rons cause the vasodilation in active muscles. Increases in (k) cardiac output distend the highly *compliant* pulmonary vessels and therefore decrease pulmonary *resistance* during running.

5. j.   The high hydrostatic pressure (55 mm Hg) in the glomerular capillaries permits *filtration* into Bowman's capsule and the low pressure in the peritubular capillaries (15 mm Hg) facilitates *reabsorption* back into the blood.

6. k.   The low hydrostatic pressure in the pulmonary capillaries (10 mm Hg) is the result of the low pressure produced by the right ventricle (27 mm Hg) and serves to keep the lungs dry (i.e., prevents *pulmonary edema*).

7. g.   This means that the heart has the highest $a\text{-}v_{O_2}$ *concentration difference* in the body and the lowest O_2 *reserve*.

8. j.   The high blood flow to the kidneys is not used to meet their metabolic needs (hence the high oxygen concentration in the renal vein), but rather to permit the removal from the blood of substantial quantities of wastes and excess electrolytes and water.

2

(1:605; 2:3-213; 3:501; 4:353; 5:1101)
Cnf 1

Which of the following circulatory systems will respond to the stimulation of its *adrenergeric sympathetic neurons* with the most intense decrease (i.e., in terms of percent of control) in its blood flow?

A. cerebral circulation
B. coronary circulation
C. cutaneous circulation
D. pulmonary circulation
E. skeletal muscle circulation

Answer and Explanation

A, B.   The cerebral and coronary circulatory systems are among the least sensitive systems in the body to the constrictor actions of norepinephrine.

C.   The cutaneous and renal vessels have a more pronounced vasoconstrictor response to the stimulation of adrenergic neurons than do the vessels of skeletal muscle and the lungs.

3

(1:544; 2:3-173; 3:482; 4:249; 5:1061)
CHpn 2

Check each of the following changes produced by a decrease in pressure in the *carotid sinus* from 90 mm Hg to 70 mm Hg:

1. a decrease in the frequency of impulses moving centrally in the glossopharyngeal nerve
2. a reflex stimulation of cardiac sympathetic neurons
3. a reflex stimulation of cholinergic postganglionic sympathetic neurons

Which one of the following best summarizes your conclusions?

A. statement 1 is correct
B. statement 2 is correct
C. statements 1 and 2 are correct
D. statements 2 and 3 are correct
E. statements 1, 2, and 3 are correct

Answer and Explanation

C.   In the healthy subject, the carotid sinus stretch receptors send progressively fewer impulses up the ninth cranial nerve as the arterial pressure decreases from 200 mm Hg to 30 mm Hg. This results in a progressively greater stimulation of adrenergic sympathetic neurons to the heart and blood vessels and a depression of the cardiac parasympathetic neurons.

E. There is no stimulation of cholinergic postganglionic sympathetic neurons during this reflex.

4

(1:594; 2:3-199; 3:485; 4:249; 5:1116)
CHRpf 2

During defecation, urination, and the lifting of heavy loads, the phenomenon of straining occurs (i.e., one utilizes the *Valsalva maneuver*). During this maneuver the following events occur:

1. a decrease in aortic pressure
2. an increase in aortic diastolic pressure
3. a second increase in aortic diastolic pressure
4. an increase in heart rate
5. an increase in intrathoracic and intraabdominal pressure

List these events in chronologic order.

A. 1-4-2-5-3
B. 2-5-1-4-3
C. 5-2-3-4-1
D. 5-2-1-4-3
E. 5-4-2-3-1

Answer and Explanation

D. In the Valsalva maneuver, the subject expires against a closed glottis and there results (5) an increased intrathoracic pressure which compresses the arteries and veins, (2) increasing *arterial pressure*. As the increase in thoracic pressure is maintained, *venous return* is impeded, the *cardiac output* decreases, and (1) the aortic pressure declines. This causes (4) a *reflex* increase in heart rate, which (3) returns (increases) the aortic pressure to normal. The maneuver ends when the subject takes an expiration.

5

(1:544; 2:3-173; 3:485; 4:249; 5:1061)
Copn 2

The total systemic *peripheral resistance* is increased in response to (check each correct statement):

1. a decreased blood volume (i.e., hemorrhage)
2. changing from a reclining to a standing position
3. hypertension
4. lifting a heavy load
5. strenuous running

Answer and Explanation

1.
2. Procedures in the healthy, conscious subject that decrease pressure (i.e., within the range of from 30 to 200 mm Hg) in the carotid sinus or aortic arch will produce a reflex increase in systemic resistance. This is due to arteriolar vasoconstriction and possibly also to an increased smooth muscle tone in arteries and veins.
3. Arterial hypertension is frequently caused by vasoconstriction, but the healthy body responds to an increased arterial pressure with vasodilation. One cause of hypertension is a malfunction of the baroreceptors owing to their loss of compliance in arteriosclerosis.
4. During this action there is the Valsalva maneuver and its associated increase in intraabdominal and intrathoracic pressures.
5. The resistance generally falls as a result of the extensive skeletal muscle vasodilation. There is also usually cutaneous vasodilation.

6

*(1:544; 2:3-176;
3:482; 4:249;
5:1061)*
CHnpo 3

The following record demonstrates an individual's *response to acetylcholine (ACh):*

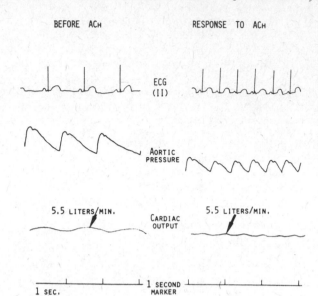

BEFORE ACh   RESPONSE TO ACh

ECG (II)

AORTIC PRESSURE

5.5 LITERS/MIN.   CARDIAC OUTPUT   5.5 LITERS/MIN.

1 SEC.   1 SECOND MARKER

On the basis of these data and your understanding of physiology, what would you conclude? Acetylcholine produced

A. an increased pulse pressure
B. an increased peripheral resistance
C. a decreased arterial pressure due to venous dilation
D. an increased systolic pressure
E. a reflex increase in heart rate

Answer and Explanation

A. The pulse pressure is reduced.
B. Peripheral resistance = pressure head/cardiac output. Since the pressure head decreased and the cardiac output remained constant, the peripheral resistance decreased.
C. Decreased arterial pressure is usually due to decreased cardiac output or decreased systemic peripheral resistance. Since changes in venous tone do not appreciably affect peripheral resistance, and the cardiac output (and, hence, the venous return of blood to the heart) is constant, the decreased arterial pressure is probably due to arteriolar vasodilation and not to any changes in the venous system.
D. Systolic pressure is peak arterial pressure. It decreases.
E. Acetylcholine has a direct slowing action on the SA node. Since the heart rate increased (i.e., there was a decreased R-R interval), apparently, the direct action of ACh is masked by a reflex stimulation of cardiac sympathetic neurons in response to the lowered arterial pressure. The receptors responsible for this reflex are in the carotid sinus and aortic arch.

7

*(1:318; 2:7-71;
3:291; 4:714;
5:1055)*
CHof 3

You are given three *catecholamine* solutions of known concentration: norepinephrine, epinephrine, and isoproterenol. Each solution is injected separately into a conscious subject at a concentration of 2 μg/kg of body weight. You can assume that, at this dose, isoproterenol stimulates only beta adrenergic receptors. What response would you obtain from the norepinephrine injection?

A. an increase in heart rate and arterial pressure that is more marked than that obtained from the other solutions

B. a decrease in heart rate and an increase in arterial pressure that is more marked than those obtained from the other solutions

C. decreases in peripheral resistance and heart rate that are more marked than those obtained with any of the other solutions

D. a decrease in peripheral resistance and an increase in cardiac output that is more marked than with any of the other solutions

E. an increase in skeletal muscle blood flow and a decrease in cutaneous blood flow

Answer and Explanation

B. Norepinephrine produces a more marked increase in the total *systemic peripheral resistance* than do the other catecholamines. It performs this function by virtue of being a strong stimulator of the *alpha receptors* of arterioles and, at these concentrations, either not a stimulator or a very weak stimulator of the beta receptors of smooth muscle. Although norepinephrine also stimulates the *beta adrenergic receptors* of the heart to increase heart rate, the increase in arterial pressure that it produces acts through the arterial baroreceptors to mask this action reflexly and produce a slowing of the heart.

```
NOREPINEPHRINE

    VASOCONSTRICTION ─────► INCREASED ARTERIAL PRESSURE

        STIMULATION OF BARORECEPTORS ────► REFLEX CARDIAC SLOWING
```

8

(1:318; 2:7-71; 3:291; 4:714; 5:907)
CHof 3

In question 7, which of the statements represents the response of a conscious individual to *isoproterenol*?

Answer and Explanation

D.

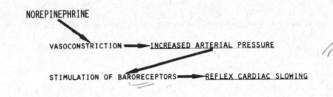

```
ISOPROTERENOL ────► VASODILATION ─────► DECREASED ARTERIAL PRESSURE

        LESS STIMULATION OF BARORECEPTORS ─────► REFLEX CARDIAC SPEEDING

        BETA RECEPTORS IN S-A NODE ─────────────► CARDIAC SPEEDING

                                    INCREASED CARDIAC OUTPUT

        BETA RECEPTORS IN THE VENTRICLES ────► INCREASED STROKE VOLUME
```

9

(1:318; 2:7-71; 3:291; 4:714; 5:907)
CHof 3

In question 7, which of the statements represents the response of an individual to epinephrine?

Answer and Explanation

E.

```
EPINEPHRINE                     DECREASED CUTANEOUS BLOOD FLOW

        ALPHA RECEPTORS ─────► CUTANEOUS VASOCONSTRICTION

                            SKELETAL MUSCLE VASOCONSTRICTION

        BETA RECEPTORS ─────► SKELETAL MUSCLE VASODILATION

                            INCREASED BLOOD FLOW TO SKELETAL MUSCLE

                            INCREASED CARDIAC OUTPUT
```

(Continued)

Epinephrine has the capacity to produce both vasoconstriction and vasodilation. At the dose injected, the skeletal muscle vasodilation effect would be expected to predominate over the skeletal muscle vasoconstriction effect. The reverse is true in the skin.

10

(1:608; 2:3-212; 3:501; 4:244; 5:1091)
Cnf 1

If a subject puts a rubber band on his arm, pulls it back, and releases it, there occurs the following sequence of events: (1) a red line, (2) a red flare, and (3) a *wheal*. Check each of the following statements that is true:

1. the red line is due to a multineuron reflex arc
2. the red flare is due to the release of histamine
3. the wheal is due to an axon reflex
4. none of these statements are true

Answer and Explanation
1. The red line is due to a direct response of the vessels to the *trauma*.
2. The flare is due to an *axon reflex*.
3. The wheal is due to an intracellular release of histamine, which causes edema.
④.

11

(1:535; 2:3-215; 3:502; 4:236; 5:100)
Cfn 1

In the heart, brain, skeletal muscle, liver, kidney, and intestine, 1 minute of *vascular occlusion* is followed by which of the following changes when the occlusion is removed (check the *two* correct answers)?

1. a period of 15 to 30 seconds during which flow progressively increases to the preocclusion value
2. a rapid (less than 2 seconds) return to the preocclusion flow
3. a period of 15 seconds or more during which the flow is markedly higher than it was prior to occlusion
4. a change in flow produced primarily by adrenergic sympathetic neurons
5. a change in flow produced primarily by cholinergic sympathetic neurons
6. a change in flow produced primarily by the accumulation of vasodilator metabolites
7. a change in flow produced primarily by local heating of the tissue

Answer and Explanation
③. This is called *reactive hyperemia*. In the heart, for example, 10 seconds of coronary artery occulusion causes, on release of the occlusion, 15 seconds during which the flow is two to four times that prior to occlusion.
⑥. Reactive hyperemia occurs in the presence of antihistaminics, in decentralized organs, and in the presence of atropine and adrenergic-blocking agents. Hypoxia, hypercapnea, acidity, and local hyperkalemia may all contribute to the vasodilation found during reactive hyperemia.

12

(1:596; 2:3-135; 3:497; 4:299; 5:1099)
CHf 1

During early ventricular systole, coronary blood flow

A. is more markedly reduced at the subepicardial surface than the subendocardial surface
B. is markedly reduced because the cusps of the aortic valve occlude the coronary arteries and therefore protect them from excess pressure
C. in the left coronary artery goes to near 0 or below
D. in the right coronary artery goes to near 0 or below
E. is well characterized by all of the above statements

Answer and Explanation

A. The occlusive force exerted on the coronary vessels during ventricular systole is greatest at the subendocardial portion. For this reason this is the part of the myocardium most prone to *ischemic damage* and subsequent infarction.

B. Eddy currents (i.e., turbulence) prevent the cusps from occluding these arteries.

Ⓒ Negative flow (back flow) is common during systole.

13

(1:623; 2:3-216; 3:508; 4:161; 5:1041)
CHfn 2

Which one of the following is most likely to occur in a healthy subject in response to *running*?

A. an increased coronary flow due to decreased cardiac adrenergic tone

B. a generalized increase in blood flow (i.e., in the kidneys, muscle, stomach, etc.) due to an increased cardiac output

C. both A and B occur during exercise

D. a decreased velocity of flow in the capillaries of the lungs

E. a decreased cardiac parasympathetic tone

Answer and Explanation

A. During exercise, there is an increased cardiac adrenergic tone to the heart.

B. During light exercise, a 300% to 400% increase in skeletal muscle blood flow and no change in renal blood flow may occur. In response to more strenuous exercise, the increased skeletal muscle flow is more marked. In other words, during exercise, at the same time that skeletal and cardiac muscle vessels are dilating, many other vessels are constricting. Renal + cutaneous

D. There is an increased velocity of flow in the pulmonary capillaries.

Ⓔ An increased heart rate during exercise is due to a decreased cardiac parasympathetic tone and an increased cardiac sympathetic tone.

14

(1:623; 2:3-164; 3:508; 4:278; 5:903)
CHNof 2

Which of the following is *not* an important mechanism for increasing the blood flow to active skeletal muscle during *exercise*?

A. an increased cardiac output

B. an increased concentration of epinephrine in the vicinity of the blood vessels of active skeletal muscle

C. an increased concentration of norepinephrine in the vicinity of the blood vessels of active skeletal muscle

D. an increased concentration of acetylcholine in the vicinity of the blood vessels of active skeletal muscle

E. the action of metabolites on skeletal muscle blood vessels

Answer and Explanation

A. The major factors regulating skeletal muscle blood flow during exercise are changes in the distribution of cardiac output initiated by vasodilation in the active skeletal muscle associated with vasoconstriction in most other areas and an increased cardiac output.

B. In strenuous exercise, epinephrine facilitates the shunting of blood to active skeletal muscle by causing vasodilation in skeletal muscle and vasoconstriction in some other areas and an increased cardiac output.

Ⓒ In exercise, there is a decreased adrenergic sympathetic tone to the blood vessels of active skeletal muscle. This decreased release of norepinephrine produces a local vasodilation and therefore a shunting of additional blood to this area.

D. The stimulation of cholinergic sympathetic neurons to skeletal muscle is probably an initial response to exercise or the anticipation of exercise.

E. Hypoxia, hypercapnea, and acidity produce skeletal muscle vasodilation. The degree to

(Continued)

which these changes occur in the vicinity of the skeletal muscle arteriole will depend on the adequacy of the response of the nervous system to exercise or the anticipation of exercise.

15

(1:623; 2:3-119; 3:508; 4:278; 5:1041)
Chcf 1

Which of the following is most likely to occur in a healthy subject in *response to strenuous running?*

A. a decrease in the ejection fraction of the ventricles
B. a decrease in the maximum dP/dt of the left ventricle
C. an increase in circulation time
D. a decrease in arteriovenous O_2 difference
E. an increase in concentration of arterial lactate

Answer and Explanation

A. In running there is a decrease in the end-systolic volume and an increase in the ejection fraction of the ventricles.
B. Catecholamines are released during exercise and have a positive inotropic action on the heart.
C. The increase in force of cardiac contraction during exercise causes an increase in velocity of blood flow.
D. The a-v_{O_2} difference increases.
E. In an exercise where the O_2 consumption is increased ninefold, the concentration of plasma lactate will double.

16

(1:623; 2:3-125; 3:508; 4:278; 5:993)
HNfq 2

In a healthy subject, running is usually associated with a decrease in the *end-systolic volume* of the right and left ventricles and with an increase in their stroke volume. The mechanism for this response is probably the following:

A. Starling's law of the heart
B. an increase in sympathetic tone to the ventricles
C. an increase in parasympathetic tone to the ventricles
D. a decrease in venous return of blood to the heart
E. an increase in pulmonary and systemic resistance

Answer and Explanation

A. The Starling law for the healthy heart states that increases in stroke work (stroke volume × mean ejection pressure) are produced by increases in end-distolic volume. The Starling mechanism, by itself, does not decrease the end-systolic volume.
B. An increase in sympathetic tone to the ventricles shifts the Starling curve to the left. In other words, there is an increase in stroke volume that produces a decrease in the end-systolic volume of the ventricles.
C. The parasympathetics play a minor role in the direct control of ventricular stroke volume. They shift the Starling curve for the atria to the right.
D. In running, there would be an increase in venous return of blood to the heart.
E. In running, there is a decrease in pulmonary and systemic resistance.

17

(1:625; 2:3-124; 3:460; 4:278; 5:1041)
Hnfa 2

A patient with a *heart transplant* is found to have no reinnervation of his heart. Check each of the following responses you expect him to have to *moderate running* and indicate the mechanism for each.

1. an increased venous return of blood to the heart
2. an increased heart rate that is more marked than that found in a normal subject during the same exercise

3. an increased stroke volume that is more marked than that found in a normal subject during the same exercise
4. a decreased sensitivity of the heart to circulating catecholamines
5. an increased stimulation of adrenergic sympathetic neurons to skeletal muscle blood vessels

Answer and Explanation
1. There is an increased venous return of blood to the heart as a result of skeletal muscle contraction and relaxation.
2. There is an increased heart rate in response to exercise in the patient with a decentralized heart, but it is a less marked increase than in the normal subject. It may be, in part, due to circulating catecholamines, to which the decentralized heart is hypersensitive.
3. The increased stroke volume is, in large part, due to an increased end-diastolic volume of the heart caused by an increased venous return of blood.
4. There is an increased sensitivity to catecholamines.
5. There is a decreased adrenergic sympathetic tone to skeletal muscle in both the above patient and in the normal subject.

18

(1:673; 2:3-207; 3:530; 4:290; 5:1112)
CRog 1

Pulmonary vascular resistance increases in response to (select the one best answer)

A. an elevated P_{O_2} in the pulmonary blood vessels
B. high altitude
C. either of the above
D. prostaglandin E_1
E. none of the above

Answer and Explanation
B. The control over the pulmonary circulation is, in many respects, different from that for the systemic circulation. Thus, hypoxia has opposite effects on systemic arterioles and pulmonary precapillary sphincters. The advantage of such a situation is that poorly ventilated parts of the lung will have some of their blood diverted to the better ventilated parts. If this did not occur we would have a situation comparable to a *right-to-left shunt* and therefore a lower saturation of the blood with O_2 in the systemic circulation.
D. This agent produces a decreased resistance to flow in both the pulmonary and systemic circulations.

19

(1:673; 2:3-207; 3:508; 4:290; 5:1041)
CRog 3

During *strenuous running*, the *cardiac output* of a subject increased fourfold, the systemic *a-v$_{O_2}$ difference* changed from 4 to 12 volumes percent, and the mean pulmonary arterial *pressure* changed from 15 mm Hg to 18 mm Hg. Check each of the conclusions that can be drawn from this typical response to exercise.

1. the increased pressure was caused by an increased pulmonary resistance to flow
2. the pulmonary artery blood had a lowered P_{O_2} during exercise
3. changes in the pulmonary artery blood P_{O_2} are, for the most part, responsible for the changes in the pulmonary peripheral resistance
4. none of the above are true

Answer and Explanation
1. Using the formula, Resistance = pressure/flow, we must conclude that resistance decreased, since flow increased fourfold and pressure 0.2-fold. In other words, the increased flow causes the increased pressure (Pressure = resistance × flow).
2. The increased a-v$_{O_2}$ differences during exercise in healthy individuals are due either solely or primarily to a decreased P_{O_2} in the systemic veins.
3. The decreased P_{O_2} of the pulmonary artery blood causes precapillary constriction. The
(Continued)

decreased resistance to flow during exercise is due to the *distention* of the vessels by the increased quantities of blood entering them from the right heart and by the high P_{O_2} characteristic of the alveoli and bronchioles during most exercises.

20

(1:673; 2:3-207; 3:530; 4:290; 5:1112)
CRo 1

Contraction of the smooth muscle of the *pulmonary circulation* occurs in response to

A. norepinephrine and epinephrine
B. serotonin and angiotensin II
C. histamine
D. none of the above
E. all of the above

Answer and Explanation

E. Pulmonary arterioles constrict in response to norepinephrine, epinephrine, and angiotensin II. Pulmonary venules constrict in response to serotonin and histamine.

21

(2:3-160; 3:464; 4:225; 5:1078)
Cqn 1

Make a list of the physiologic characteristics and functions served by the *systemic veins* and then answer the following question: The systemic veins differ from the systemic arteries in that the systemic veins (check each correct statement)

1. contain a much greater blood volume
2. are much more compliant
3. have a more variable blood volume
4. do not receive a sympathetic innervation
5. in the arms and thighs contain valves

Answer and Explanation

1. In a 70-kg man, the systemic veins (3440 ml of blood) contain over three times as much blood as systemic arteries (1000 ml).
2.
3. During hemorrhage and water loading, the volume of the veins changes more than that of the arteries. Because of this, the systemic veins are an important *volume buffering system* for the arteries.
4. It has been suggested that increasing adrenergic sympathetic tone to the veins is a mechanism for decreasing venous compliance and shifting blood to the arterial side of the circulation.
5. The only arterial valves are the aortic and pulmonary valves. These serve to separate the arterial blood from that in the ventricles.

22

(1:677; 2:3-202; 3:530; 4:290; 5:1108)
CRoqn 2

The *pulmonary circulation* differs from the systemic circulation in that in the pulmonary circulation (check each correct statement)

1. the veins do not contain an important reservoir of blood for the heart
2. the arterial system is more distensible
3. adrenergic sympathetic neurons are less important in controlling arteriolar and precapillary resistances
4. the capillaries have a higher transmural pressure
5. the arteries serve as more important blood reservoirs
6. obstruction of blood flow is much more likely to cause a retrograde ventricular failure

Answer and Explanation

1. Both the pulmonary arteries and veins serve as important *reservoirs* of blood that the

'left heart can call upon. The blood volume in the pulmonary system of an adult may vary from 200 ml to 1000 ml.

②.

③. In the pulmonary circulation, unlike the systemic circulation, the capillaries represent the major source of *resistance*, and many of the arterioles are almost devoid of smooth muscle. Apparently, the main effect elicited by the stimulation of *adrenergic sympathetic neurons* to the pulmonary blood vessels is to decrease the *compliance* of arteries and veins (i.e., to stimulate their smooth muscles).

4. If this is the case, *pulmonary edema* is present.

⑤. Almost as much blood volume lies in the pulmonary arteries as in the pulmonary veins.

6. One would have to have an obstruction of over two-thirds of the pulmonary vascular bed before a right ventricular failure would develop. Left ventricular failure is 30 times more common than right ventricular failure.

23

(1:672; 2:3-204; 3:530; 4:290; 5:1108)
Copv 2

During running the cardiac output doubles. What changes would you expect in the *pulmonary system*?

Pulmonary Vascular Resistance	Pulmonary Arterial Pressure
A. a small decrease	a large increase
B. a large decrease	a small increase
C. a large decrease	a small decrease
D. a small increase	a large increase
E. a large increase	a large increase

Answer and Explanation

B. A doubling of the cardiac output is usually associated with a 2 to 5 mm Hg increase in pulmonary artery pressure. This represents approximately a 40% reduction in pulmonary resistance. The large compliance of the pulmonary circuit results in large decreases in resistance in response to increases in right heart output. In the systemic circuit, on the other hand, adrenergic sympathetic tone to the arterioles is a much more important factor in the control of resistance.

24

(1:672; 2:3-190; 3:530; 4:290; 5:1108)
CHpo 2

Systemic arterial pressure in the adult is approximately six times that of *pulmonary arterial pressure* because (select the one best statement)

A. left ventricular stroke volume is greater than right ventricular stroke volume
B. systemic blood volume exceeds pulmonary blood volume
C. systemic resistance exceeds pulmonary resistance
D. pulmonary compliance exceeds systemic compliance
E. intraabdominal pressure exceeds intrathoracic pressure

Answer and Explanation

A. Right and left ventricular stroke volumes are approximately equal.
C. Pressure = resistance × flow.
Since the flow from the left ventricle approximately equals the flow from the right ventricle, resistance becomes the determining factor.

25

(1:987; 2:3-169;
3:247; 4:259;
5:1059)
CpEa 2

A sustained raised systemic *arterial pressure* (hypertension) may result from (check each correct answer)

1. a decreased peripheral resistance
2. a decreased secretion of prostaglandin E_1
3. an increased concentration of blood histamine
4. excessive secretion of ACTH (adrenocorticotropic hormone)

Answer and Explanation

1. An increased resistance tends to increase arterial pressure.
2. This prostaglandin is a vasodilator. It has been suggested that the hypertension that occurs after removal of both kidneys (renoprival hypertension) is due to decreased levels of circulating prostaglandins.
3. In anaphylactic shock, there are increased concentrations of *histamine* or a histamine-like substance in the blood.
4. *ACTH* facilitates the release of *cortisol*, which increases the sensitivity of the body to *norepinephrine* and *epinephrine*. Cortisol may also cause Na^+ and water retention by the body. In *Cushing's syndrome*, plasma cortisol values are elevated, and about 85% of the individuals with this malady are hypertensive.

26

(1:420; 2:3-216;
3:478; 4:243;
5:1054)
Cocb 2

Check each of the following statements about *bradykinin* that is true:

1. it is formed by sweat glands, salivary glands, and the pancreas
2. it, like angiotensin, is formed from circulating globulins
3. it, like histamine, produces vasodilation
4. it, like histamine, produces smooth muscle contraction in the viscera and is capable of attracting leukocytes to an area

Which one of the following statements best summarizes your conclusions?

A. statements 1, 2, 3, and 4 are true
B. statements 1, 2, and 3 are true
C. statements 1, 2, and 4 are true
D. statements 1 and 3 are true
E. statements 2 and 3 are true

Answer and Explanation
A.

27

(1:628; 2:3-184;
3:510; 4:332;
5:1081)
Cbqa 2

A patient suffers a *blood loss* over a period of 20 minutes. At the end of this period, his arterial pressure has changed from 100 to 70 mm Hg and his heart rate from 70 to 140/min. His hematocrit is 30% and his skin is cold. What other changes have occurred (check each correct statement)?

1. a decreased capillary hydrostatic pressure
2. a decreased interstitial fluid volume
3. an increased plasma colloid osmotic pressure
4. an increased total systemic resistance

Answer and Explanation
1, 2, 4.

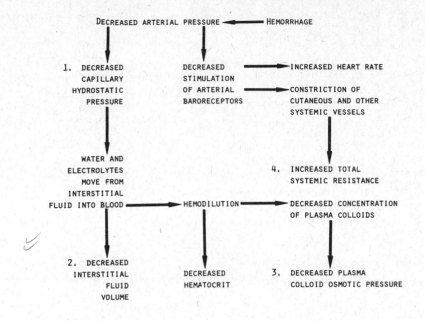

DECREASED ARTERIAL PRESSURE ◄——— HEMORRHAGE

1. DECREASED
   CAPILLARY
   HYDROSTATIC
   PRESSURE

DECREASED
STIMULATION
OF ARTERIAL
BARORECEPTORS

INCREASED HEART RATE

CONSTRICTION OF
CUTANEOUS AND OTHER
SYSTEMIC VESSELS

WATER AND
ELECTROLYTES
MOVE FROM
INTERSTITIAL
FLUID INTO BLOOD ——— HEMODILUTION ———

4. INCREASED TOTAL
   SYSTEMIC RESISTANCE

DECREASED CONCENTRATION
OF PLASMA COLLOIDS

2. DECREASED
   INTERSTITIAL
   FLUID
   VOLUME

DECREASED
HEMATOCRIT

3. DECREASED PLASMA
   COLLOID OSMOTIC PRESSURE

28

*(1:628; 2:3-184;
3:510; 4:332;
5:1081)*
Cbqa 2

Some other responses to the hemorrhage discussed in question 27 include (check each correct answer)

1. a decreased glomerular filtration rate
2. a decrease in venous tone
3. sodium retention
4. a decreased secretion of antidiuretic hormone (ADH)
5. an increased production of angiotensin II (A-II)

Answer and Explanation
1, 3, 5.

DECREASED ARTERIAL PRESSURE ◄——— HEMORRHAGE

1. DECREASED
   GLOMERULAR
   FILTRATION
   RATE

DECREASED
STIMULATION
OF
ARTERIAL
BARORECEPTORS

INCREASED
RELEASE OF
RENIN BY
KIDNEYS

5. INCREASED
   PRODUCTION
   OF A-I, A-II,
   ALDOSTERONE

4. INCREASED
   RELEASE OF
   ADH BY POSTERIOR
   PITUITARY GLAND

3. NA
   RETENTION
   BY KIDNEYS
   (HYPERNATREMIA)

2. INCREASED
   VENOUS TONE

DECREASED URINE VOLUME
AND INCREASED BLOOD VOLUME

29

What other set of baroreceptors respond to hemorrhage in a manner similar to the arterial baroreceptors?

Answer and Explanation

The *cardiopulmonary receptors* respond to a decreased stretch by sending fewer impulses to their sensory neurons and this results in increases in systemic resistance, venous tone, renin secretion, antidiuretic hormone secretion, and possibly cardiac output.

30

(1:718; 2:6-66; 3:482; 4:520; 5:1067)
Cbga 3

Thirty percent of the blood volume of an anesthetized dog was removed. This was followed by a marked increase in peripheral resistance and a decrease in arterial pressure to 40 mm Hg. Next, the *ninth cranial nerve* was severed bilaterally, and the *arterial pressure* decreased to 20 mm Hg. This second decrease in the arterial pressure was probably due to the destruction of

A. parasympathetic motor neurons
B. sympathetic motor neurons
C. sensory neurons from the carotid sinus
D. sensory neurons from the carotid bodies
E. sensory neurons from volume receptors

Answer and Explanation

A. Parasympathetic motor neurons in the *glossopharyngeal nerve* play essentially no role in the control of arterial pressure.
B. There are no sympathetic neurons in this nerve.
C. At pressures as low as 40 mm Hg, the aortic *pressoreceptor* message level to the brain is near 0. At higher arterial pressure levels, decentralization of the carotid sinus causes an elevated arterial pressure.
D. Under the conditions of the experiment, there should be an intense stimulation of these *hypoxia*-sensing receptors that will not only affect the medullary respiratory centers but will also spill over into its neighbor, the *vasomotor center*. When this stimulation to the vasomotor center is lost, arterial pressure will decrease. Under resting conditions, the *carotid bodies* play little or no role in the control of blood pressure.

31

(1:318; 2:9-64; 3:291; 4:714; 5:907)
Cofa 3

A patient who has lost 30% of his blood volume can be successfully treated if that blood volume is returned to him within the first hour, but he may die if this transfusion is delayed 3 or 4 hours. It has been suggested that this condition is due in part to (1) a baroreceptor-induced reflex vasoconstriction in the viscera, (2) visceral ischemia, (3) deterioration of the capillary barrier in the GI tract, and (4) entrance of bacteria and toxins into the GI capillaries. Which of the following agents would be most effective in relieving the GI ischemia?

A. epinephrine, because it increases the cardiac output and decreases the peripheral resistance
B. the related catecholamine, norepinephrine
C. the beta adrenergic stimulating agent, isoproterenol
D. acetylcholine
E. an acetylcholine blocking agent for smooth muscle and the heart, such as atropine

Answer and Explanation

A, B. Epinephrine and norepinephrine cause vasoconstriction in the GI tract and therefore may not relieve its ischemia.
C. It will, by increasing cardiac output, tend to maintain arterial pressure while decreasing resistance to flow in the GI tract. Unfortunately, like all *catecholamines*, it may produce dangerous arrhythmias of the heart.

D. It will lower peripheral resistance to such an extent that there is no pressure left to maintain what little blood flow there is already present. It, unlike isoproterenol, does not elevate cardiac output.

E. Acetylcholine is not the cause of the problem in hypovolemia.

32

(1:425; 2:4-78; 3:432; 4:99; 5:1139)
Cbfa 3

A patient has suffered a severe *hemorrhage* at least 4 hours before admission to the hospital. She is given numerous transfusions of blood but is apparently bleeding into the alimentary, respiratory, urinary, and auditory tracts. The laboratory reports that her *prothrombin time* is infinitely long. How would this bleeding condition best be treated? By administering intravenously

A. sodium citrate until the prothrombin time is near normal
B. the antivitamin K drug, dicumarol, until the prothrombin time is near normal
C. heparin until the prothrombin time is near normal
D. serotonin
E. thrombin

Answer and Explanation

A. Sodium *citrate* would produce a *hypocalcemia* and a resultant hyperreflexia that might be fatal. It is most often used as an *anticoagulant* for the storage of blood outside the body.

B. One of the reasons for this problem may be a disseminated intravascular *coagulation*, which depletes the body of prothrombin as fast as it is added to the blood. *Dicumarol* interferes with prothrombin production by the liver.

C. *Heparin* is an anticoagulant that interferes with the formation of thrombin. In this case, it would be used to prevent prothrombin and fibrinogen consumption and would be withdrawn when the prothrombin concentration was at a more effective level.

D. The vasoconstriction produced by serotonin would be counterproductive.

E. This would cause the production of clots at or near the point of injection rather than at the points where bleeding is occurring.

33

(1:628; 2:2-91; 3:510; 4:332; 5:1282)
Cbqa 2

One week after a 70-kg subject loses 1000 ml of *blood*

A. the plasma volume is still below normal
B. the intestinal absorption of iron is greater than before hemorrhage
C. glomerular filtration is higher than before hemorrhage
D. the concentration of reticulocytes in the blood is below normal
E. there is a greater than normal cutaneous vasoconstriction

Answer and Explanation

A. The plasma volume, but not the blood volume, is back to normal within a day or two. Factors that increase the plasma volume include (1) a decrease in the capillary hydrostatic pressure, (2) an increase in plasma concentrations of renin, aldosterone, and antidiuretic hormone, (3) an increase in water and salt intake, (4) conservation of water and salt by the kidneys.

B. After hemorrhage, there is an increase in production of erythrocytes and, therefore utilization of iron for the formation of hemoglobin. The removal of iron from the plasma results in an increase in absorption of iron from the intestinal tract.

C. Glomerular filtration should be near normal at this time.

D. The reticulocyte is a precursor to the erythrocyte and increases its concentration in the blood during increases in red blood cell production.

E. The cutaneous vascular tone should be back to normal by now.

34

(1:536; 2:3-166;
3:479; 4:238;
5:1069)
HCN 2

What are the roles played by the *spinal cord, medulla oblongata, hypothalamus,* and *cerebral cortex* in the control of the cardiovascular system? After you have answered this question, compare your answer with those for the following three questions. Which of the following statements best characterizes the role of the *thoracic spinal cord* in the regulation of the heart and blood vessels?

A. it is devoid of efferent neurons controlling the heart
B. its efferent neurons to blood vessels are dependent upon impulses from the brain for stimulation
C. its efferent neurons are part of a simple reflex arc that includes afferent neurons that enter the cord without first going to the brain
D. statements A and B are correct
E. statements A and C are correct

Answer and Explanation

A. Preganglionic sympathetic neurons to the heart originate in the intermediolateral cell column of the thoracic cord.
B. Investigators have blocked the cervical cord in the dog with cocaine and have noted only a small decrease in arterial pressure. Transection of the cervical cord in human beings and dogs, on the other hand, causes a marked decrease in arterial pressure. In time, however, adrenergic sympathetic tone returns, the arterial pressure rises, and many of the *cardiovascular responses to limb movement and pain* return.
C. There are also reflex connections with the arterial baroreceptors and chemoreceptors that are in the brain.

35

(1:536; 2:3-166;
3:480; 4:238;
5:1069)
CHN 2

Which one of the following statements best characterizes the role of the *medulla oblongata* in the regulation of the heart and blood vessels?

A. it contains a vasomotor center on the right side of the floor of the fourth ventricle which controls blood vessels on both the ipsilateral and contralateral sides of the body
B. it contains a vasomotor center in which all descending neurons affecting the control of the blood vessels synapse
C. statements A and B are correct
D. it contains a cardioinhibitory center that, unlike the vasomotor center, is little influenced by impulses from arterial baroreceptors
E. all of the above statements are correct

Answer and Explanation

A. There is extensive *crossover* between the vasomotor centers on the right and left sides of the medulla.
B. Many of the axons from internuncial neurons in the hypothalamus descend to preganglionic sympathetic neurons in the thoracic cord without synapsing in the brain stem.
D. The vascular and cardiac centers of the brain are strongly influenced by impulses coming from the *arterial baroreceptors.*

36

(1:536; 2:3-167;
3:480; 4:242;
5:1071)
CHNp 2

Check each of the following statements that accurately characterizes the *hypothalamus.*

1. it is somewhat less sensitive to central nervous system depressants, such as the barbiturates, than the respiratory centers of the medulla
2. it functions as a synaptic link with the cerebral cortex and, as such, does not play a role in the control of heart rate and arterial pressure in the resting, unexcited subject
3. its destruction prevents increases in heart rate and peripheral resistance in response to decreases in carotid sinus distention

4. its destruction delays increases in stroke volume and heart rate resulting from exercise
5. it plays an important role in temperature regulation
6. it produces hormones and controls the release of hormones by the pituitary

Answer and Explanation

1. *Barbiturates* depress hypothalamic function and thereby cause changes in heart rate and arterial pressure at concentrations which have little or no effect on the hindbrain and the upper spinal cord.
2. It is not only an important link with the *cortex* but also an important influence on the vascular and cardiac centers of the medulla and their response to impulses from the *aortic arch* and *carotid sinus*.
3. It is through the *corticohypothalamic* pathway that we produce cardiovascular changes in anticipation of stresses such as *exercise*, examinations, etc.
4. Warming the blood to the hypothalamus causes *cutaneous vasodilation*.
5. The hypothalamus produces (1) a number of *releasing factors* (growth hormone-releasing factor, for example) which control pituitary secretion and (2) some *hormones* which are released by the pituitary (antidiuretic hormone, for example.)

37

(1:318; 2:3-114; 3:176; 4:715; 5:896)
Hn 1

Identify the response of the heart to the stimulation of vagal parasympathetic neurons to structures 1, 2, and 3 in the following diagram. (*Hint:* Structure 2 represents the atrial myocardium.)

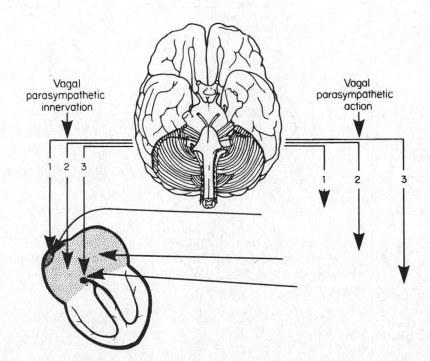

Answer and Explanation

1. A decreased *heart rate* or an ectopic beat (i.e., *vagal escape*)
2. A negative *inotropic* action on atrium
3. Atrioventricular conduction *delay or block*

Identify structures 2, 4, and 8 in the following diagram. Structures 2 and 4 contain important cardiovascular centers. Structure 8 represents an important group of peripheral receptors.

(1:536; 2:3-174; 3:481; 4:241; 5:1070)
CHNp 1

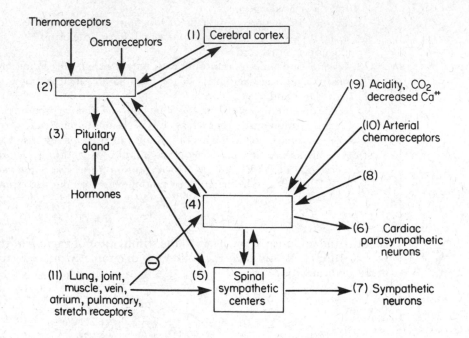

Answer and Explanation

2. *Hypothalamus* in the diencephalon
4. *Medulla* oblongata
8. Arterial stretch receptors in right and left carotid sinus and in aorta (also called pressoreceptors and *baroreceptors*)

(1:623; 2:3-124; 3:460; 4:249; 5:1039)
Hna 2

Contrast the response of a normal conscious subject and a subject without cardiovascular reflexes to (1) changing from a reclining to a standing position, (2) changing from a standing to a running condition, and (3) the intravenous injection of norepinephrine.

Answer and Explanation

	Conscious, Normal Subject	Subject without Cardiovascular Reflexes
1. Change from a reclining to a standing position	*Reflex* increase in heart rate and total systemic peripheral resistance (carotid sinus reflex)	*Orthostatic hypotension* (decreased carotid artery pressure causes inadequate cerebral blood flow)

2. Change from a standing to a running condition	*Anticipatory* increase in heart rate and cardiac output and a local vasodilation	No anticipatory changes
	Marked increases in *heart rate*	Small increases in heart rate
	During *exercise* the increases in heart rate and cardiac output are maintained, in part, by adrenergic sympathetic neurons	During exercise the increases in *stroke volume* and cardiac output are maintained, in part, by an increased end-diastolic volume (Frank-Starling mechanism)
3. Response to intravenous norepinephrine	Increased total systemic resistance Increase in arterial *pressure* Decreased heart rate and *cardiac output* (carotid sinus reflex)	Increased total systemic resistance Exaggerated increase in arterial pressure Increased heart rate and cardiac output (direct action on the heart)

(1:1055; 2:7-62; 3:311; 4:944)
CEKpq 1

Identify items 1, 2, 3, and 4 in the following diagram. Item 1 is a physical change in certain arteries. Items 2, 3, and 4 are molecules produced in the body.

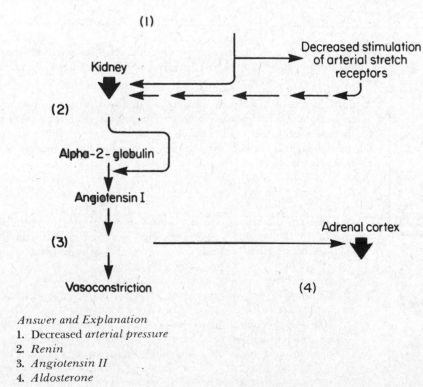

Answer and Explanation
1. Decreased *arterial pressure*
2. *Renin*
3. *Angiotensin II*
4. *Aldosterone*

41

(1:1055; 2:5-90;
3:309; 4:437;
5:1621)
EKCq 1

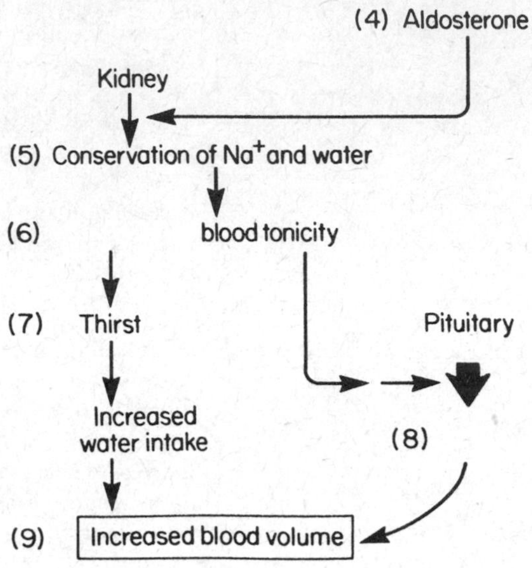

Identify the change (*increase* or *decrease*) in blood tonicity that is represented by item 6 and identify the molecule that is released by the pituitary (item 8) in the diagram below.

(4) Aldosterone

Kidney

(5) Conservation of Na^+ and water

(6)   blood tonicity

(7) Thirst   Pituitary

Increased
water intake   (8)

(9) | Increased blood volume |

Answer and Explanation
6. Increased blood *tonicity*
8. *Antidiuretic hormone* (ADH)

Questions 42 through 45 consist of two main parts: a statement and a reason for the statement. Choose the best answer for each of these questions from the options **A** through **E** below and mark your choice in the space provided.

A. both the statement and the reason are true and are related as cause and effect (i.e., as indicated)
B. both the statement and the reason are true but are not related as cause and effect
C. the statement is true but the reason is false
D. the statement is false but the reason is an accepted fact or principle
E. both the statement and the reason are false

42

(1:596; 2:3-137;
3:189; 4:944;
5:1100)
CHnf 2

Statement
_____ The stimulation of cardiac sympathetic neurons causes an increased coronary blood flow because

Reason
cardiac sympathetic neurons release a catecholamine that stimulates alpha adrenergic receptors.

Answer and Explanation
B. *Catecholamines*, by stimulating beta$_1$ adrenergic receptors, cause an accumulation of *metabolites* which in turn causes *coronary dilation*. Epinephrine and norepinephrine also stimulate the alpha receptors of the coronary arterioles. Although this action alone causes a constriction, the metabolic action of these agents, in the absence of beta blockage, predominates, and there is a net decrease in coronary resistance.

43

*(1:586; 2:3-112;
3:458; 4:163; 5:993)*
HWMq 3

Statement
_____ Increases in heart rate (from 60 to 100 per minute) and end-diastolic volume each cause increases in contractility because

Reason
an increase in contractility is always associated with a shift of the Frank-Starling curve to the left.

Answer and Explanation
D. Increases in heart rate do cause increases in *contractility* (+ inotropic action). This is sometimes called the *treppe* or staircase phenomenon. Increases in ventricular *end-diastolic volume*, like positive inotropic agents and procedures, cause increases in ventricular stroke work but, by definition, do not affect inotropicity. In other words, a change in end-diastolic volume moves you to a different spot on a single *Frank-Starling* curve, and a change in contractility or inotropicity moves you to a different curve. It has only been during the past 2 decades that physiologists and clinicians have recognized the importance of the observation that there is a family of Frank-Starling curves for each of the heart's chambers. See the illustrations in question 39, page 95.

44

*(1:504; 2:3-118;
3:460; 4:510; 5:997)*
HWpf 2

Statement
_____ A doubling of left ventricular stroke volume will have the same effect on the O_2 consumption of the heart as a doubling of left ventricular mean ejection pressure because

Reason
either change will cause a doubling of left ventricular stroke work.

Answer and Explanation
D. Increases in ventricular stroke volume resulting from increases in ventricular *end-diastolic volume* cause increases in *efficiency*, whereas increases in *mean ejection pressure* do not cause increases in efficiency.

45

(1:590; 2:3-175)
Cp 2

Statement
_____ When one changes from a reclining to a standing position, there is a reflex increase in heart rate because

Reason
when one stands, there is an increase in the pressure in the arteries of the thigh and leg.

Answer and Explanation
B. The increased heart rate is due to the decreased pressure in the *carotid sinuses* and the aortic arch.

11
Capillaries and Sinusoids

1

(1:517; 2:4-88; 3:463; 4:358; 5:1085)
Cd 1

What one *structural feature* do all blood capillaries have in common?

A. absence of intracellular fenestrations in the endothelial cells
B. presence of intracellular fenestrations
C. a discontinuous endothelium
D. a continuous basement membrane
E. they remain patent in the healthy subject

Answer and Explanation
A. The blood capillaries of the heart, skeletal muscle, lung, nervous system, dermis, subcutaneous and adipose tissue, and the placenta are devoid of a fenestrated endothelium.
B. Fenestrations are found in the capillaries of the glands, renal glomerulus, and vagina.
C. The cerebral capillaries apparently have a continuous endothelium with no apparent intercellular space between one endothelial cell and the next.
D. This helps to prevent overdistention of blood capillaries.
E. At any one point in time, most of the blood capillaries in the body are collapsed. It has been noted that during strenuous exercise, the number of *patent capillaries* in an active muscle may be five times that in the resting state.

2

(1:517; 2:4-88; 3:463; 4:358; 5:1085)
Cbd 1

Which structures in the body are normally devoid of *blood capillaries*?

A. cartilaginous plates of long bones
B. lenses of the eyes
C. the cartilaginous plates and the lens
D. corneas
E. dermis of the skin

Answer and Explanation
C.
D. They are sparsely concentrated in the cornea.

3

(1:528; 2:4-88;
3:464; 4:370;
5:1085)
Cd 1

Select the one statement below that best characterizes *lymph capillaries*. Lymph capillaries

A. have a smaller diameter than blood capillaries
B. are less permeable than blood capillaries
C. have no endothelial lining
D. have a discontinuous basement membrane
E. are found only in the liver, spleen, bone marrow, lymph nodes, and gastrointestinal tract

Answer and Explanation
A. The diameter of patent blood capillaries is about 7 μm and that for lymph capillaries may exceed 14 μm.
B. Albumin passes much more freely through lymph capillaries.
D.
E. They are found throughout the body adjacent to the blood capillaries (i.e., in skeletal muscle, the heart, the dermis of the skin, etc.).

4

(1:617; 3:463;
4:350; 5:1085)
Cd 1

Select the one statement below that best characterizes the *sinusoids*. The sinusoids

A. have a smaller diameter than lymph capillaries
B. are not found in skeletal muscle
C. have a continuous endothelial lining
D. have a continuous basement membrane
E. are less permeable than lymph capillaries

Answer and Explanation
B. Most of the sinusoids of the body are in the liver, bone marrow, and spleen.
C. Some of the spaces between the cells that line their lumen exceed 1 μm.
D. The sinusoids of the bone marrow appear to lack any basement membrane. The sinusoids of the spleen have a discontinuous basement membrane.

5

(1:522; 2:4-489;
3:472; 4:359; 5:8)
Cdg 2

Carbon dioxide moves readily from the extravascular space into the blood. The mechanism or mechanisms responsible for the permeability of skeletal muscle capillaries to CO_2 is

A. the lipid solubility of CO_2
B. pinocytosis
C. diapedesis
D. endothelial fenestrae
E. active transport

Answer and Explanation
A. Carbon dioxide moves through the *capillary barrier* more readily than do gases of comparable or smaller size (O_2 for example).
B. Particles with a radius of less than 0.03 μm can be transported across the endothelial lining by *microphagocytosis*. This is a process that is unimportant for the transport of gases, hexoses, and water, because it is incapable of moving large quantities of any molecule. It is important in moving large molecules to which the capillary is impermeable and which are important in trace amounts (vitamin B_{12}, for example).
C. This is a mechanism by which neutrophils and other wandering cells pass the endothelial barrier.
D. The endothelium of the capillaries of skeletal muscle is devoid of fenestrae.

6

(1:526; 2:9-52; 3:491; 4:347; 5:1787)
Cdb 2

If radioactive urea were injected into the right fermoral artery, the time it would take for it to reach equilibrium in the various interstitial fluids would vary. The *interstitial fluid* into which it would <u>move most slowly</u> would be that found in the

A. brain
B. heart
C. left leg
D. right leg
E. kidney

Answer and Explanation

A. With the notable exception of CO_2, most substances do not move readily between the blood and the intercellular fluid of the brain. There is said to be a *blood-brain barrier*. Part of the reason for this barrier is the processes of the astrocytes that surround the cerebral capillaries. Urea in the blood takes more than 5 hours to come into equilibrium with the intercellular fluid of the brain. In the case of the other tissues mentioned, this occurs in less than an hour.

7

(1:823; 2:5-3; 3:13; 4:391; 5:1150)
Cd 2

A substance was injected intravenously and found to be distributed through 30% of the body water. It probably

A. did not pass freely through the blood capillaries
B. was distributed uniformly throughout the body water
C. did not enter the cells of the body
D. was not excreted or utilized in the body
E. was excluded from the cerebrospinal fluid

Answer and Explanation

C. Since approximately 7% of the body water is plasma and 21% is interstitial fluid, it seems reasonable to assume that if a substance is restricted to 30% of the body water, it is not entering the body cells to an important degree, since the body's cells contain approximately 72% of the body water.

8

(1:522; 2:3-205; 3:473; 4:367; 5:1089)
Cdp 2

The following data were collected from a skeletal muscle capillary:

Capillary hydrostatic pressure:	30 mm Hg
Tissue hydrostatic pressure:	2 mm Hg
Effective osmotic pressure:	23 mm Hg
Capillary colloid osmotic pressure:	25 mm Hg
Tissue colloid osmotic pressure:	2 mm Hg

The *filtration pressure* in this capillary was

A. greater than 23 mm Hg pushing fluid out of the capillary
B. between 17 and 23 mm Hg pushing fluid out of the capillary
C. between 9 and 16 mm Hg pushing fluid out of the capillary
D. between 1 and 8 mm Hg pushing fluid out of the capillary
E. negative, i.e., there was a net influx into the capillary

Answer and Explanation

D. If you let a plus (+) represent a pressure moving fluid out of and a minus (−) represent a pressure moving fluid into a capillary, then:

Effective *hydrostatic pressure* = 30 − 2 = + 28 mm Hg.

Effective *osmotic pressure* = 2 − 25 = − 23 mm Hg.

Filtration pressure = 28 − 23 = + 5 mm Hg.

9

(1:522; 2:4-92;
3:472; 4:363;
5:1089)
Cdva 2

A patient exhibits swelling of the ankles and a bloated abdomen and has a history of *malnutrition*. The bloated abdomen is probably due to

A. increased intestinal gas
B. slow, chronic abdominal hemorrhage
C. increased capillary hydrostatic pressure
D. increased capillary colloid osmotic pressure
E. decreased capillary colloid osmotic pressure

Answer and Explanation
E. Vitamin deficiency, amino acid deficiency, caloric deficiency, liver damage, or excessive loss of plasma proteins in the urine can produce a decreased capillary colloid osmotic pressure, which causes an increased movement of water from the plasma and into the abdomen and other intercellular spaces. The other suggestions are either incorrect or highly unlikely.

10

(1:522; 2:4-91;
3:472; 4:363;
5:1087)
Cnopq 2

Blood flow ($\dot{Q}$), venous oxygen concentration (Cvo_2), and *interstitial fluid volume* (ISF) were measured in the leg. The metabolism of the structures in the leg was kept constant as was the arterial Po_2 and pressure. Under these circumstances either an increased venous pressure or a decreased vasoconstrictor tone to the leg would cause:

A. an increased $\dot{Q}$
B. an increased Cvo_2
C. an increased ISF
D. an increased $\dot{Q}$ and Cvo_2
E. an increased $\dot{Q}$ and ISF

Answer and Explanation
C. Blood flow through the capillaries is directly related to the *perfusion pressure* (arterial pressure-venous pressure) and inversely related to the *resistance* (i.e., the degree of vasoconstriction). Increases in capillary hydrostatic pressure produce increases in ISF and *lymph flow* and are caused by increases in venous pressure or arteriolar dilation:

	$\dot{Q}$	Cvo_2	ISF
Increased venous pressure:	decrease	decrease	increase
Decreased vasoconstrictor tone:	increase	increase	increase

11

(1:528; 2:4-89;
3:173; 4:367;
5:1089)
Con 2

The *loss of fluid* by the systemic blood capillaries in a healthy subject is

A. more than the gain of fluid by the capillaries
B. usually increased by a local arteriolar dilation
C. both A and B are correct
D. decreased by the stimulation of adrenergic sympathetic neurons to the veins
E. increased by the dilation of postcapillary sphincters

Answer and Explanation
C. The *lymphatic system* returns the liquid lost from the capillaries back to the circulation. Arteriolar dilation permits more of the arterial pressure to be transferred to the capillaries.
D. The importance of adrenergic sympathetic neurons to veins is not fully understood. We do know, however, that under experimental conditions, their stimulation can increase smooth muscle tone, decrease venous compliance, and increase venous pressure. Increases in venous pressure increase the capillary hydrostatic pressure.
E. This decreases capillary hydrostatic pressure.

12

(1:528; 2:7-45;
3:474; 4:375;
5:1511)
CKHE 2

Which one of the following is least likely to cause *edema* in the adult?

A. nephritic syndrome
B. congestive heart failure
C. liver malfunction due to chronic alcoholism
D. mastectomy
E. administration of thyroxine to a hypothyroid patient

Answer and Explanation

E. When hypothyroidism develops in the adult, it is called myxedema (Gk.*myxa*, mucus, + *oidema*, swelling). In this condition, there is a high incidence of edema in the (1) face (79%), (2) eyelids (90%), and (3) periphery (55%). The edema of the skin is, for the most part, due to the accumulation of a variety of protein-carbohydrate complexes (i.e., colloids) in the intercellular space and can be relieved by the administration of *thyroxine.*

13

(1:522; 2:4-93;
3:474; 4:375;
5:1158)
CdKH 3

What are the mechanisms responsible for producing *edema* in the conditions listed in A through D of question 12?

Answer and Explanation

A. In the nephritic syndrome, a loss of albumin in the urine causes a reduction of the capillary *colloid osmotic pressure.*
B. In congestive heart failure, there is an increased venous pressure that causes an increased capillary *hydrostatic pressure.*
C. In liver malfunction, there may be a decreased production of albumin and therefore a decreased capillary *colloid osmotic pressure.*
D. In mastectomy, the *lymphatic drainage* from the arms is markedly decreased.

14

(1:529; 2:2-91;
3:473; 4:824; 5:30)
CfDE 2

The blood capillaries and systemic veins are more important than the *lymph capillaries and ducts* in the transportation of

A. cholesterol and stearic acid from the intestine
B. glucose from the intestine
C. cholesterol, stearic acid, and glucose from the intestine
D. albumin from the intercellular space of the intestine
E. renin from its site of production in the kidney

Answer and Explanation

A. Most of the long-chain fats enter the circulation via the lymphatic system rather than the *hepatic portal system.*
B. Glucose and most other nutrients move readily into the blood capillaries of the intestine.
D. The lymphatic system moves a quantity of albumin approximately equivalent to half of the total plasma albumin into the circulatory system each day. This consists of albumin that has escaped from the blood through the blood capillaries. It moves more readily from the perivascular space into the lymph capillaries than into the blood capillaries because of the greater permeability of the lymph capillaries.

15

(1:529; 2:4-88;
3:473; 4:370;
5:1091)
Cfb 2

The *lymph ducts* differ from the veins in that the lymph ducts

A. are devoid of similunar valves
B. contain a fluid with a higher velocity of flow
C. contain a fluid almost devoid of glucose
D. contain a fluid devoid of leukocytes
E. have none of the above characteristics

Answer and Explanation

A. The *semilunar valves* in the veins and lymph ducts facilitate the transport of blood and lymph.

B. It has been estimated that in the adult the *thoracic duct* moves approximately 4 liters of lymph into the blood per day and the inferior vena cava moves over 7000 liters per day.

C. Thoracic duct lymph in the dog averages 124 mg of glucose/100 ml.

D. In the rat, the thoracic duct carries approximately 10^9 cells (i.e., *erythrocytes* and *leukocytes*) to the blood per day.

Ⓔ.

16

(1:521; 2:6-19; 3:473; 4:358; 5:1088)
Cdf 2

In a healthy individual two substances, f and g, are found to have a lower concentration on the venular side of a capillary than the arteriolar side. Their concentration in the blood changes as they pass through the capillary in the following manner.

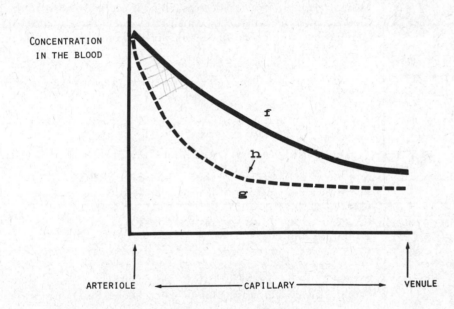

What valid conclusions can you make from this figure?

A. substance f is blood flow-limited (i.e., perfusion-limited)
B. substance g is diffusion-limited
C. both A and B are correct
D. substance g could be O_2
E. all of the above are correct

Answer and Explanation

A. Substances in the blood (f in this case) that do not reach equilibrium with the tissues in their passage through the capillary are diffusion-limited.

B, Ⓓ. Substance g has reached equilibrium at point h. Oxygen has a concentration curve similar to g and is flow-limited. The delivery of substance g to the tissues can be increased by increasing blood flow to those tissues.

17

(1:520; 2:3-25; 3:467; 4:367; 5:1026)
Cfo 2

A local *constriction* of the cutaneous arterioles of the hand can cause a local (select the one best answer)

A. increase in cutaneous capillary flow
B. increase in capillary transmural pressure
C. both of the above
D. complete closure of the capillary lumen
E. increase in lymph flow

Answer and Explanation
D. Constriction of precapillary sphincters, metarterioles, and arterioles lying between arteries and capillaries may so lower the pressure in the capillaries to cause the loss of their lumen. The intraluminal, hydrostatic pressure at which vessels collapse is called their *critical closing pressure* and may vary from 10 to 60 mm Hg in the skin.

18

(1:830; 2:5-25; 3:565; 4:403; 5:1167)
Cp 2

Most *arterioles* produce a decrease in *hydrostatic pressure* in the capillaries with which they are in series when they constrict. Name a specific arteriole that has the opposite action (i.e., produces an increased capillary pressure when it constricts).

Answer and Explanation
When the renal *efferent arteriole* constricts it causes an increased pressure in the glomerular capillaries.

19

(1:695; 2:4-44; 3:533; 4:509; 5:1732)
Rgb 2

As blood passes through the capillary beds of skeletal muscle, there is

A. a net flux of Cl^- out of the red blood cell (RBC)
B. an increased affinity of hemoglobin for O_2
C. both of the above
D. an increased affinity of hemoglobin for H^+
E. both A and D

Answer and Explanation
A. There is a net flux of HCO_3^- out of and of Cl^- into the RBC.
B. There is a decreased affinity of hemoglobin for O_2 owing to the increased H^+ and CO_2 concentration.
D. *Deoxyhemoglobin* has a greater affinity for H^+ than does oxyhemoglobin.

12
Extravascular Fluid Systems

1

(2:9-52; 3:489; 4:383; 5:1218)
NCfpa 1

Check each of the following statements concerning *cerebrospinal fluid* (CSF) that is true:

1. over 30% of the cerebrospinal fluid is formed by the dural sinuses
2. over 30% of the cerebrospinal fluid is formed by the choroid plexi
3. normally, occlusion of the jugular veins causes a prompt (less than 12 seconds) rise in CSF pressure in the subarachnoid space at the level of the fourth lumbar vertebra
4. The arachnoid villi that project into the superior sagittal sinus tend to facilitate the movement of fluid into this venous sinus but also tend to prevent filtration from the venous sinus into the subarachnoid space
5. CSF reduces the weight of the central nervous system

Answer and Explanation
1. CSF is absorbed into the blood at the *dural sinuses.*
2. The major sites for the formation of CSF are probably the *choroid plexi* in the ventricles of the brain, the glial elements in the ependymal wall of the brain, and the cerebral blood vessels.
3. In the adult, the *subarachnoid space* extends to approximately S-2. Jugular occlusion should cause an increased CSF pressure throughout the subarachnoid space and ventricles of the brain, since it decreases absorption of CSF at the dural sinuses but has little effect on formation. If occlusion does not promptly increase CSF pressure throughout the system, this is called a positive *Queckenstedt-Stokey sign* of pathology.
4. These villi collapse when CSF pressure exceeds cerebral vein pressure.
5. The brain and cord weigh about 1500 g outside the body but inside the body are *buoyed up* so that their weight is reduced to 50 g.

2

(2:9-50; 3:489; 4:383; 5:1225)
NCpi 2

Check one of the following statements about *cerebrospinal fluid* that is true:

A. when CSF pressure increases by more than 5 mm Hg its formation ceases
B. CSF tends to produce an exaggerated transmural pressure in the blood vessels of the brain
C. CSF is a more poorly buffered solution than blood
D. CSF has approximately the same concentration of Ca^{++} and K^+ as plasma
E. all of the above statements are true

(Continued)

Answer and Explanation

A. CSF formation is not dependent upon a high filtration pressure. It is formed by an *active process*.

B. CSF tends to maintain a fairly constant *transmural pressure* during the Valsalva maneuver, during headward *acceleration* and *deceleration*, and when one changes from a horizontal to a vertical position. For example, in the standing position, while *gravity* is tending to pull blood caudad, it is also tending to pull CSF caudad. In this way, CSF tends to maintain the *patency* of the cerebral vessels. During headward deceleration the CSF tends to prevent *rupture* of the cerebral blood vessels.

C. Plasma has 300 times the concentration of proteins as CSF and therefore is a better buffered solution.

D. CSF has a lower concentration of K^+ and Ca^{++} than plasma (0.62 times the concentration of K^+ and 0.49 times the concentration of Ca^{++}).

3

(2:9-53; 4:384; 5:1238)
Cpa 2

Obstruction of the *cerebral aqueduct* causes

A. no important change in CSF pressure in the lateral ventricles of the brain because there are alternate pathways between the first, second, and third ventricles and the fourth ventricle

B. an important decrease in the CSF pressure in the lateral ventricles

C. a cessation of CSF production in the lateral ventricles

D. both B and C

E. an important increase in CSF pressure in the lateral ventricles

Answer and Explanation

E. When the cerebral aqueduct is blocked, the *CSF pressure* may change from a normal value of 100 mm H_2O (7 mm Hg) to one of 160 mm H_2O (12 mm Hg). In the fetus or newborn child, this may affect brain development and cause *hydrocephalus*. In the adult, it may cause brain ischemia.

4

(1:97; 2:8-67; 3:109; 4:386; 5:1240)
Cspd 2

Intraocular fluid (aqueous humor) differs from cerebrospinal fluid in that intraocular fluid

A. has a protein concentration similar to that in plasma

B. does not cause a serious increase in pressure when reabsorption is decreased

C. is an ultrafiltrate of plasma

D. serves no important nutritive function

E. contains a higher concentration of ascorbic acid

Answer and Explanation

A. Intraocular fluid has a lower *protein* concentration than either plasma or CSF.

B. The pressure may rise from a normal value of 20 mm Hg to one in excess of 80 mm Hg. Pressures as high as 40 mm Hg can cause *blindness*.

C. It, like CSF, is formed by active transport.

D. It is an important source of nutrient for the *lens*.

E. It contains 18 times the concentration of *ascorbic acid* found in plasma and 12 times the concentration found in CSF.

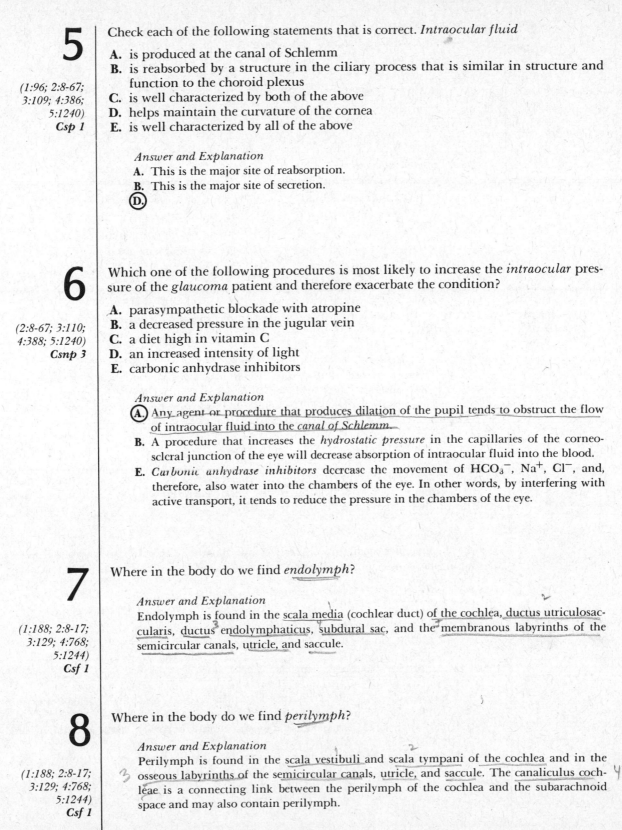

5

*(1:96; 2:8-67;
3:109; 4:386;
5:1240)*
Csp 1

Check each of the following statements that is correct. *Intraocular fluid*

A. is produced at the canal of Schlemm
B. is reabsorbed by a structure in the ciliary process that is similar in structure and function to the choroid plexus
C. is well characterized by both of the above
D. helps maintain the curvature of the cornea
E. is well characterized by all of the above

Answer and Explanation
A. This is the major site of reabsorption.
B. This is the major site of secretion.
(D.)

6

*(2:8-67; 3:110;
4:388; 5:1240)*
Csnp 3

Which one of the following procedures is most likely to increase the *intraocular* pressure of the *glaucoma* patient and therefore exacerbate the condition?

A. parasympathetic blockade with atropine
B. a decreased pressure in the jugular vein
C. a diet high in vitamin C
D. an increased intensity of light
E. carbonic anhydrase inhibitors

Answer and Explanation
(A.) Any agent or procedure that produces dilation of the pupil tends to obstruct the flow of intraocular fluid into the *canal of Schlemm.*
B. A procedure that increases the *hydrostatic pressure* in the capillaries of the corneoscleral junction of the eye will decrease absorption of intraocular fluid into the blood.
E. *Carbonic anhydrase inhibitors* decrease the movement of HCO_3^-, Na^+, Cl^-, and, therefore, also water into the chambers of the eye. In other words, by interfering with active transport, it tends to reduce the pressure in the chambers of the eye.

7

*(1:188; 2:8-17;
3:129; 4:768;
5:1244)*
Csf 1

Where in the body do we find *endolymph*?

Answer and Explanation
Endolymph is found in the scala media (cochlear duct) of the cochlea, ductus utriculosaccularis, ductus endolymphaticus, subdural sac, and the membranous labyrinths of the semicircular canals, utricle, and saccule.

8

*(1:188; 2:8-17;
3:129; 4:768;
5:1244)*
Csf 1

Where in the body do we find *perilymph*?

Answer and Explanation
Perilymph is found in the scala vestibuli and scala tympani of the cochlea and in the osseous labyrinths of the semicircular canals, utricle, and saccule. The canaliculus cochleae is a connecting link between the perilymph of the cochlea and the subarachnoid space and may also contain perilymph.

9

(1:407; 2:8-17; 3:135; 4:451; 5:1226)
Csdc 1

Which of the fluids listed below normally have *protein concentrations* of less than 3 mg/ml (*Hint*: Plasma has a concentration of 72 mg of protein/ml)?

A. cerebrospinal fluid
B. intraocular fluid
C. endolymph
D. perilymph
E. amniotic fluid
F. urine

Answer and Explanation

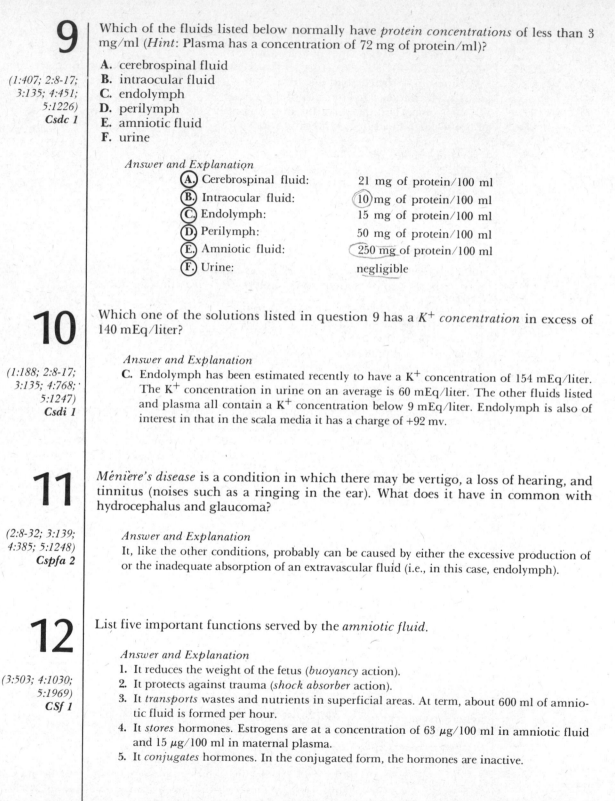

(A.) Cerebrospinal fluid:	21 mg of protein/100 ml
(B.) Intraocular fluid:	(10) mg of protein/100 ml
(C.) Endolymph:	15 mg of protein/100 ml
(D.) Perilymph:	50 mg of protein/100 ml
(E.) Amniotic fluid:	250 mg of protein/100 ml
(F.) Urine:	negligible

10

(1:188; 2:8-17; 3:135; 4:768; 5:1247)
Csdi 1

Which one of the solutions listed in question 9 has a *K^+ concentration* in excess of 140 mEq/liter?

Answer and Explanation

C. Endolymph has been estimated recently to have a K^+ concentration of 154 mEq/liter. The K^+ concentration in urine on an average is 60 mEq/liter. The other fluids listed and plasma all contain a K^+ concentration below 9 mEq/liter. Endolymph is also of interest in that in the scala media it has a charge of +92 mv.

11

(2:8-32; 3:139; 4:385; 5:1248)
Cspfa 2

Ménière's disease is a condition in which there may be vertigo, a loss of hearing, and tinnitus (noises such as a ringing in the ear). What does it have in common with hydrocephalus and glaucoma?

Answer and Explanation

It, like the other conditions, probably can be caused by either the excessive production of or the inadequate absorption of an extravascular fluid (i.e., in this case, endolymph).

12

(3:503; 4:1030; 5:1969)
CSf 1

List five important functions served by the *amniotic fluid*.

Answer and Explanation

1. It reduces the weight of the fetus (*buoyancy* action).
2. It protects against trauma (*shock absorber* action).
3. It *transports* wastes and nutrients in superficial areas. At term, about 600 ml of amniotic fluid is formed per hour.
4. It *stores* hormones. Estrogens are at a concentration of 63 μg/100 ml in amniotic fluid and 15 μg/100 ml in maternal plasma.
5. It *conjugates* hormones. In the conjugated form, the hormones are inactive.

13
Hemostasis

1

(1:411; 2:4-78; 3:430; 4:92; 5:1137)
Cbf 1

List four important responses of the body that prevent *blood* loss after the rupture of a small blood vessel.

Answer and Explanation
1. Vasoconstriction.
2. Formation of insoluble *fibrin* threads.
3. Formation of a platelet *plug*.
4. An increased perivascular pressure associated with the production of a *hematoma*.

2

(1:431; 2:3-230; 3:431; 4:92; 5:1137)
Cbf 1

Which one of the following is released by *blood platelets* during hemorrhage and tends to produce *vasoconstriction*?

A. serotonin
B. histamine
C. thrombosthenin
D. accelerator globulin
E. bradykinin

Answer and Explanation
A.
B. *Histamine* produces vasodilation in the normal systemic arteriole.
C. *Thrombosthenin* is a contractile substance in the platelet that causes clot retraction.

3

(1:433; 2:4-78; 3:431; 4:92; 5:1137)
Cbf 1

In questions 1 and 2, we have noted that *platelets* prevent bleeding by (1) forming a white thrombus (i.e., a plug) and (2) releasing a vasoconstrictor. What other important functions do platelets serve?

Answer and Explanation
3. Through the release of platelet factors, they initiate and accelerate the formation of a *fibrin clot*.
4. They are essential for *clot retraction*.
5. Possibly, they are also deposited in the capillary membrane, where they contribute to its integrity.

4

Which one of the following statements is most correct? A *procoagulant* not normally circulating in the plasma is

A. prothrombin
B. fibrinogen
C. antihemophilic factor (factor VIII)
D. Ac-globulin (factor V or proaccelerin)
E. none of the above

(1:417; 2:4-80; 3:431; 4:93; 5:1138)
Cbf 1

Answer and Explanation
E. All of the factors listed are procoagulants found in plasma.

5

Dicumarol is a drug that impairs the utilization of *vitamin K* by the liver. Dicumarol therapy, therefore, would decrease the plasma concentration of which one of the following *procoagulants*?

A. prothrombin
B. fibrinogen
C. antihemophilic factor (factor VIII)
D. Ac-globulin (factor V)
E. none of the above

(1:414; 2:4-87; 3:432; 4:101; 5:1141)
Cbfa 1

Answer and Explanation
A. Avitaminosis K causes either a decreased plasma concentration or a decreased production of the following factors during coagulation:

- prothrombin
- serum prothrombin conversion accelerator (SPCA, factor VII, autoprothrombin I)
- plasma thromboplastin component (PTC, factor IX, Christmas factor, antihemophilic factor B, autoprothrombin II)
- Stuart-Prower factor (factor X, autoprothrombin III)

These are all procoagulants that are either produced by the liver or whose precursors are produced by the liver.

6

Which one of the following combinations of substances in plasma will cause the production of a *clot*?

A. prothrombin, accelerator globulin, antihemophilic factor, platelet factor 3, fibrinogen
B. thrombin and fibrinogen
C. prothrombin, accelerator globulin, tissue extract, Ca^{++}
D. accelerator globulin, antihemophilic factor, platelet factor 3, Ca^{++}, fibrinogen
E. prothrombin, accelerator globulin, fibrinogen, Ca^{++}

(1:415; 2:4-80; 3:431; 4:94; 5:1141)
Cbfc 2

Answer and Explanation
B. Thrombin catalyzes the conversion of fibrinogen to a monomer, fibrin. Fibrin monomers undergo end-to-end and side-to-side polymerization to form extensive networks of fibrin threads.

7

*(1:418; 2:4-80;
3:431; 4:93; 5:1141)*
Cbfc 2

Indicate what substance should be added to the plasma solutions in parts A, C, D, and E of question 6 in order to form a clot.

Answer and Explanation

A. All of the factors listed in question 6, with the exception of fibrinogen, are primarily important in that they facilitate the conversion of prothrombin to thrombin. In solution A, the addition of Ca^{++} (as $CaCl_2$, for example) causes the formation of a clot.

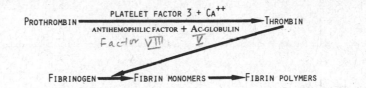

C. The addition of *fibrinogen* causes the formation of a clot in this solution.
D. The addition of *prothrombin* causes the solution to clot.
E. The addition of *tissue extract* (i.e., thromboplastin) causes the formation of a clot.

8

*(1:429; 2:4-84;
3:431; 4:92; 5:1137)*
Cbfc 3

Rheomacrodex (low molecular weight *dextran*) is administered to a patient and found to have no effect on his recalcification plasma *clotting time* or one-stage *prothrombin time*, but does increase his *bleeding time*. Which of the following statements is the most likely explanation of this action? Rheomacrodex

A. inhibits platelet aggregation
B. inhibits rouleau formation
C. inhibits fibrin polymerization
D. inhibits the action of thrombin
E. facilitates the activation of factor X

Answer and Explanation

A. Both the recalcification plasma clotting time and the prothrombin time are performed on plasma (i.e., in the absence of platelets). The fact that these two tests yield normal results is indicative of a normal clotting system. The fact that the bleeding time is prolonged is indicative of a failure in some other aspect of the hemostatic mechanism.
B. There is no evidence to indicate that the tendency of red cells to form a rouleau affects the bleeding time.
E. Factor X (Stewart factor) catalyzes the formation of thrombin.

9

*(1:427; 2:4-84;
3:431; 4:92; 5:1138)*
Cbfc 2

Which one of the following statements is correct? *Serum* does not contain

A. prothrombin
B. plasma thromboplastin component (PTC, factor IX)
C. Ca^{++}
D. serum prothrombin conversion accelerator (SPCA, factor VII)
E. any of the above

Answer and Explanation

A. Serum is the solution released from a blood clot during clot retraction. Normally, it is devoid of prothrombin, fibrinogen, and factors V (Ac-globulin), VIII (antihemophilic factor), and XI (plasma thromboplastin antecedent: PTA). Apparently, these substances are consumed in the clotting reaction.
B, C, D. PTC, Ca^{++}, and SPCA all either catalyze or facilitate the formation of thrombin without being completely consumed. Factors X, XII, and XIII are also found in serum.

10

(1:415; 2:4-82;
3:431; 4:93; 5:1142)
Cbfc 1

Contrast the roles of plasma *transglutaminase* (factor XIII) and *plasmin* in hemostasis.

Answer and Explanation

Plasma transglutaminase (*fibrin-stabilizing* factor, FSF), in the presence of thrombin, changes from an inactive to an active form. Active plasma transglutaminase facilitates the conversion of fibrin-s (soluble in a 5 molar urea solution) to fibrin-i (insoluble in 5 molar urea). Plasmin (fibrinolysin) facilitates the digestion of fibrin and fibrinogen and the formation of anticoagulants. It is important in bringing about the *dissolution of a blood clot.*

11

(1:425; 2:4-83;
3:432; 4:98; 5:1142)
Cbfc 1

What substances facilitate the conversion of *profibrinolysin* to fibrinolysin?

Answer and Explanation

Tissue enzymes (cytofibrinokinase), bacterial enzymes (streptokinase and staphylokinase), and certain substances found in urine (urokinase) facilitate this chemical change.

12

(1:428; 2:4-83;
3:432; 4:98; 5:1142)
Cbc 2

You have a patient in whom you wish to prolong the *Lee-White clotting time* for the next 10 hours. What would be the procedure of choice?

A. an intravenous drip of a heparin solution
B. an intravenous injection of sodium citrate
C. the administration of dicumarol
D. any of the above would be effective—one would make a choice on the basis of cost, availability, and facilities
E. intravenous vitamin K

Answer and Explanation
A.

13

(1:428; 2:4-86;
3:432; 4:101;
5:1141)
Cbc 2

Why is the administration of *sodium citrate* and *dicumarol* inappropriate in question 12?

Answer and Explanation

• *Sodium citrate* (a chelating agent) removes Ca^{++} from solution. In doing this, it produces a hyperexcitability that can lead to a laryngeal spasm and death due to asphyxiation. Injections of sodium citrate that produce no change in the clotting time can cause serious changes in excitability.

• *Dicumarol* exerts its anticoagulant action by decreasing the production of prothrombin. It, unlike heparin, would take many hours before it increased the coagulation time.

14

*(1:428; 2:4-86;
3:432; 4:101;
5:1141)*
Cbc 2

Under what circumstances is (1) *sodium citrate* the anticoagulant of choice and (2) *dicumarol* the anticoagulant of choice?

Answer and Explanation
1. Sodium citrate is used to prevent clotting when blood is stored outside the body. It is cheaper than heparin. When small volumes of citrated blood are transfused into a patient, no supplemental calcium is necessary.
2. Dicumarol is used when you wish to increase the Lee-White clotting time in a patient over a long period of time and an immediate effect is not necessary. It is less expensive than heparin and more convenient to use, since it is effective when taken once a day by mouth.

15

*(1:434; 2:4-83;
3:430; 4:101;
5:1141)*
Cbca 3

Check each of the following statements that is true:

1. all coagulopathies (deficiencies in one of the 13 clotting factors) cause an increased Lee-White whole-blood clotting time
2. all coagulopathies cause an increased one-stage prothrombin time
3. all thrombocytopathies (platelet dysfunctions) are associated with a prolonged clot retraction time
4. a prolonged Lee-White whole-blood clotting time is usually associated with a prolonged bleeding time
5. a prolonged prothrombin time is always associated with a prolonged Lee-White time

Answer and Explanation
All of the above statements are false:
1. Deficiencies in fibrin-stabilizing factor (factor XIII) and serum prothrombin conversion accelerator (SPCA, factor VII) do not cause a prolonged *Lee-White clotting time*. In factor VII deficiency, however, the prothrombin time is prolonged.
2. Deficiencies in factors VIII, IX, XI, XII, and XIII do not produce a prolonged *prothrombin time*. This does not mean that all of these procoagulants are unimportant in the activation of prothrombin but that in a prothrombin time procedure an *excess of thromboplastin* is added to the sample. This excess will mask the action of the above factors.
3. In hereditary, hemorrhagic thrombasthenia, *clot retraction* is either absent or prolonged. In von Willebrand's syndrome (an inherited thrombocytopathy), clot retraction is normal.
4. In most coagulopathies the *bleeding time* is normal. A prolonged bleeding time with a normal whole-blood clotting time and prothrombin time is usually caused by a thrombocytopathy.
5. In fibrinogen, prothrombin, and factor V, VII, and X deficiencies, the prothrombin time is prolonged, but the Lee-White clotting time usually is not prolonged.

14
Blood

1

(1:407; 2:4-5;
3:429; 4:208;
5:1127)
Cbpa 1

Which one of the following statements about *albumin* is true?

A. less than 10% is degraded each month
B. it has a molecular weight greater than that for the gamma globulins
C. it is responsible for most of the osmotic pressure of plasma
D. in liver disease and nephrosis, it approaches, in plasma, the concentration of the plasma globulins
E. at a pH of 7.4 it is a positively charged particle

Answer and Explanation
A. Approximately 6% to 10% of the albumin in plasma is degraded each day.
B. The molecular weight for albumin is approximately 69,000 and that for gamma globulin is 156,000.
C. The osmotic pressure of plasma is 5776 mm Hg. Albumin normally is responsible for 16.4 mm Hg pressure and the other colloids about 9 mm Hg.
D. Normally, the albumin-globulin (A-G) ratio is 2:1. In liver disease, albumin production decreases. In nephrosis, the loss of albumin increases (albuminuria). Since these conditions are not associated with a similar decrease in globulin, but may cause an increased plasma globulin concentration, the A-G ratio may change from 2:1 to 1:1.
E. Albumin is negatively charged at a pH of 7.4.

2

(1:407; 2:4-5;
3:429; 4:208;
5:1565)
Cbc 2

The *plasma proteins* function to (1) destroy certain foreign materials (antibodies), (2) control body activity (hormones), (3) prevent bleeding (procoagulants), and (4) lyse clots (profibrinolysin). What other functions do they serve?

Answer and Explanation
5. They prevent *edema* by causing the plasma osmotic pressure to be higher than the perivascular osmotic pressure.
6. They provide one-sixth of the *buffering* capacity of the blood.
7. They *prevent excretion and destruction* of hormones, vitamins, iron, lipids, drugs, etc. The plasma proteins do this by forming complexes with these substances. For example, a 6-year-old girl with a transferrin (a beta$_1$ globulin) deficiency had a plasma iron-binding capacity of 15 μg/100 ml (330 μg/100 ml is normal) and a disappearance time for iron of 5 minutes (70 to 140 minutes is normal).
8. They *decrease the potency* of agents (thyroxine, for example) by forming protein complexes that serve as reservoirs but are not themselves active.

3

(1:408; 2:4-95; 3:429; 4:74; 5:1127)
Cbc 1

Which one of the following statements concerning *antibodies* is incorrect?

A. they are all proteins
B. they are produced by the ribosomes in the cells of the spleen and lymph nodes
C. some cause lysis of antigen-containing cells
D. some cause precipitation of an antigen
E. they have a half-life of less than 24 hours

Answer and Explanation
B. It is the *plasma cell* in these organs that produces the antibodies.
E. The average biological *half-life* is about 13 days in the adult.

4

(1:408; 2:4-95; 3:417; 4:76; 5:1133)
Cbca 2

Which one of the following statements concerning *antibodies* is incorrect?

A. the fetus in the 8th month of pregnancy has about one-half the antibody-producing potential the adult has
B. antibody production is more rapid and more marked after the second exposure to an antigen than after a first exposure 1 month earlier
C. a high antibody titer is usually associated with a high lymphocyte count
D. antibody production is depressed by adrenocorticotrophic hormone (ACTH), radiation, and nitrogen mustards
E. antigen-antibody reactions are associated with the release of histamine

Answer and Explanation
A. At birth, the child has little or no antibody-producing potential.
B. A first exposure to an antigen results in a maximum antibody titer in 8 to 12 days. A second exposure results in a maximum titer that is higher than occurred initially and is seen in 5 to 9 days.
C. Small lymphocytes produce and carry antibodies.
D. ACTH facilitates the release of cortisol, which, in turn, depresses antibody production and decreases the number of circulating lymphocytes.
E. The release of histamine accounts for some of the symptoms found in allergic reactions.

5

(1:410; 2:4-104; 3:417; 4:76; 5:1570)
Cb 1

Which one of the following statements about *lymphocytes* is incorrect?

A. they are produced by the thymus, red bone marrow, spleen, and lymph nodes
B. their concentration in the blood falls abruptly after removal of the thymus gland in the adult, and the immune reaction is disturbed
C. they are capable of changing into plasma cells which secrete antibodies
D. they constitute 20% to 40% of the leukocytes
E. they do not perform an important phagocytic function

Answer and Explanation
B. The thymus is smaller and of less importance in the adult than in the child. In the child, the thymus is an important source of lymphocytes. It has been suggested that in the newborn mouse, the thymus is essential for the development of immune responses, but these observations have not been confirmed on pathogen-free animals. In the fetus, lymphocyte precursors are changed by the *thymus* to T lymphocytes and these are changed by the liver and spleen to B lymphocytes. Both T and B lymphocytes migrate to the lymph nodes and bone marrow, where they function in the child and adult in cellular and humoral immunity.

6

(1:409; 3:417; 4:67; 5:1133)
Cba 1

Which one of the following statements concerning the *monocyte* is incorrect?

A. it is more common in the blood than the eosinophil and basophil
B. it is produced in the adult by the bone marrow and lymph nodes
C. unlike the neutrophil, it is rich in lipase
D. unlike the neutrophil, it does not accumulate outside the circulation in an area of inflammation
E. it is not classified as a granulocyte

Answer and Explanation
A. Monocytes constitute 4% to 8% of the leukocytes, eosinophils 1% to 3%, and basophils 0% to 1%.
(D.) Both rapidly accumulate at a site of infection, the neutrophils predominating initially. The neutrophils disintegrate within about 3 days and then the monocytes and lymphocytes predominate.

7

(1:409; 2:4-65; 3:416; 4:67; 5:1133)
Cb 1

Which one of the following statements is correct? The *neutrophil* 15 days

A. is the second most numerous leukocyte in the blood
B. is not actively phagocytic in the bloodstream
C. has a life span of about 120 days RBC
D. acts as a reservoir for vitamin B$_{12}$
E. is formed in the adult mainly in the spleen and lymph nodes

Answer and Explanation
A. The neutrophil represents 50% to 70% of the *leukocytes* in the blood.
B. It is *phagocytic* both in and out of the blood. In the blood it phagocytoses bacteria, small insoluble particles, and fibrin.
C. Its *life span* is less than 2 weeks.
(D.)
E. It is formed mainly in the red *bone marrow*.

8

(1:409; 2:4-66; 3:476; 4:65; 5:1133)
Cb 1

Which one of the following statements is incorrect? *Neutrophilic granulocytes*

A. contain histamine
B. are in the blood at a concentration of 200,000 to 500,000/mm³
C. are attracted to a site of tissue damage by chemical agents such as histamine
D. contain lysosomes
E. increase in concentration during exercise

Answer and Explanation
A. It has been suggested that the release of histamine by neutrophils serves to increase the circulation to an area (vasodilation) and serves as a chemical signal to bring to an area of infection more neutrophils (chemotaxis).
(B.) There are normally 3000 to 6000 neutrophils per mm³ of blood.

9

(1:409; 2:4-66; 3:476; 4:67; 5:1133)
CbHa 1

Which one of the following statements is incorrect? The concentration of *neutrophils* in the blood

A. rises following a cardiac infarction (necrosis of an area in the heart)
B. rises whenever the lymphocyte count rises
C. rises whenever an infection is associated with the formation of pus
D. falls in response to hypoxia
E. falls during vitamin B$_{12}$ deficiency

Answer and Explanation

A. The products of tissue damage facilitate the movement of neutrophils from reservoirs into the blood.

B. The movement of lymphocytes and neutrophils into the blood can be independently controlled. *Adrenocorticotropic hormone* will, for example, increase the number of circulating neutrophils while it decreases the number of circulating *lymphocytes and eosinophils.*

C. *Pus* consists primarily of dead neutrophils.

D. *Hypoxia* also causes an increase in erythrocyte production as a result of an increase in *erythropoietic activity* in the bone marrow and possibly also as a result of a decrease in *leukopoiesis.*

10

(1:409; 2:4-66; 3:416; 4:65; 5:1133)
Cb 1

There is still a great deal that we do not know about the function of the *leukocytes.* Which one of the following perspectives is least likely to be true?

A. basophils are granulocytes that contain histamine and heparin in their meta-chromatic granules

B. eosinophils phagocytose the antigen-antibody complex

C. the eosinophil concentration in the blood increases during allergic reactions

D. eosinophils are sites for the manufacture of antibodies

E. neutrophils have multilobed nuclei

Answer and Explanation

D.

11

(1:408; 2:4-50; 3:426; 4:86; 5:1132)
Cb 1

Which one of the following statements is incorrect?

A. antigens are not always proteins

B. in some cases, antigens require the presence of a plasma factor (complement) before they will combine with an antibody

C. the body can form antibodies against substances with a molecular weight less than 10,000

D. the antigens responsible for the A and B blood groups are unconjugated proteins   *mucopolysacchrides*

E. the antigen responsible for Rh positivity has been called, by Fisher, the D factor

Answer and Explanation

A. *Antigens* are usually proteins, polysaccharides, or combinations.

C. Characteristically, the body forms antibodies against substances with a molecular weight in excess of 10,000, but molecules with a molecular weight of less than this can combine with an antigen and, in this way, stimulate the production of antibodies that will react against the low molecular weight substance (called a hapten) when it is not in combination with another molecule.

D. They are mucopolysaccharides.

12

(1:425; 2:4-55; 3:429; 4:95)
Cbca 2

Blood for transfusion is stored at 4°C for periods up to 21 days. It usually has had sodium citrate (prevents clotting), citric acid (reduces pH), and dextrose (additional source of nutrient) added to it. Which one of the following changes would not occur in the blood after 7 days of *storage*?

A. a decreased concentration of dextrose and an increased concentration of lactic acid

B. a decreased concentration of plasma K^+

C. an increased prothrombin time

(Continued)

D. a decreased concentration of SPCA (factor VII), antihemophilic factor (AHF, factor VIII), and PTC (factor IX)

E. a decreased concentration of platelets

Answer and Explanation

A. The *dextrose* will change from a plasma concentration of 350 mg/100 ml to 300 mg/100 ml, and the lactic acid will change from 20 mg/100 ml to 70 mg/100 ml.

B. The *plasma K^+* will change from 3.5 mEq/liter to 12 mEq/liter. Other signs of deterioration of cellular function include an increased *plasma hemoglobin* (from 0 to 10 to 25 mg/100 ml), plasma *inorganic phosphate* (1.8 to 4.5 mg/100 ml), and plasma *ammonia* (50 to 260 μg/100 ml).

C. The *prothrombin time* increases by 50%.

D. The plasma concentrations of the following decrease: *SPCA* (−11%), *AHF* (−62%), *PTC* (−18%).

E. The *platelet count* decreases 52%.

13

(1:408; 2:4-50; 3:426; 4:86)
Cb 2

Which one of the following statements is correct? An individual with *blood group A_1*

A. will not develop plasma agglutinins to type B blood in the absence of exposure to a B agglutinogen

B. will never agglutinate the erythrocytes from a group A_1 donor

C. usually will agglutinate the erythrocytes from a group O donor

D. usually will agglutinate the erythrocytes from a group AB donor

E. usually will agglutinate the erythrocytes from a type A_2 donor

Answer and Explanation

A. The anti-B *agglutinin* is always found in the plasma of individuals with blood group A_1, A_2, and O. There is no evidence it develops in response to the exposure of the individual to a B *agglutinogen*.

B. If the A_1 recipient is Rh-negative and has previously received Rh-positive whole blood, he may agglutinate Rh-positive A_1 cells. Physicians, by using the ABO classification, can reduce transfusion deaths to 2 per 1000 transfusions. By using, in addition, the Rh classification, they can still further reduce transfusion deaths.

C. The group O donor's erythrocytes lack the A and B agglutinogens.

D. The A_1 recipient contains anti-B agglutinins.

E. The A_1 recipient generally does not contain anti-A_2 agglutinins.

14

(1:408; 2:4-50; 3:426; 4:86)
Cb 2

Why is it safer to *transfuse* type A blood into a type AB recipient than to transfuse type AB blood into a type A recipient?

Answer and Explanation

Generally the donor's *agglutinins* become so diluted in the recipient's plasma that they are incapable of causing agglutination of the recipient's erythrocytes. When large quantities of blood are transfused, clumping of both the donor's and the recipient's erythrocytes may occur.

15

(1:408; 2:407; 3:426; 4:86)
Cb 2

Which one of the following statements is correct? An individual with *blood group* B

A. has the most common blood type

B. cannot be the biologic father of a child with an O blood type

C. cannot be the biologic father of a child with an A blood type

D. cannot be the biologic father of a child with an AB blood type if the mother is type O

E. all of the above statements are correct

Answer and Explanation
A. The *most common blood type* among Caucasians (45%), Negroes (48%), and Orientals (36%) is group O. Group A ranks second (41%, 27%, 28%), group B third (10%, 21%, 23%), and group AB fourth (4%, 4%, and 13%).
B. An individual with a group B *phenotype* can have a BB or BO *genotype*.
C. A child with an A phenotype can have either an AA or AO genotype.
D. A child with an AB phenotype must have an AB genotype. If its father is type B, its mother must have an A or AB phenotype.

16

(1:409; 2:4-50; 3:429; 4:87; 5:1132)
Cb 2

Which one of the following represents the most potentially dangerous situation?
A. an Rh-positive mother who is bearing her second Rh-negative child
B. an Rh-negative mother who is bearing her second Rh-positive child
C. an Rh-positive mother who is bearing her first Rh-negative child
D. an Rh-negative mother who is bearing her first Rh-positive child
E. the transfusion of 1 unit of O-negative blood into an A-positive recipient

Answer and Explanation
B. The fetus is incapable of producing dangerous quantities of antibodies against its mother's cells. On the other hand, an Rh-negative mother can produce dangerous quantities of antibodies against the cells of an Rh-positive fetus. An Rh-negative mother who has not been previously exposed to Rh-positive cells will have no agglutination problem during a first Rh-positive pregnancy. Only 1 in 42 have problems in their second pregnancy and only 1 in 12 in their fifth pregnancy. In other words, the incidence of *erythroblastosis fetalis* will depend on the number of exposures of the mother to Rh-positive cells, and this will depend, in part, on the number of pregnancies and the adequacy of the placental barrier.
E. This is a safe procedure.

17

(1:409; 2:4-50; 3:429; 4:87)
Cb 2

What is the most appropriate treatment for a newborn child (blood group A) suffering from *erythroblastosis fetalis?*
A. simultaneous bleeding and transfusion with A-positive blood
B. simultaneous bleeding and transfusion with A-negative blood
C. simultaneous bleeding and transfusion with either A-positive or A-negative blood
D. transfusion with type O blood
E. transfusion with type AB blood

Answer and Explanation
A. The newborn with hemolytic disease has two problems: (1) too many antibodies to his Rh-positive cells and (2) Rh-positive cells. Transfusion with Rh-positive blood tends to dilute the antibodies but does nothing about the number of Rh-positive cells.
B. Since the Rh-positive infant shows little or no reaction against Rh-negative cells, this is the procedure of choice.
D. Type O-negative would be an acceptable transfusion.
E. Transfusion of the infant with type AB blood would be counterproductive.

18

(1:408; 2:4-50;
3:427; 4:86;)
Cb 2

The absence of anti-A and anti-Rh *agglutinins* in the plasma means the subject is

A. A-positive or AB-positive
B. A-negative or AB-negative
C. A-positive, AB-positive, A-negative, or AB-negative
D. Rh-positive
E. type O

Answer and Explanation

A, B, Ⓒ. In the majority of the cases, both the Rh-positive and Rh-negative individual will lack anti-Rh factors, since this is a factor that usually occurs in an Rh-negative individual in response to the transfusion of Rh-positive blood. The blood type B or O individual inherits the anti-A agglutinin.

19

(1:693; 2:3-16;
3:538; 4:512)
Rbc 1

In which of the following is *carbonic anhydrase* (CA) found in highest concentration?

A. blood
B. erythrocytes
C. neutrophilic leukocytes
D. skeletal muscle fibers
E. interstitial fluid of skeletal muscle

Answer and Explanation

B. $CO_2 + H_2O \overset{CA}{\underset{CA}{\rightleftharpoons}} H_2CO_3 \rightleftharpoons H^+ + HCO_3^-.$

Since the erythrocyte contains most of the CA found in the blood, it is in the erythrocyte that most of the above reactions occur. Carbonic anhydrase is also found in high concentration in the stomach's acid-secreting cells and in renal tubular cells.

15
The Erythrocyte

1

(2:4-20; 3:414;
4:1037; 5:1958)
GCSb 1

During the 3rd, 4th, and 5th months of pregnancy, *fetal hemoglobin* is produced primarily in which one of the following structures?

A. blood islands
B. red bone marrow
C. spleen
D. liver
E. lymph nodes

Answer and Explanation
D.

2

(1:408; 2:4-20;
3:414; 4:57;
5:1127)
GCb 1

Check each of the following statements that is true. In a healthy 40-year-old male, *red blood cells are being formed in the*:

1. femur
2. tibia
3. vertebrae
4. sternum
5. rib

Which one of the following best summarizes your conclusions?

A. statement 1 is true
B. statements 1 and 2 are true
C. statement 3 is true
D. statements 2 and 4 are true
E. statements 3, 4, and 5 are true

Answer and Explanation
A, B. The femur and tibia stop producing erythrocytes at about *age 20.*
E.

3

(1:408; 2:4-23;
3:422; 4:58;
5:1130)
Cb 1

The principal site of production of the *erythropoietic factor* (i.e., erythrogenin) is thought to be the

A. red bone marrow
B. spleen
C. lymph nodes
D. kidney
E. small intestine

Answer and Explanation
   D. It is proposed that hypoxia acts to increase the release of REF (renal erythropoietic factor) from the kidney and a globulin from the liver. The REF acts in the blood on the globulin to change it to erythropoietin. The erythropoietin is apparently essential for the maintenance of normal red cell production. It is destroyed in the liver and has a half-life of about 5 hours.

4

(1:408; 2:4-41;
3:423; 4:56;
5:1127)
b 2

Which one of the following statements concerning the *red blood cell* (RBC) and *hemoglobin* (Hb) is correct? Under healthy conditions the blood contains approximately

A. 5×10^6 RBC per ml and 15 g of Hb per ml
B. 5×10^6 RBC per ml and 0.15 g of Hb per ml
C. 5×10^6 RBC per ml and 0.15 mg of Hb per ml
D. 5×10^6 RBC per mm^3 and 0.15 mg of Hb per ml
E. 5×10^6 RBC per mm^3 and 0.15 g of Hb per ml

Answer and Explanation
   A, B, C. There are between 4.3 and 6.0×10^9 RBC per ml and 0.15 g of Hb per ml. Since 1 ml of blood weighs between 1.050 and 1.060 g, it cannot contain 15 g of Hb per ml.
   (E.)

5

(1:688; 2:6-6;
3:520; 4:492;
5:1697)
Rqpm 2

The following data are collected during a study on a patient: Total volume of air expired in 5 minutes: 100 liters at standard temperature, pressure, and dryness (STPD)

Pressure:	760 mm Hg
O_2 in expired gas:	17% (STPD)
CO_2 in expired gas:	2.5% (STPD)
O_2 in inspired gas:	20% (STPD)

What would the partial *partial pressure* of O_2 be in the expired gas at STPD?

A. 99 mm Hg
B. 109 mm Hg
C. 119 mm Hg
D. 129 mm Hg
E. 139 mm Hg

Answer and Explanation
   D. Partial pressure of O_2:

$$P_{O_2} = (\%O_2/100) \times (\text{pressure}) = (17/100) \times (760 \text{ mm Hg}) = 129.2 \text{ mm Hg}.$$

6

(1:681; 3:520;
4:494; 5:1344)
R Wgm 2

What would the *oxygen consumption* of the patient in question 5 be?

A. 0.2 liters/min
B. 0.4 liters/min
C. 0.6 liters/min
D. 0.8 liters/min
E. 1.0 liters/min

Answer and Explanation

C. O_2 consumption $= \dfrac{(\% \ O_2 \ \text{inspired} - \% \ O_2 \ \text{expired})}{100} \times (\text{volume expired/min})$

$= \dfrac{(20 - 17)}{100} \times (100 \ \text{liters/5 min}) = 0.6 \ \text{liters/min}.$

7

*(1:682; 2:1-5;
3:222; 4:514;
5:1357)*
R Wgm 2

What would the respiratory exchange ratio, R, for the patient in question 6 be?

A. 1.02
B. 0.97
C. 0.92
D. 0.87
E. 0.82

Answer and Explanation

E. R = (CO_2 production)/(O_2 consumption).

CO_2 production $= \dfrac{(\% \ CO_2 \ \text{expired} - \% \ CO_2 \ \text{inspired})}{100} \times (\text{volume expired/min})$

$= \dfrac{(2.5 - 0)}{100} \times (100 \ \text{liters/5 min}) = 0.5 \ \text{liters/min}.$

R = (0.5 liters/min)(0.6 liters/min) = 0.83.

In doing this problem, it is assumed the inspired CO_2 concentration is approximately 0%. Since it is usually less than 0.4% this is a good approximation for this problem.

8

*(1:681; 2:4-44;
3:520; 4:494;
5:693)*
CRbg 2

A subject is breathing 50% O_2 and 50% N_2 at a barometric *pressure* of 760 mm Hg and room temperature. Which one of the following statements is incorrect?

A. $P_{N_2} = P_{O_2}$ in the air
B. $P_{N_2} < 370$ mm Hg in the aortic blood
C. $P_{O_2} > P_{N_2}$ in the aortic blood
D. the O_2 content of the aortic blood exceeds the N_2 content of the aortic blood
E. the P_{O_2} in the aortic blood will be directly related to the atmospheric pressure

Answer and Explanation

B. The inspired air will be diluted with H_2O vapor (6.2% of the alveolar air; 47 mm Hg) and the residual volume of air in the respiratory system. The dilution of the air with H_2O will reduce the P_{N_2} from 380 mm Hg to 356.5 mm Hg [380 − (47/2 mm Hg)]. The rest of the dilution will produce only minor changes in the P_{N_2}.
C. The P_{N_2} will be greater than the P_{O_2}. This is because the inspired air will be diluted with an O_2 poor residual volume.
D. Hemoglobin markedly increases the O_2 content of the arterial blood.
E. In a healthy subject, the aortic P_{O_2} = (% concentration in alveoli/100) (atmospheric pressure).

9

*(1:687; 2:4-43;
3:533; 4:507;
5:1721)*
CRbg 2

When the P_{O_2} of the blood is 100 mm Hg, how much O_2 is dissolved in the blood as O_2?

A. less than 5% of the total O_2 that enters the blood
B. less than 20%
C. less than 40%
D. less than 60%

(Continued)

E. more than 60%

Answer and Explanation
A. At this P_{O_2}, each 100 ml of blood will contain 19.5 ml of O_2 combined with *hemoglobin* and 0.3 ml dissolved as free O_2.

10

(1:687; 2:4-44; 3:534; 4:507; 5:1402)
Rbgm 2

When the P_{O_2} of the blood changes from 100 to 600 mm Hg, there is
A. a sixfold increase in the O_2 content of the blood
B. less than a 1% increase in the O_2 content of the blood
C. a twofold increase in the O_2 content of the blood
D. a sixfold increase in the quantity of dissolved O_2
E. a sixfold increase in the quantity of dissolved O_2 and a twofold increase in the O_2 content of the blood

Answer and Explanation
D. At a P_{O_2} of 100 mm Hg, the major carrier of O_2 in the blood is about 97.5% saturated. Therefore, much of the increase in blood O_2 content gained by increasing the P_{O_2} of the blood to 600 mm Hg will be gained by increasing the quantity of dissolved O_2.

P_{O_2}	100 mm Hg	600 mm Hg
Dissolved O_2 in blood:	0.3 ml/100 ml*	1.8 ml/100 ml
O_2 combined with Hb:	19.5 ml/100 ml	20.8 ml/100 ml

ml of O_2 per 100 ml of blood

11

(1:688; 2:4-44; 3:534; 4:508; 5:1725)
CRgb 2

If one increases the *alveolar* P_{O_2} from 75 mm Hg to 100 mm Hg, the systemic artery blood will carry

A. one-third more O_2
B. one-third less O_2
C. 5% more O_2
D. 5% less O_2
E. 20% more O_2

Answer and Explanation
C. The relationship between the quantity of O_2 carried by *hemoglobin* (ordinate) and the P_{O_2} (abscissa) yields an "S"-shaped curve. Since the hemoglobin is about 94% saturated with O_2 at a P_{O_2} of 75 mm Hg, the amount of additional O_2 that will be carried by the blood in healthy subjects above this alveolar P_{O_2} is relatively small.

12

(1:688; 2:4-44; 3:534; 4:508; 5:1725)
CRgb 2

If one decreases the plasma P_{O_2} from 50 mm Hg to 30 mm Hg, the blood from a normal subject will carry

A. one-third more O_2
B. one-third less O_2
C. 5% more O_2
D. 5% less O_2
E. 20% more O_2

Answer and Explanation
B. Usually, between a P_{O_2} of 10 and 50 mm Hg, we are dealing with the steepest part of the *oxyhemoglobin dissociation curve*.

Plasma P_{O_2}	10 mm Hg	30 mm Hg	50 mm Hg
Whole blood O_2:	2.73 ml*	11.49 ml	16.85
Dissolved O_2:	0.03 ml	0.09 ml	0.15 ml
Hemoglobin O_2:	2.70 ml	11.40 ml	16.70 ml
Hemoglobin saturation:	13.5%	57%	83.50%

*ml of O_2 at STPD per 100 ml of blood

13

(1:689; 2:4-44;
3:533; 4:509;
5:1725)
RCgbc 2

Curve H in the following diagram represents a curve for normal arterial blood in a resting subject breathing room air.

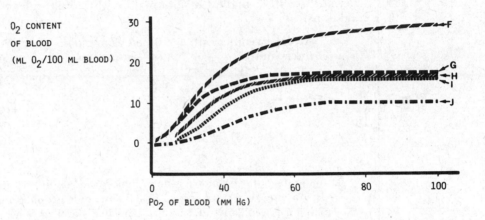

Which of the following statements is correct?

A. curve F is for arterial blood from a subject who is hyperventilating
B. curve G is for arterial blood from a subject with anemia
C. curve I is for arterial blood from a subject with anemia
D. curve I is for blood leaving an exercising muscle
E. curve J is for blood leaving an exercising muscle

Answer and Explanation

A. *Hyperventilation* can markedly increase arterial P_{O_2}, but not O_2 content at a particular P_{O_2}.

B, C. Curve J is from an individual with *anemia*. In anemia the problem is a low O_2 content at each P_{O_2}.

D. Metabolically active muscle produces heat, CO_2, and H^+. Each of these *metabolites* shifts the curve to the right at P_{O_2}'s below 80 mm Hg.

14

(1:689; 2:4-44;
3:534; 4:509;
5:1725)
Rgb 1

Check each of the following statements that is true. The concentration of O_2 in ml of O_2/100 ml of blood held at a P_{O_2} of 40 mm Hg is increased by

1. an increased plasma pH from 7.4 to 7.6
2. an increased plasma P_{CO_2} from 40 to 80 mm Hg
3. an increased body temperature from 38 to 40°C

Which one of the following best summarizes your conclusions?

(Continued)

A. statement 1 is true
B. statement 2 is true
C. statement 3 is true
D. statements 1 and 3 are true
E. statements 2 and 3 are true

Answer and Explanation
A. A decrease in *acidity* shifts the PO_2–hemoglobin saturation curve to the left.
B. C. Increases in PCO_2 and increases in *temperature* will decrease the amount of O_2 carried by the blood.

15

(1:687; 2:4-44; 3:533; 4:504; 5:1402)
RCbg 2

Arterial PO_2 will be highest in a normal healthy adult under which of the following conditions?

	Alveolar ventilation (liters/min)	Blood hemoglobin concentration (g/dl)	PO_2 of inspired air (mm Hg)
A.	8	8	90
B.	7	8	300
C.	7	18	150
D.	4	8	90
E.	4	21	150

Answer and Explanation
B. In the absence of pulmonary shunts, blood hemoglobin concentration plays no role in determining the PO_2 of arterial blood. Since only about 2% of the blood bypasses the pulmonary capillaries in a normal individual, the effect of hemoglobin on arterial PO_2 is negligible. On the other hand, a low hemoglobin may cause a reduction in venous PO_2. B is a better answer than A because the high PO_2 in B more than compensates the somewhat lower ventilation.

16

(1:690; 2:4-44; 3:534; 4:510; 5:1729)
CRbc 2

The P_{50} for blood is the PO_2 at which the *hemoglobin* is 50% saturated with O_2. Under resting conditions, it is about 26 mm Hg. Check each of the following statements that is true.

1. 2,3-diphosphoglycerate (DPG) increases the P_{50}
2. DPG shifts the oxygen hemoglobin dissociation curve to the left
3. DPG decreases O_2 liberation in the systemic capillaries
4. fetal hemoglobin has a greater affinity for binding O_2 than does adult hemoglobin A because fetal hemoglobin contains less DPG
5. acidosis, hypercapnea, and hypoxia decrease the concentration of DPG
6. 1 mol of deoxygenated hemoglobin A binds 1 mol of DPG
7. thyroid hormones, growth hormone, and androgens increase the P_{50}

Answer and Explanation
1. Hemoglobin, at a high DPG level, will be 50% saturated with O_2 at a PO_2 greater than 26 mm Hg.
2, 3. DPG, by shifting the curve to the right, increases the amount of O_2 liberated from the blood at the PO_2 found in the perivascular spaces of the systemic capillaries.
4.
5. They increase the concentration of DPG.
6.
7. These hormones increase the concentration of DPG.

17

*(1:688; 2:4-42;
3:551; 4:535;
5:1848)*
RCbga 1

Cyanosis is caused by

A. an increased concentration of reduced hemoglobin
B. a decreased concentration of hemoglobin
C. a decreased concentration of oxyhemoglobin
D. carbon monoxide poisoning
E. hypoxia

Answer and Explanation

A. Cyanosis is first noted when the concentration of reduced hemoglobin exceeds 5 g/100 ml of blood.

B, C. Iron deficiency *anemia*, as well as other forms of anemia, can produce marked decreases in the concentration of *hemoglobin* and *oxyhemoglobin* without producing blueness.

D. Carbon monoxide poisoning, as well as anemia, tends to prevent cyanosis by decreasing the concentration of reduced hemoglobin.

E. When hypoxia is caused by anemia, KCN, or CO, there is generally no cyanosis.

18

*(1:690; 2:4-43;
3:557; 5:1724)*
Rgba 2

Check each of the following statements that is true. *Nitrites* have the following effects:

1. They reduce the ferric iron in hemoglobin to ferrous iron
2. They increase the O_2 holding capacity of hemoglobin
3. They may produce hypoxia.

Which one of the following best summarizes your conclusions?

A. statement 1 is true
B. statement 2 is true
C. statement 3 is true
D. statements 1 and 3 are true
E. statements 2 and 3 are true

Answer and Explanation

A. Nitrites oxidize the ferrous iron to ferric iron.

B. Nitrites decrease the O_2 holding capacity of hemoglobin. This represents the major problem. The *methemoglobin* formed by the oxidation of the ferrous iron is an ineffective O_2-carrying system.

C.

19

*(1:681; 2:4-48;
3:556; 4:511;
5:1693)*
Rgba 2

A healthy subject breathing air contaminated with *CO* (0.1% of the total volume) at a pressure of 760 mm Hg will have an arterial *plasma*

A. P_{CO_2} that exceeds the venous plasma P_{CO_2}
B. P_{CO_2} that approximately equals the venous plasma P_{CO_2}
C. P_{CO} that exceeds the arterial plasma P_{O_2}
D. P_{O_2} that exceeds the arterial plasma P_{N_2}
E. P_{N_2} that exceeds the arterial plasma P_{O_2}

Answer and Explanation

A, B. Since CO_2 is added to the blood in the systemic capillaries, the systemic vein blood will have a higher P_{CO_2} than the systemic artery blood.

C, D. Under steady state conditions the partial pressure of gases in the alveolus will equal the partial pressure of gases in the systemic artery blood. In other words, the arterial P_{O_2} will exceed the P_{CO} and P_{CO_2} and be less than the P_{N_2}.

E.

20

(1:691; 2:6-15;
3:556; 4:511;
5:1920)
Rgba 2

If the conditions in question 19 are maintained, the subject dies. This is caused by

A. the irritant action of CO on the respiratory membrane
B. the direct action of CO upon the enzymes of the cell
C. CO interfering with the action of carbonic anhydrase
D. CO blocking the action of erythropoietin
E. CO combining with hemoglobin

Answer and Explanation

B. This is the mode of action of KCN, but not CO.

C, D, (E.) CO, under the conditions specified, combines with 50% of the O_2 receptor sites of *hemoglobin* and, in this way, halves the O_2 carrying capacity of the blood.

21

(1:691; 2:6-23;
3:552; 4:535;
5:1708)
Rgba 1

In which of the following conditions is the *arterial* P_{O_2} reduced?

A. anemia
B. KCN poisoning
C. pulmonary hypoventilation
D. CO poisoning
E. moderate exercise

Answer and Explanation

A. The arterial P_{O_2} is the concentration of O_2 in plasma expressed in mm Hg. It is not controlled by the concentration of hemoglobin or the number of red blood cells.

B. Cyanide prevents the cell from utilizing O_2. It does not affect the arterial P_{O_2}.

(C.) In the healthy subject, the P_{O_2} of the blood leaving the lungs is equal to that in the alveoli.

D. Carbon monoxide ties up O_2 binding sites on hemoglobin. It does not affect the arterial P_{O_2}.

E. In the healthy subject, arterial P_{O_2} does not change appreciably during moderate exercise. Venous P_{O_2}, on the other hand, is lowered.

22

(1:682; 2:6-13;
3:551; 4:535;
5:1721)
Rgba 2

Anemia and *hypoxic hypoxia* affect the concentration of gases in the aortic blood. Indicate the consequences of these two conditions by completing the following table. In answering this question, do not be concerned with the compensatory responses to the anemia and hypoxic hypoxia.

	Normal values	Changes (+, −, 0±)	
		Anemia	Hypoxic hypoxia
O_2 content:	_____ ml/100 ml	_____	_____
O_2-carrying capacity of Hb:	20.8 ml/100 ml	_____	_____
O_2 saturation of Hb:	_____ %	_____	_____
P_{O_2} in aorta:	_____ mm Hg	_____	_____

Answer and Explanation

	Normal values	Changes (+, −, 0±)	
		Anemia	Hypoxic hypoxia
O_2 content:	20.3 ml/100 ml	−	−
O_2-carrying capacity of Hb:	20.8 ml/100 ml	−	0
O_2 saturation of Hb:	97.5%	0	−
P_{O_2} in aorta:	100 mm Hg	0	−

The numerical data given above are approximations.

$$O_2 \text{ content} = \frac{\% \text{ saturation} \times \text{capacity}}{100}$$

$$= \frac{97.5 \times 20.8}{100} = 20.3 \text{ ml } O_2/100 \text{ ml blood.}$$

In anemia, the concentration of available hemoglobin is reduced. Therefore, the O_2 content and carrying capacity of the blood are reduced. In severe anemia, there may also be a reflex increase in ventilation. In hypoxic hypoxia (pulmonary congestion, for example), the aortic Po_2 is reduced. The decrease in O_2 *content* and *saturation* may be quite minor because of the relative flatness of the oxygen dissociation curve at O_2 partial pressures above 70 mm Hg. Initially, the O_2-carrying capacity of the blood is little affected, but later there may be an increased hematocrit. There will also be a reflex increase in ventilation.

23

(1:692; 2:1724; 3:556; 4:535; 5:1724)
Rgba 2

Which one of the following changes in the blood of the right atrium is most likely to be seen in *anemia*?

A. an increase in O_2-carrying capacity
B. a decrease in hemoglobin saturation
C. an increase in O_2 content
D. an increase in Po_2
E. none of the above

Answer and Explanation
(B.) A decreased *Hb saturation* with O_2 is not seen in the blood of the systemic arteries of the patient with uncomplicated anemia, but since the cells of the body continue to withdraw large quantities of O_2 from the blood, it will be seen in the systemic veins because of the reduced O_2 *content* in the systemic artery blood.
C, D. In the anemic patient, there is a reduced right arterial blood O_2 content and partial pressure.

24

(2:4-36; 3:425; 4:63)
Rgba 2

Which one of the following statements is incorrect?

A. an individual who is homozygous for hemoglobin-S has hemoglobin that is insoluble at a low Po_2
B. an individual who is heterozygous (i.e., has one gene for Hb-S and one for Hb-A) has only normal hemoglobin
C. an individual who has the sickle cell trait usually has some Hb-A in his red cells
D. 10% of the U.S. black population contains the sickle cell trait
E. in sickle cell anemia, the P_{50} is increased

Answer and Explanation
A. At a low Po_2, Hb-S crystallizes within the erythrocyte to cause it to change from a flexible, biconcave to a fragile, rigid, crescent-shaped structure that tends to cause local *ischemia*.
(B.) Half of this individual's *hemoglobin* will be *Hb-S* and half *Hb-A*.
C. He is heterozygous for Hb-S. Since his erythrocytes usually contain combinations of Hb-S and Hb-A, they become less rigid during hypoxia and less likely to occlude the microcirculation than the erythrocytes from an individual with *sickle cell anemia*.

25

(1:693; 2:5-12; 3:536; 4:509; 5:1735)
Rbc 2

When the pH of the blood changes from 7.5 to 6.5, 31.74 mEq of H^+ combine with blood proteins. It is estimated that 87% of this *buffering action* is due to

A. the plasma proteins because of the buffering capacity of albumin
B. the plasma proteins because of the buffering capacity of the globulins
C. hemoglobin, because it is in higher concentration in the blood than the plasma proteins
D. hemoglobin, because it has a greater buffering capacity per gram
E. hemoglobin, because it is in higher concentration and has a greater buffering capacity

Answer and Explanation

E. There are normally about 3.9 g of *plasma protein* and 15 g of *hemoglobin* per 100 ml of blood.

26

(1:693; 2:6-16; 3:538; 4:512; 5:1733)
Rgb 1

Approximately 90% of the CO_2 that enters the blood is carried as

A. dissolved CO_2
B. carbonic anhydrase
C. carbamino CO_2
D. carbonic acid
E. bicarbonate

Answer and Explanation

E. In the presence of the catalyst, *carbonic anhydrase*, CO_2 and H_2O are rapidly converted to H_2CO_3. The H_2CO_3 ionizes to H^+ and HCO_3^-. These reactions occur in areas where CO_2 is being produced. The reverse occurs where CO_2 is being lost (i.e., in the lungs).

27

(1:693; 2:6-16; 3:538; 4:492; 5:1696)
Rbdg 3

In arterial blood, the concentration of dissolved O_2 is 0.3 ml/100 ml of blood and the concentration of *dissolved CO_2* is 2.4 ml/100 ml. Why is the concentration of dissolved CO_2 so much higher?

A. in the alveolus, the concentration of CO_2 is higher than the concentration of O_2
B. in the lungs, the diffusing capacity (ml of gas moved from the alveolus to the blood per minute per mm Hg difference in partial pressure) for CO_2 is greater than that for O_2
C. most of the O_2 that enters the blood is combined with hemoglobin
D. in plasma, the solubility (ml of gas dissolved/ml of solution) of CO_2 is greater than that for O_2
E. in blood, CO_2 is in equilibrium with HCO_3^-

Answer and Explanation

This question is concerned with mechanisms (causes and effects). Statements B through E are true, but only one answer is the cause of the higher concentration of CO_2. Only one is causally related to the higher concentration of CO_2.

A. CO_2 constitutes about 5% and O_2 about 13% of the *alveolar air*.
B. In normal subjects, the blood is in the pulmonary capillaries long enough to come into equilibrium with the alveolar air. On the other hand, when there is a *diffusion* block (pulmonary edema, for example) the lower rate of diffusion for O_2 may result in a hypoxia not associated with hypercapnea.
C. An equilibrium between alveolar O_2, dissolved O_2, and oxyhemoglobin is reached in the pulmonary capillaries, and therefore, the affinity of hemoglobin for O_2 does not affect the concentration of dissolved O_2.

(D.) CO_2 is about 20 times more soluble in plasma than O_2.

E. The equilibrium between CO_2 and HCO_3^- is reached before the blood leaves the pulmonary capillaries.

28

(1:693; 2:6-16; 3:538; 4:512; 5:1733)
CRbg 1

In questions 26 and 27, we noted that some of the CO_2 that enters the blood is carried by the blood as CO_2 (dissolved) and that some is changed to (HCO_3^-). Into what other form is the CO_2 changed?

Answer and Explanation

In each 100 ml of venous blood, there is approximately 47.1 ml of CO_2 carried as bicarbonate, 3.3 ml as carbamino-CO_2, and traces as H_2CO_3. Most of the *carbamino-CO_2* is formed by the *hemoglobin*, but some is formed by the *plasma proteins*:

$$\text{Protein-NH}_2 + CO_2 \rightleftharpoons \text{protein-NH·COOH}.$$

29

(1:694; 2:6-17; 3:538; 4:512; 5:1735)
Rbgp 3

If you compare the *Po_2-blood O_2 curve* with the *Pco_2-blood CO_2 curve* (Po_2 and Pco_2 are on the abscissa), you find that

A. both are relatively flat at partial pressures above 60 mm Hg

B. at a given partial pressure, you have approximately the same gas content for O_2 and CO_2

C. between a partial pressure of 10 mm Hg and 100 mm Hg, both curves have a straight-line relationship between pressure and content

D. increases in Po_2 shift the CO_2 curve to the right and increases in Pco_2 shift the O_2 curve to the right

E. increases in acidity shift the two curves to the left

Answer and Explanation

A. The CO_2 curve is not flat at a Pco_2 above 60 mm Hg.

B. At a Pco_2 of 40 mm Hg you have 49 ml of CO_2 being carried by each 100 ml of blood as HCO_3^-, carbamino-CO_2, etc. At a Po_2 of 40 mm Hg you have 15 ml of $O_2/100$ ml of blood.

C. Neither curve has a straight-line relationship at these pressures.

(D.) Carbaminohemoglobin ($HbCO_2^-$) and acid hemoglobin (HHb) have less affinity for O_2 than hemoglobin (Hb^-). Oxyhemoglobin (HbO_2^-) and acid hemoglobin (HHb) have less affinity for O_2 than hemoglobin (Hb^-).

E. Acidity shifts the two curves to the right.

30

(1:695; 2:6-16; 3:538; 4:512; 5:1732)
Rbgd 2

When CO_2 is removed from the blood in the *lungs*, which one of the following events also occurs? There is a net

A. efflux of HCO_3^- from the erythrocyte (RBC)

B. efflux of Cl^- from the erythrocyte

C. influx of CO_2

D. efflux of O_2

E. efflux of O_2 and influx of CO_2

Answer and Explanation

A. *Bicarbonate ions* diffuse into the RBC as its *carbonic anhydrase* catalyzes the formation of CO_2 and H_2O from H_2CO_3.

(B.) Associated with the influx of negatively charged bicarbonate ions is the efflux of negatively charged chloride ions. This is called the *chloride shift* and serves to maintain a relatively constant transmembrane potential.

(Continued)

C. In the lungs, the RBC is producing CO_2 and the CO_2 is diffusing out into the plasma.

D. Since the Po_2 in the alveolus is greater than the Po_2 in the venous blood entering the pulmonary capillaries, the Po_2 of the plasma will be initially greater than the Po_2 of the RBC cytoplasm. Therefore, there is an influx of O_2.

31

(1:10; 2:1-18; 3:21; 4:44; 5:10)
dgie 2

Check each of the following statements concerning the *cell membrane* of the *human erythrocyte* that is true. It is more *permeable* to

1. Cl^- than K^+
2. K^+ than Na^+
3. K^+ than urea
4. O_2 than CO_2
5. O_2 than nitrous oxide
6. glucose than glycerol

Answer and Explanation

1. Although the *diameters* of hydrated Cl^- and K^+ are similar (3.86 *vs.* 3.96 A), the Cl^- moves through the membrane much more rapidly. The *permeability coefficient* (in $cm/sec/cm^2$) for each of these particles will vary markedly from one cell to the next. Generally, however, the distribution of *charges* in the cell membrane favors the diffusion of negatively charged particles over positively charged particles.

2. Hydrated K^+ has a smaller diameter than hydrated Na^+ (3.96 *vs.* 5.12 A). On the other hand, muscle and nerve cells, during their initial response to stimulation, have a higher permeability to Na.

3. Hydrated K^+ has a larger diameter than urea (3.96 *vs.* 3.6 A), but in this case, more important than differences in size is the fact that the permeability of a cell membrane to a substance is inversely related to the *hydrophilic* nature of that substance. This is due to the lipid (*hydrophobic*) character of the membrane. Hydrophilic substances such as Na^+, K^+, *sugars, and amino acids* penetrate more slowly than *urea, water, and glycerol*. The *permeability coefficients* for water (10^{-2}), urea (10^{-4}), and Cl^- (10^{-4}) are markedly above those for K^+ (10^{-8}) and Na (10^{-10}).

4. CO_2 diffuses on an average of 20.3 times faster than O_2 through membranes.

5. Nitrous oxide moves on an average of 13.9 times faster.

6. Glycerol is smaller and less hydrophilic than glucose.

PART FIVE
RESPIRATION

16
Ventilation

1

(1:668; 2:6-45;
3:521; 4:476;
5:1684)
RMf 1

Normal *inspiration* results from

A. a decreased intrapleural pressure
B. an increased alveolar pressure
C. both of the above
D. depression of the thorax
E. relaxation of the diaphragm

Answer and Explanation
A.

2

(1:668; 2:6-44;
3:521; 4:476;
5:1681)
Rp 2

During the initial phase of *inspiration* in a healthy subject at rest,

A. intrapulmonary pressure rises
B. intraabdominal pressure rises
C. intrapulmonary and intraabdominal pressures rise
D. there is less muscular effort than during the initial phase of expiration
E. the larynx is elevated

Answer and Explanation
A, (B.) The contraction of the *diaphragm* decreases intrathoracic pressure and increases intraabdominal pressure.
D. It is primarily elastic *recoil* that is responsible for expiration in a subject at rest.
E. Elevation of the *larynx* would occlude the glottis and prevent inspiration.

3

(1:668; 2:6-36;
3:523; 4:476;
5:1682)
Rg 2

Which one of the following statements is true?

A. the respiratory bronchioles and trachea contain cartilage that maintains their patency
B. dry air at 0°C that is inspired through the nasal cavity does not reach body temperature and 100% humidity until after it reaches the lobar bronchus
C. the nasal and pharyngeal epithelium and trachea are devoid of functional cilia
D. during quiet breathing, little or no mucus (less than 1 ml per hour) is produced in the respiratory tract

(Continued)

E. during a quiet inspiration, there is a submaximal contraction of the diaphragm and external intercostal muscles

Answer and Explanation

A. *Cartilage* is found from the trachea through the small bronchi. It is not normally found below the bronchi (i.e., in the bronchioles and alveoli).

B. Inspired dry air at −10°C has reached body *temperature* and 100% *humidity* by the time it leaves the nasal cavity.

C. Mucus is being continuously carried by *cilia* from the nasal cavity and tracheobronchial tree toward the hypopharynx, where it is swallowed

D. Approximately 432 g of *mucus* are produced by the respiratory epithelium per hour [(5.4 g/m²/hr) × (80 m²)].

(E.)

4

(1:668; 2:6-36; 3:523; 4:476; 5:1682)
RMN 3

A patient is taken to surgery and given a general anesthetic and muscle relaxant. An anesthetist is responsible for maintaining an appropriate depth of anesthesia. Under these conditions, what is most likely responsible for each of the following events?

A. a movement inward during inspiration of the intercostal tissue due to a paralysis of the external intercostal muscles

B. a movement inward during inspiration of the intercostal tissue due to a contraction of the internal intercostal muscles

C. a movement inward during inspiration of the intercostal tissue due to a relaxation of the external and a contraction of the internal intercostal tissues

D. an abdominal breathing due to the paralysis of the diaphragm

E. an abdominal breathing due to the contraction of the abdominal muscles

Answer and Explanation

(A.) B, C. The *external intercostal muscles* serve to elevate the ribs and stabilize the intercostal space during inspiration. Their paralysis, in this case, is probably due to the muscle relaxant. The sucking inward of the soft tissue is due to the negative intrathoracic pressure during inspiration.

D, E. Abdominal breathing during anesthetization is usually due to a loss of *abdominal muscle tone*, which produces a pronounced distention of the abdomen during inspiration.

5

(1:668; 2:6-36; 3:525; 4:476; 5:1681)
RfM 2

Under resting conditions, an individual inspires about 6 liters of air per minute. When his ventilation exceeds 100 liters per minute, inspiration will be produced not only by the contraction of the diaphragm and external intercostal *muscles* but also by the contraction of

A. the scalene, sternomastoid, and trapezius muscles

B. the scalene, sternomastoid, and transversus abdominus muscles

C. the sternomastoid, transversus abdominus, and trapezius muscles

D. the transversus abdominus, trapezius, and scalene muscles

E. the transversus abdominus, trapezius, scalene, and sternomastoid muscles

Answer and Explanation

(A.) The *scalene and sternomastoid* muscles elevate the rib cage, and the *trapezius* muscle stabilizes the head so that contraction of the sternomastoid muscle does not move the head and thus dampen its action on the rib cage.

B, C, D, E. The *transversus abdominus* muscle compresses the abdominal contents and pushes the diaphragm cephalad.

6

*(1:644; 2:6-43;
3:524; 4:481;
5:1685)*
RqGa 2

The *residual volume* for the lungs

A. is the volume of air that remains in the lungs after expiring the resting tidal volume of air
B. is the volume of air remaining in the lungs after inspiring the resting tidal volume of air
C. is generally greater at age 75 than at age 45
D. is less than 0.5 liter in the adult
E. increases in atelectasis

Answer and Explanation
A, B. It is the volume of air left in the lungs after a maximal expiration.
C. D. In *aging*, there is a loss of the lungs' elastic pull on the thorax and a resultant increase in residual volume. At age 45, the residual volume averages 1.48 liters, at age 65 it is 1.72 liters, and at age 75 1.92 liters.
E. *Atelectasis* is a collapse of the alveoli.

7

*(1:644; 2:6-10;
3:524; 4:481;
5:1685)*
Rqf 1

Which of the following *formulae* is correct?

A. Vital capacity = inspiratory reserve volume + expiratory reserve volume
B. Dead air space = resting tidal volume + residual volume
C. Alveolar minute ventilation = (respiratory rate) × (tidal volume – dead air space)
D. Vital capacity = inspiratory reserve volume + resting tidal volume + expiratory reserve volume + residual volume
E. Inspiratory reserve volume = vital capacity – resting tidal volume

Answer and Explanation
A. Vital capacity = inspiratory reserve volume + resting tidal volume + expiratory reserve volume.
B. Dead air space = tidal volume – the volume of fresh air that enters the respiratory portion of the respiratory system during an inspiration.
C. D, E. Inspiratory reserve volume = (vital capacity) – (expiratory reserve volume + resting tidal volume).

8

*(1:644; 2:6-7;
3:549; 4:485;
5:1697)*
Rfmw 3

An athlete at rest has a cardiac output of 4 liters per minute, an oxygen consumption of 0.25 liter per minute, and a pulmonary ventilation of 5 liters per minute. During running, his cardiac output increases to 20 liters per minute and his oxygen consumption to 3 liters per minute. Approximately what would his *pulmonary ventilation* be?

A. 5 liters/min
B. 20 liters/min
C. 60 liters/min
D. 144 liters/min
E. 200 liters/min

Answer and Explanation
C. (Pulmonary ventilation at rest) × $\left(\dfrac{O_2 \text{ consumption during exercise}}{O_2 \text{ consumption before exercise}} \right)$ =

(5 liters of air/min) × $\left(\dfrac{3 \text{ liters of } O_2/\text{min}}{0.25 \text{ liters of } O_2/\text{min}} \right)$ =

60 liters of air/min = pulmonary ventilation during exercise.

Under these circumstances, the arterial P_{O_2} would be expected to stay fairly constant during exercise, and therefore, there would be a direct relationship between the increase in O_2 consumption and the increase in pulmonary ventilation. Changes in cardiac output are not needed for this calculation.

9

(1:644; 2:6-12; 3:524; 4:484; 5:1797)
Rqm 2

A subject is asked to breathe air through a tube. During the experiment the following data are collected:

Tidal volume:	500 ml
Alveolar P_{CO_2}:	40 mm Hg
P_{CO_2} of expired gas:	30 mm Hg

What is the respiratory *dead air space* of this individual?

A. less than 100 ml
B. between 100 and 150 ml
C. between 150 and 200 ml
D. between 200 and 250 ml
E. more than 250 ml

Answer and Explanation

B. $\dfrac{\text{Dead space volume}}{\text{Tidal volume}} = (\text{alveolar } P_{CO_2} - \text{expired } P_{CO_2})/\text{alveolar } P_{CO_2}.$

$\dfrac{\text{Dead space volume}}{500 \text{ ml}} = (40 \text{ mm Hg} - 30 \text{ mm Hg})/40 \text{ mm Hg}.$

Dead space volume $= (0.25) \times (500 \text{ ml}) = 125 \text{ ml}.$

In the above equation, you are assuming that the P_{CO_2} of the inspired air is approximately 0 mm Hg. Since the dead air space in a healthy individual with a tidal volume of 500 ml varies from 100 to 200 ml, it would seem that the tube through which this subject breathes contributes little to his dead space.

10

(1:644; 3:529; 4:484; 5:1697)
Rq 2

The anatomic *dead space*

A. is the same at the functional residual capacity and at total lung capacity
B. represents approximately 5% of the functional residual capacity
C. both of the above are true
D. represents approximately 5% of the total lung capacity
E. and the physiologic dead space are determined by different methods but yield values in both normal and pathologic states that agree to within 10%

Answer and Explanation

A. It is larger at the total lung capacity because the more negative intrapleural pressure in this state causes dilation of the bronchi and bronchioles.

B. Dead space $= \dfrac{(150 \text{ ml} \times 100)}{(3000 \text{ ml})} = 5\%$ of functional residual capacity.

D. Dead space $= \dfrac{(180 \text{ ml} \times 100)}{(6000 \text{ ml})} = 3\%$ of total lung capacity.

E. The physiologic dead space may be considerably greater than the anatomic dead space in certain pathologic conditions.

11

(1:644; 2:6-8;
3:529; 4:483;
5:1697)
Rfp 2

What might account for a decreased arterial Po_2 in a subject with a constant CO_2 production and *respiratory minute volume*?

A. a decreased functional residual capacity
B. a decreased respiratory rate and tidal volume
C. an increased respiratory rate and a decreased tidal volume
D. a decreased respiratory rate and an increased tidal volume
E. an increased respiratory rate and tidal volume

Answer and Explanation

B. Since respiratory minute volume = tidal volume × respiratory rate, you cannot have a constant minute volume associated with a decrease of both tidal volume and rate.

C. A decrease in alveolar ventilation causes a decrease in the arterial Po_2. It will be associated with no change in respiratory minute volume when there is an increased "dead space minute volume" (= dead space × respiratory rate). See the formula for alveolar ventilation in question 7C.

12

(1:661; 2:6-12;
3:528; 4:485;
5:1706)
Rgq 2

A subject, starting at midinspiration, takes as deep an inspiration of 100% O_2 as possible and then steadily exhales maximally while the N_2 content of the gas is continuously recorded. The following record summarizes the results.

Which one of the following statements best characterizes these results?

A. the gas at J is primarily alveolar gas
B. the gas at K is only dead space gas
C. the gas at L is primarily dead space gas
D. the volume of air moved during phase M is the dead space
E. the high N_2 concentration in phase M is because the subject is expiring gas that comes primarily from the upper portions of the lungs

Answer and Explanation

A. The gas at J is the first gas expired and the only gas containing no N_2. It is only from the *dead space*.

B. This is the second phase of expiration. The gas is a mixture of dead space gas (0% N_2) and alveolar gas.

(Continued)

C. This is the third phase and represents *alveolar gas*.

E. In this study, the first gas inspired into the alveoli was from the dead space and therefore contained the highest concentration of N_2. Most of it moved into the upper portions of the lungs. During phase M, many of the airways in the lower portions have closed because of their low transmural pressures. Therefore most of the last expired air comes from these N_2-rich upper lungs. They are also responsible for the slight slope seen in phase L.

13

(1:661; 2:6-12; 3:528; 4:485; 5:1706)
Rgq 2

Match W, X, Y, and Z in the record in question 12 with the appropriate volumes listed below.

A. _____ closing volume
B. _____ dead space volume
C. _____ residual volume
D. _____ vital capacity

Answer and Explanation
A. X. The *closing volume* is the lung volume above residual volume at which the lower parts of the lung begin to close during expiration.
B. W. The dead space is the volume from the beginning of expiration to a vertical line drawn at the half-way point of the second phase (K).
C. Y. The *residual volume* is the air left in the lung at the end of a maximum expiration.
D. Z. The *vital capacity* is the maximum volume of air that can be expired after a maximum inspiration.

14

(1:662; 2:6-53; 3:525; 4:478; 5:1687)
Rqpq 2

The following pressure-volume loops were obtained from subjects X and Y during quiet breathing at a rate of 14/min.

What conclusion can you draw from these loops? Subject X has the

A. higher pulmonary compliance
B. higher tidal volume
C. higher pulmonary compliance and tidal volume

D. lower pulmonary compliance
E. lower pulmonary compliance and tidal volume

Answer and Explanation
C. The *compliances* (C) are represented by the slopes of lines F-G (subject X) and F-H (subject Y): slope = (liters)/(cm H_2O) = C. The height of point G (subject X) and H (subject Y) equal the *tidal volumes*.

15

(1:649; 2:6-37; 3:554; 4:477; 5:1683)
Rqpa 3

A unilateral *pneumothorax* causes which one of the following changes?

A. an increase in pulmonary residual volume
B. a collapse of the chest wall inward
C. a drecrease in the intrapleural pressure
D. an increase in resting tidal volume
E. a shift of the mediastinum toward the normal side

Answer and Explanation
A. A pneumothorax may result from a ruptured alveolus, an incision through the chest wall, or an injection of air through the chest wall. When it occurs, the lungs, owing to their elastic characteristics, recoil and the residual volume is reduced.
B. The elastic characteristics of the lungs are such that at the end of either a resting or maximal expiration they are pulling the chest wall inward. Usually, pneumorthorax causes the chest wall to obtain a size similar to that found at the end of a normal resting inspiration. The presence of air in the intrapleural space would dampen the inward pull of the lung on the chest wall and tend to make the wall expand rather than collapse.
C. The *intrapleural pressure* would move away from its subatmospheric values of between −2 and −6 mm Hg toward atmospheric pressure.
D. The pneumothorax would cause a dampening of the pressure changes during inspiration as a result of the presence of a distensible fluid (i.e., air) in the intrapleural space.
E. There will be a lower intrathroacic pressure on the normal side that will tend to pull the *mediastinum* toward that side.

16

(1:643; 2:6-38; 3:525; 4:478; 5:1687)
Rqpm 2

In young, healthy men the compliance for both the pulmonary system and the thoracic cage is 0.2 liter/cm of H_2O. What is the *compliance* of the respiratory system? (Assume that the respiratory system consists of only the thoracic cage and the pulmonary system.)

Answer and Explanation
These two compliances are parallel. Therefore, the total compliance (c_T) is:

$$\frac{1}{c_T} = \frac{1}{c_1} + \frac{1}{c_2} = \frac{1}{0.2} + \frac{1}{0.2} = \frac{2}{0.2} = \frac{1}{0.1}.$$

c_T = 0.1 liter/cm of H_2O.

17

(1:643; 2:6-38; 3:525; 4:478; 5:1687)
Rqpa 2

How do you calculate *specific compliance*?

Answer and Explanation

$$\text{Specific compliance} = \frac{\text{Compliance}}{\text{Lung volume}} = \frac{\text{change in volume/change in pressure}}{\text{lung volume}}.$$

18

(1:662; 2:6-38;
3:525; 4:478;
5:1687)
Rqpa 3

In which one of the following conditions is there most likely to be an increased respiratory *compliance* and a decreased *specific compliance*?

A. obesity and pregnancy
B. diffuse alveolar fibrosis
C. alveolar edema
D. aging (between 45 and 90 years)
E. decreased production of surfactant

Answer and Explanation

A. Obesity and pregnancy decrease compliance by interfering with an increase in the cephalad-caudad dimension of the thorax.
B. In this condition, compliance may change from a normal value of 165 ml/cm of H_2O to 10 ml/cm of H_2O.
C. In alveolar edema there is a decreased residual volume and decreased compliance.
D. In aging there is an increased residual volume which causes the increased compliance and an infiltration of the lung with collagen which causes the decreased specific compliance.
E. Loss of surfactant production causes an increased surface tension at the air-liquid interface of the alveoli and a decreased compliance.

19

(1:646; 2:6-38;
3:526; 4:477;
5:1681)
Rqpa 2

During thoracic surgery the *lungs collapse*. Prior to the final closure of the chest, the surgeon applies air under pressure to the mouth in order to expand the lungs. Under what conditions would the least amount of pressure be required?

A. the first 50 ml of expansion with the external nares open
B. the third 50 ml of expansion with the external nares open
C. the first 50 ml of expansion with the external nares closed
D. the third 50 ml of expansion with the external nares closed
E. the first 50 ml of expansion in the absence of surfactant production

Answer and Explanation

A, B. Passing air into the mouth with the *external nares* open will not expand the lung.
C. The pressure needed to open the *collapsed alveoli* is greater than that needed to distend the lung once the alveoli are patent.
D.
E. The absence of *surfactant* makes the lung less compliant.

20

(1:647; 2:6-39;
3:526; 4:477;
5:1973)
GRqa 2

The aspiration of *amniotic fluid* by transabdominal amniocentesis is proving a useful diagnostic tool. When the ratio of *lecithin to sphingomyelin* in the amniotic fluid

A. is greater than 2, the fetus is immature and, if delivered, will probably suffer severe emphysema (dilation of alveoli)
B. is greater than 2, the fetus is immature and, if delivered, will probably suffer severe atelectasis (collapse of alveoli)
C. is less than 1, the fetus is immature and, if delivered, will probably suffer severe emphysema
D. is less than 1, the fetus is immature and, if delivered, will probably suffer severe atelectasis
E. is greater than 2, the fetus should be delivered by cesarean section

Answer and Explanation
D. At 26 weeks of gestation, the lecithin/sphingomyelin ratio is less than 1, but lecithin production has begun, and the lung is beginning to produce the dipalmitoyl lecithin-containing substance, surfactant. At 35 weeks, there is a rapid jump in lecithin production, but sphingomyelin concentration stays constant. As the ratio increases from less than 1 to greater than 2, the incidence of death due to respiratory distress syndrome (RDS) or hyaline membrane disease decreases. At a ratio of 1, there is a survival of about 10%. At a ratio greater than 2, there are no deaths due to RDS.

21

(1:64; 2:6-39; 3:526; 4:477; 5:1681)
Gqa 2

Which one of the following statements is correct? *Surfactant*

A. is distributed homogeneously throughout the liquid that covers the alveolar epithelium
B. is usually at a high concentration in the lung of the premature infant
C. causes the surface tension exerted on small alveoli to be less than that exerted on large alveoli
D. increases surface tension
E. is not formed by alveolar cells

Answer and Explanation
A. It is localized at the air–liquid interface.
B. Its production begins late in fetal life, and the premature infant is frequently deficient in surfactant. This deficiency causes the atelectasis seen in *hyaline membrane disease of the newborn*.
C. It is a detergent-like substance that lowers surface tension. As the alveolus enlarges, the concentration of surfactant at the surface becomes less, and the surface tension therefore goes up.
D. It decreases surface tension.
E. It is formed by cells lining the alveoli.

22

(1:641; 2:6-45; 3:521; 4:477; 5:1684)
Rp 2

The *alveolar pressure* is

A. higher (less negative) than the intrapleural pressure during inspiration
B. higher (less negative) than the intrapleural pressure during expiration
C. higher (less negative) than the intrapleural pressure during inspiration and expiration
D. lower (more negative) than the intrapleural pressure during expiration
E. lower (more negative) than the intrapleural pressure during inspiration and expiration

Answer and Explanation
C.

23

(1:649; 2:6-39; 3:521; 4:477; 5:1681)
Rp 2

What mechanism or mechanisms are responsible for the *alveolar pressure* being higher than the intrapleural pressure during expiration?

Answer and Explanation
The *surface tension* exerted by the liquid lining the alveolar lumen and the *recoil* characteristics of the alveolar wall tend to increase the pressure in the alveolus during the relaxation of the diaphragm.

24

*(1:646; 2:6-37;
3:526; 4:532;
5:1830)*
Raof 1

In *emphysema,*

A. a forced expiration causes a collapse of some of the bronchioles
B. a forced expiration is essential for a normal tidal volume
C. there is a decreased resistance to air flow on inspiration
D. there is an increased resistance to air flow on inspiration
E. there is a tendency for alveoli to collapse

Answer and Explanation
   A.

25

*(1:642; 2:6-37;
3:526; 4:532;
5:1830)*
RaoM 1

In *emphysema,* there is an increased tendency for bronchioles to collapse during a forced expiration. This is due to

A. a decrease in the elasticity of the lungs
B. a loss of collaginous tissue from the lungs
C. a failure of bronchiolar chondroblasts
D. excessive tone in the bronchiolar smooth muscle
E. hypersensitivity to epinephrine

Answer and Explanation
(A.) In a normal subject, the ability of the lungs to recoil during expiration helps to keep a higher pressure in the lumen of the bronchiole than there is in the peribronchiolar space. In emphysema, the loss of lung elasticity causes an increase in the residual volume of the lungs and bronchiolar occlusion during a forced expiration.
C. Bronchioles do not contain cartilage, and cartilage-forming cells do not play a role in the patency of bronchioles.
D. This is more characteristic of allergic reactions than emphysema.
E. Epinephrine produces a decrease in bronchiolar muscle tone.

26

*(1:699; 2:6-49;
3:528; 4:500;
5:1714)*
Rq 1

Which one of the following statements best characterizes the pattern of *ventilation* in the lungs during quiet breathing?

A. surfactant keeps each region of the lung equally distended and ventilated
B. gravity in the erect individual keeps the base of the lung (inferior portion) more poorly expanded and ventilated than the apex
C. gravity in the erect individual keeps the base of the lung (inferior portion) more poorly expanded and better ventilated than the apex
D. gravity in the erect individual keeps the base of the lung (inferior portion) more expanded and ventilated than the apex
E. gravity in the erect individual keeps the base of the lung (inferior portion) more expanded and less ventilated than the apex

Answer and Explanation
A. *Surfactant* does serve to reduce overdistention, but there is good evidence for uneven ventilation of the alveoli.
(C.) The intrapleural pressure at the base of the lung is less negative, and therefore, lung volume in this region is also less than found in the upper portions of the lungs. The smaller volume of each inferior alveolus will mean that it will distend more than the larger cephalad alveoli during inspiration and, therefore, be better ventilated.

27

(1:663; 2:6-53; 3:527; 4:480; 5:1687)
RpfW 1

How does one calculate the *work* performed by the respiratory system during a single inspiration?

Answer and Explanation
Integrate intrapleural pressure and change in lung volume during inspiration:

Work (dyne-cm) = pressure (dynes/cm^2) × volume (cm^3)

28

(1:663; 2:6-53; 3:531; 4:480; 5:1687)
RpfW 2

Check each of the following factors that would increase the *respiratory minute work.*

1. airway constriction
2. disappearance of turbulence
3. decreased density of the inspired gas
4. increased compliance of the lungs
5. increased tidal volume

Answer and Explanation
1. This increases airway resistance and therefore minute work.
2. Turbulence increases the respiratory minute work.
3. An increase in density increases the respiratory minute work.
4. A decrease in compliance increases the respiratory minute work.
5. See answer 27.

29

(1:663; 2:6-45; 3:531; 4:531; 5:1686)
Roqa 2

Name a pathologic state in which the *work of inspiration* is increased because of (1) an increased *resistance*. Name another pathologic state where the work of breathing is increased because of (2) lung *stiffness*.

Answer and Explanation
1. In asthma, bronchitis, and other allergic reactions, the airway resistance is increased during inspiration and expiration. In emphysema the major problem is an increased resistance during expiration.
2. In diffuse pulmonary fibrosis, there is a decrease in lung compliance. There is also an increased inspiratory work in obesity, the late stages of pregnancy, and kyphoscoliosis. All of these conditions are examples of restrictive ventilatory problems. In each, the *timed vital capacity* is less than normal.

30

(1:664; 2:6-53; 3:527; 4:531; 5:1689)
Roqa 2

What is the most *efficient* method for the asthmatic to use during respiration?

A. hyperventilation
B. higher respiratory rate and lower tidal volume than the healthy subject
C. lower respiratory rate and higher tidal volume than the healthy subject
D. a respiratory rate and volume comparable to that of a healthy person
E. a respiratory rate and volume comparable to that for a person with a reduced lung compliance

Answer and Explanation
A. During quiet breathing, 1.5% of the O_2 consumption occurs in the respiratory muscles. A sevenfold increase in ventilation causes 3% of the O_2 consumption to occur in respiratory muscles.

(Continued)

B. This is the most efficient ventilation for an individual with a restrictive ventilatory insufficiency (lung fibrosis, obesity, etc.).

C. The total work performed during inspiration equals (1) the *work used to overcome elastic forces* plus (2) the *work used to overcome resistance*. The asthmatic can maintain a normal alveolar ventilation and keep his efficiency optimal under the circumstances by increasing tidal volume and decreasing respiratory rate.

31

(1:681; 2:6-74; 3:520; 4:492; 5:1853)
Rgpm 2

A mountain climber has been resting for an hour near the summit of Mount Everest (barometric pressure = 247 mm Hg). Under these circumstances

A. at the end of inspiration, the air in her bronchioles will have a Po_2 of less than 45 mm Hg
B. at the end of inspiration, the air in her bronchioles will have a PH_2O of less than what is found at sea level (i.e., less than 35 mm Hg)
C. statements A and B will both be true
D. her arterial blood will have a lower pH than what is found at sea level
E. all of the above statements will be true

Answer and Explanation

A, B. This air should differ from the air outside the body in that it is at body temperature and has a PH_2O of 47 mm Hg (i.e., at a humidity of 100%):

$$Po_2 = (247 - 47 \text{ mm Hg}) (0.20) = 40 \text{ mm Hg}.$$

D. The *hypoxia* that occurs at this altitude should cause a greater ventilation than would be found at sea level and therefore a lower arterial CO_2 and H^+ concentration in the arterial blood.

32

(1:682; 3:222; 4:514)
Rd 2

The *respiratory exchange ratio* (R)

A. decreases during strenuous running
B. decreases during metabolic acidosis
C. decreases during both of the above
D. increases during hyperventilation
E. increases in individuals who shift from a high carbohydrate diet to a high fat diet

Answer and Explanation

A, B, E. R = (CO_2 expired)/(O_2 take-up by lungs and blood). It is increased by increases in ventilation and increases in *respiratory quotient* (RQ). During (A) exercise, (B) metabolic acidosis, and (D) *hyperventilation* there is an increased ventilation which produces a depletion of CO_2 stores without a comparable movement of O_2 into the body. (E) Under steady-state conditions the body's R and RQ are equal. Carbohydrate catabolism moves the RQ toward 1 and fat catabolism moves it toward 0.7.

D.

33

(1:736; 3:532; 4:486)
REb 1

Which one of the following functions is not performed by the *lungs*?

A. preventing emboli from moving through the lungs to the aorta
B. fibrinolysis
C. activation of angiotensin I to angiotensin II
D. removal from the blood of epinephrine, vasopressin, and oxytocin ✗ EVO
E. removal from the blood of adenine nucleotides, serotonin, and norepinephrine

Answer and Explanation

D. The lungs remove none of the substances listed in D.

Questions 34 and 35 consist of two main parts: a statement and a reason for the statement. Choose the best answer for each of these questions from the options A through E below and mark your choice in the space provided.

A. both the statement and the reason are true and are related as cause and effect (i.e., as indicated)
B. both the statement and the reason are true but are not related as cause and effect
C. the statement is true but the reason is false
D. the statement is false but the reason is an accepted fact or principle
E. both the statement and the reason are false

34

(1:736; 2:6-88; 3:531; 4:487; 5:1678)
Rq 1

Statement

_____ When a fine dust is inhaled, most of the dust particles with a diameter of 5 μm will be removed from the inspired air before they reach the alveoli because

Reason

the mucociliary system in the respiratory tract is an effective means of trapping these particles and transporting them to the esophagus, where they are swallowed.

Answer and Explanation
A. Most particles larger than 10 μm are removed in the *nose*, whereas particles smaller than 2 μm may reach the *alveoli*, where they will be removed by *phagocytosis*.

35

(1:685; 2:6-10; 3:521; 4:495; 5:1697)
Rqf 2

Statement

_____ If the production of CO_2 by a healthy subject is doubled, the respiratory tidal volume is doubled, and the respiratory rate stays constant the arterial P_{CO_2} will stay constant because

Reason

a doubling of alveolar ventilation associated with a doubling of CO_2 production will maintain the arterial P_{CO_2} constant.

Answer and Explanation
D. A doubling of the *tidal volume* will cause a decreased P_{CO_2} because it causes more than a doubling of *ventilation*. For example, if the tidal volume goes from 500 ml to 1000 ml, the respiratory rate is 20/min, and the dead space is 150 ml, the alveolar ventilation goes from 7000 ml/min to 17,000 ml/min, a 2.4-fold increase in ventilation:

Alveolar ventilation = (respiratory rate) $\times$ (tidal volume − dead space):

$(20) \times (500 - 150) = 7000$ ml/min at a tidal volume of 500 ml,

$(20) \times (1000 - 150) = 17,000$ ml/min at a tidal volume of 1000 ml.

17
Diffusion and Perfusion

1

(1:704; 2:6-3;
3:522; 4:498;
5:1680)
RCd 1

List the structures (or *barriers*) through which O_2 must diffuse in passing from the alveolar lumen to hemoglobin.

Answer and Explanation
Surfactant containing liquid → alveolar membrane → basement membrane → capillary endothelium → plasma → erythrocyte membrane.

2

(1:706; 2:6-27;
3:529; 4:498;
5:1697)
RCdq 2

The physiologist, in discussing the flow of particles (Na^+, CO_2, etc.), uses the terms *conductance* (g) and *diffusing capacity* (D). Write a formula that defines each of these terms.

Answer and Explanation
$$D = \frac{\text{flow}}{\text{driving pressure}} = \frac{\text{ml/min}}{P_1 - P_2 \text{ in mm Hg}} = g.$$

3

(1:703; 2:6-26;
4:498; 5:1390)
RCdg 2

List five factors that determine the quantity of gas that will diffuse through a *barrier*.

Answer and Explanation
1. Surface area available for diffusion: A
2. Thickness of barrier: T
3. Molecular weight of diffusing particle: M
4. Solubility of particle in medium: S
5. Driving pressure: $P_1 - P_2$

4

(1:703; 2:6-24;
4:498; 5:1697)
RCdg 3

Using the parameters listed in question 3, write a formula that defines the volume of a gas that will diffuse through a *barrier* per unit time.

Answer and Explanation
$$\dot{Q} = \dot{V} \propto \left(\frac{A}{T}\right) \times \left(\frac{S}{\sqrt{M}}\right) \times (P_1 - P_2).$$

The value of $\dot{V}$ for a given gas is also dependent upon the character of the media through which it must diffuse. We have already noted that diffusion from blood through the capillaries of the brain is slower than that from blood through the capillaries of skeletal muscle.

5

(1:706; 2:6-25;
4:500; 5:1711)
Rdgp 3

A series of gas mixtures is inhaled by a healthy subject. Which one of the following gases would *diffuse* most slowly from the lungs into the blood?

A. CO_2 at a P_{CO_2} of 60 mm Hg
B. CO at a P_{CO} of 0.5 mm Hg
C. O_2 at a P_{O_2} of 130 mm Hg
D. O_2 at a P_{O_2} of 150 mm Hg
E. nitrous oxide at a P_{N_2O} of 0.3 mm Hg

Answer and Explanation
B.

6

(1:703; 2:6-80;
4:500; 5:1696)
Rdgp 3

In question 5, *CO* at a partial pressure head ($= P_1 - P_2$) greater than that for nitrous oxide is noted to diffuse less rapidly into the blood. Its molecular weight is less than that for nitrous oxide. What was responsible for the slower diffusion of CO?

Answer and Explanation
CO is 0.047 times as soluble in body fluids as N_2O.

$$\dot{V} \propto \left(\frac{A}{T} \right) \times \left(\frac{S}{\sqrt{M}} \right) \times (P_1 - P_2).$$

7

(2:6-80; 3:560;
4:555; 5:1697)
Rpgd 2

Name a common gas that, like CO, is *poorly soluble* in body fluids and is responsible for most of the problems that occur in divers who are rapidly *decompressed*.

Answer and Explanation
N_2.

8

(2:6-80; 3:559;
4:556; 5:1924)
Rpgd 3

If an individual who is *scuba diving* 66 ft below the surface of the sea is exposed to a barometric pressure of 3 atmos, what will be the P_{N_2} in the inspired tracheal air of an individual who is 132 ft below the surface of the sea and is breathing air from his tank?

A. less than 2500 mm Hg
B. 2600 mm Hg
C. 2800 mm Hg
D. 3000 mm Hg
E. over 3100 mm Hg

Answer and Explanation
D. Barometric pressure at 132 ft = (1 atmos) + $\left(\dfrac{132 \text{ ft}}{33 \text{ ft/atmos}} \right)$ = 5 atmos.

P_{N_2} = (0.8) × [(5 atmos × 760 mm Hg/atmos) − 47 mm Hg] = 3002 mm Hg.

Note that in this solution (1) an allowance is made for the dilution of the inspired air with water vapor (PH_2O = 47 mm Hg), (2) the assumption is made that 80% of the air in the tank is N_2, and (3) the conclusion is drawn that each 33 ft below the surface of the sea adds 1 atmos of pressure to the subject and the air he breathes.

9

(2:6-80; 3:559; 4:554; 5:1927)
Rdgp 2

If an individual who is breathing air has been scuba diving for 10 minutes at a depth of 132 ft and rapidly rises to the surface, she will generally experience no ill effects. If an individual has been breathing air for 540 minutes at a depth of 90 ft, it is recommended that she go through a period of 720 minutes of *decompression*. If she does not, she may experience any one of a number of symptoms characterized as the *bends* (skin rash, pain in the extremities, substernal distress, spastic paralysis, unconsciousness, death). How do you explain these observations?

Answer and Explanation

In the study, there are two important variables: (1) the PN_2 and (2) the duration of exposure to that PN_2. The subject who is most susceptible to the bends has a *longer exposure* to the elevated PN_2. This longer exposure provides the time necessary for N_2, which *diffuses slowly*, to move into the perivascular tissues. It necessitates a longer period of decompression for N_2 to diffuse out of the body. It has been estimated that it takes 1 hour for a subject breathing 100% O_2 to lose 90% of her N_2, and 3 hours for her to lose 95% of her N_2. If decompression is too rapid, N_2 comes out of solution and forms N_2 *bubbles* in the tissues and sometimes in the blood.

10

(1:704; 2:6-25; 3:529; 4:505; 5:1710)
RCdg 2

Under what circumstances, in the healthy subject, will the Po_2 of the blood leaving a *pulmonary capillary* be lower than the Po_2 of the alveolus served by that capillary (select the one best answer)?

A. breathing air with a high Po_2
B. performing an exercise in which the cardiac output is tripled
C. breathing air with a high Pco_2
D. all of the above
E. none of the above

Answer and Explanation

E. Under resting conditions the blood stays in the pulmonary capillary for about 0.75 second and acquires the alveolar Po_2 during the first 0.3 second. In exercise the cardiac output increases and the blood passes through the pulmonary capillaries more rapidly. When the cardiac output is increased threefold or less, there still is an equilibrium reached between the blood Po_2 and the alveolar Po_2 before the blood leaves the lungs. In other words, in the lung of the healthy subject, *oxygen transfer* is seldom if ever *diffusion limited*.

11

(1:705; 2:6-25; 3:529; 4:493; 5:1711)
RCdg 1

Which one of the following gases in the healthy subject is *diffusion limited* (select the one best answer)?

A. CO_2
B. CO
C. both of the above
D. O_2
E. nitrous oxide

Answer and Explanation

B. The quantity of CO that is carried from the lungs does not increase with increases in pulmonary blood flow. In other words, its transfer is not *perfusion limited* but is diffusion limited. CO_2, O_2, and nitrous oxide transfer in the healthy subject, on the other hand, is not diffusion limited but is perfusion limited.

12

*(1:706; 2:6-27;
3:529; 4:498;
5:1709)*
RCdg 1

A subject inspires a mixture of gases containing CO and holds his breath for 10 seconds. It is calculated that during the 10 seconds when the subject holds his breath the alveolar P_{CO} is 0.5 mm Hg and the CO uptake is 25 ml/min. What is the *diffusing capacity* for CO?

Answer and Explanation

$$D_{CO} = \frac{I_{CO}}{P_1 CO - P_2 CO} = \frac{25\ ml/min}{(0.5 - 0)\ (mm\ Hg)} = 50\ ml/min/mm\ Hg.$$

In this procedure, it is assumed that $P_2 CO$ (CO pressure in pulmonary capillary blood) is 0.

13

*(1:705; 2:6-25;
3:530; 4:531;
5:1713)*
Rdag 2

A patient suffered from *pulmonary congestion* as a result of left ventricular failure. Which one of the following would be most consistent with the finding of a moderate diffusion block in this patient?

A. lowered pulmonary vein pressure
B. elevated P_{CO_2} in systemic arteries
C. elevated P_{O_2} in systemic arteries
D. lowered P_{O_2} in systemic arteries
E. elevated P_{CO_2} and P_{O_2} in systemic arteries

Answer and Explanation
A. An increased pressure in the pulmonary system is causing pulmonary edema and, hence, is causing the diffusion block.
B. CO_2 diffuses through barriers much more readily than does O_2. In fact, the lowered arterial P_{O_2} produced in this condition can cause a reflex increase in alveolar ventilation and, thereby, a decrease in alveolar and arterial P_{CO_2}. On the other hand, a severe diffusion block can cause an elevated arterial P_{CO_2}.
C, D. O_2 diffuses through barriers about one-twentieth as readily as CO_2. Characteristic of a moderate diffusion block, therefore, is a hypoxia not associated with hypercapnea.

14

*(1:703; 2:6-24;
3:551; 4:535;
5:1710)*
Radb 3

A patient is noted to be *cyanotic*. You suspect that she has either (1) a reduced diffusing capacity or (2) a true right-to-left shunt due to a tetralogy of Fallot. Which of the following features distinguishes these two types of cyanosis?

A. in (1) only, there will be an abnormally low arterial P_{O_2}
B. in (1) only, there will be an abnormally low alveolar P_{CO_2}
C. in (1) only, the patient will remain cyanotic when breathing 100% O_2
D. in (2) only, there will be an abnormally low arterial P_{O_2}
E. in (2) only, the patient will remain cyanotic when breathing 100% O_2

Answer and Explanation
A. A low arterial P_{O_2} is characteristic of both conditions.
B. The alveolar P_{CO_2} may be the same in both conditions. A reduced diffusing capacity for O_2 may be associated with a normal diffusing capacity for CO_2.
C. In diffusion blocks, it is possible to deliver normal quantities of O_2 to the blood by increasing the P_{O_2} of the inspired air.
D, E. Breathing 100% O_2 does not appreciably increase the O_2 content of the blood leav-

(Continued)

ing the lung of a normal patient or one with a tetralogy of Fallot. The reason for this is that in a normal individual or a patient with a tetralogy, most of the O_2 is carried in combination with hemoglobin and the hemoglobin leaving the lung is over 95% saturated with O_2 when this individual is breathing room air. On the other hand, one can overcome a diffusion block by increasing the PO_2 of the inspired air.

15

(1:699; 2:6-32; 3:500; 4:531; 5:1714)
CRba 2

In the erect subject, the *perfusion of the lung* via the pulmonary vessels is

A. uniform throughout the lung
B. increased during positive pressure artificial respiration if the alveolar pressure is kept above atmospheric pressure
C. both of the above are true
D. greatest in the superior part of the lung
E. greatest in the inferior part of the lung

Answer and Explanation
B. An increased alveolar pressure tends to occlude pulmonary vessels and, therefore, decrease the perfusion of the lung.
(E.) In the erect individual, the weight of the blood tends to make the arterial, capillary, and venous pressure in the most inferior part of the lung about 23 mm Hg higher than in the most superior part of the lung. This contributes to a smaller resistance to flow in the inferior part of the lung than in the superior part.

16

(1:698; 2:6-32; 3:528; 4:501; 5:1710)
CRgb 2

In the healthy, erect adult, the *arterial blood* has a PO_2 that is about 4 mm Hg lower than that found in the *mixed alveolar air*. What is the mechanism or what are the mechanisms responsible for this?

A. some of the alveoli with a low alveolar PO_2 receive a greater perfusion with blood than alveoli with a high PO_2
B. the thebesian veins empty venous blood into the left ventricle
C. bronchial vessels empty venous blood into the pulmonary vein
D. all of the above contribute to the lower arterial PO_2
E. the diffusion characteristics of the lung

Answer and Explanation
A. The alveoli in the superior part of the lung may have a PO_2 of 130 mm Hg, while those in the inferior part may have a PO_2 of 90 mm Hg. The inferior alveoli have a greater blood flow.
(D.)
E. O_2 transfer is not diffusion limited in the healthy adult.

17

(1:699; 2:6-32; 3:528; 4:500; 5:1714)
CRgb 3

Which one of the following statements best characterizes the *alveoli* of the healthy lung in an erect subject? The alveoli

A. have their highest ventilation/blood perfusion ratio in the superior part of the lung
B. have their highest ventilation/blood perfusion ratio in the inferior part of the lung
C. all have a ventilation/blood perfusion ratio of 1
D. in the inferior part of the lung usually have a lower ventilation than those in the superior part of the lung
E. statements B and D are both correct

Answer and Explanation

A. As one goes from the superior to the inferior part of the lung, there is an increase in ventilation and blood perfusion but a decrease in *ventilation/blood perfusion ratio*. In other words, the ventilation increases less than the blood perfusion.

18

(1:698; 2:6-30; 3:554; 4:500; 5:1714)
CRgba 2

What happens when the *ventilation/perfusion ratio* of a lung unit decreases? The alveoli in that unit develop a

A. higher P_{O_2}
B. lower P_{N_2}
C. higher P_{O_2} and lower P_{CO_2}
D. higher P_{CO_2}
E. higher P_{N_2} and higher P_{O_2}

Answer and Explanation

D. There will be a higher P_{CO_2} and a lower P_{O_2} in the alveoli.

19

(1:694; 2:6-18; 3:539; 4:513; 5:1735)
CRgba 2

A number of patients with an exaggerated *ventilation/perfusion inequality* have a lowered arterial P_{O_2} but a normal arterial P_{CO_2}. Why is this (select the one best answer)?

A. CO_2 diffuses more rapidly than O_2
B. CO_2 is more soluble in the blood than O_2
C. both of the above
D. the O_2 and CO_2 dissociation curves have different shapes
E. there is a greater pressure head for CO_2 than O_2 between the alveoli and the pulmonary capillary blood

Answer and Explanation

A. In the healthy subject, the blood, before it leaves the pulmonary capillary, has come into equilibrium with the O_2 and CO_2 in the alveoli it serves (i.e., CO_2 and O_2 are not diffusion limited in this case). Therefore, the speed of diffusion is not an important factor.

D. Since the *CO_2-dissociation curve* is not "S" shaped (i.e., not plateaued) but is closer to a straight line, alveoli with a high ventilation/perfusion ratio (low P_{CO_2}) will compensate for alveoli with a low ventilation/perfusion ratio (high P_{CO_2}). This is not true for O_2. Alveoli with a P_{O_2} of 90 and 180 mm Hg add almost the same amount of O_2 to the blood.

E. The pressure head for CO_2 is about 6 mm Hg and that for O_2 is about 60 mm Hg.

20

(1:706; 2:6-27; 3:529; 4:498; 5:1709)
RCdq 2

The following data were obtained from a patient:

Arterial P_{O_2}:	100 mm Hg
Right atrial P_{O_2}:	40 mm Hg
Diffusing capacity for O_2:	30 ml/min/mm Hg
O_2 consumption:	240 ml/min

What is the patient's average P_{O_2} difference between his alveolar gas and his alveolar capillary blood?

Answer and Explanation

$$\textit{Diffusing capacity} \text{ (ml/min/mm Hg)} = \frac{\text{flow (ml/min)}}{\text{driving pressure (mm Hg)}}.$$

(Continued)

$$\text{Driving pressure} = \frac{\text{flow}}{\text{diffusing capacity}} = \frac{240 \text{ ml/min}}{30 \text{ ml/min/mm Hg}} = 8 \text{ mm Hg.}$$

Questions 21 through 24 consist of two main parts—a statement and a reason for the statement. Choose the best answer for each of these questions from the options A through E below and mark your choice in the space provided.

A. both the statement and the reason are true and are related as cause and effect (i.e., as indicated)
B. both the statement and the reason are true but are not related as cause and effect
C. the statement is true but the reason is false
D. the statement is false but the reason is an accepted fact or principle
E. both the statement and the reason are false

21

(1:698; 2:6-28; 3:528; 4:498; 5:1714)
CRqba 2

Statement
_____ If the ventilation to a lobule of the lung is decreased but the blood flow stays constant, the blood leaving the lobule will have an increased pH because

Reason
under these conditions alveolar P_{CO_2} will increase.

Answer and Explanation
D. The increased P_{CO_2} causes a decreased pH.

22

(1:705; 2:6-28; 3:529; 4:507; 5:1714)
RCdq 2

Statement
_____ If the velocity of flow through the pulmonary capillaries of a healthy subject is doubled, the arterial P_{CO_2} will increase by more than 10% because

Reason
an increased velocity of flow through a capillary decreases the period that each ml of blood is exposed to the capillary surface during each circuit.

Answer and Explanation
D. Threefold increases in *velocity of flow* do not change either the arterial P_{O_2} or the P_{CO_2}. These gases, unlike CO, are not, in the healthy individual, *diffusion* limited.

23

(1:690; 2:4-44; 3:534; 4:510; 5:1729)
RCbq 2

Statement
_____ An elevation in the concentration of 2,3-diphosphoglycerate (DPG) in the red blood cell (RBC) causes an increase in the P_{50} of the blood because

Reason
an increase in DPG shifts the O_2-dissociation curve to the right.

Answer and Explanation
A. DPG will increase in response to chronic hypoxia, thyroid hormones, growth hormone, androgens, and alkalosis. Its concentration is low in the fetus and newborn.

24

(1:687; 2:4-44;
3:534; 4:508;
5:1735)
RCbq 2

Statement

_____ A healthy individual breathing 100% O_2 will have an arterial P_{O_2} of less than 120 mm Hg because

Reason

the O_2 content of the arterial blood is increased by less than 20% when you change from breathing 20% O_2 to breathing 100% O_2.

Answer and Explanation

D. Breathing 100% O_2 will cause a fivefold increase in arterial P_{O_2}. On the other hand, it will increase the amount of O_2 carried by *hemoglobin* by less than 3%.

18
Control of Respiration

1

*(1:710; 2:6-54;
3:540; 4:516;
5:1765)*
RNa 2

Which one of the following procedures will cause an immediate *cessation of respiration*?

A. transection of the cord at C-6
B. transection of the cord at C-2
C. transection between the medulla and pons
D. all of the above procedures cause apnea
E. none of the above procedures causes apnea

Answer and Explanation
A, (B.) Somatic efferent neurons originating in *C-3, 4, and 5* pass down the right and left *phrenic nerves* to initiate the contraction of the diaphragm. They are not dependent upon impulses originating below C-5 for their stimulation but are dependent upon impulses traveling down the cord from the *medulla*.

2

*(1:715; 2:6-57;
3:544; 4:517;
5:1751)*
RN 2

Which one of the following statements is correct? The medullary *inspiratory center*

A. responds to mild increases in P_{CO_2} and acidity in the cerebrospinal fluid (CSF) of the 4th ventricle of the brain by increasing the number of inspirations per minute
B. responds to mild increases in P_{CO_2} and acidity in the cerebrospinal fluid (CSF) of the 4th ventricle of the brain by increasing the depth of inspiration
C. responds as stated in both A and B
D. when stimulated, stimulates the medullary expiratory center
E. none of the above are true

Answer and Explanation
(C.)
D. The inspiratory and expiratory centers are mutually inhibitory, i.e., the stimulation of one causes the inhibition of the other.

3

(1:710; 2:6-55; 3:540; 4:516; 5:1749)
RngM 1

The healthy heart has its *pacemaker* in the sinoatrial node. The stimuli for the contraction of the diaphragm originate

A. in the floor of the 4th ventricle of the brain
B. in the pneumotaxic center
C. in the apneustic center
D. in chemoreceptors
E. in stretch receptors

Answer and Explanation
A. There is an area in the medulla that sends out volleys of stimuli to the neurons of the phrenic nerves. This area continues to fire periodically when isolated from most of the rest of the body. When this area is depressed, it may become dependent upon impinging stimuli from peripheral chemoreceptors and elsewhere for continued function.

4

(1:711; 2:6-58; 3:541; 4:517; 5:1751)
RNMa 2

A cat has its ninth and tenth cranial nerves severed bilaterally, and a section is made in the pons just above the *apneustic center*. How do these procedures modify the respiratory pattern? They cause

A. decerebrate rigidity
B. long inspiratory gasps
C. both of the above
D. apnea
E. eupnea

Answer and Explanation
C.

5

(1:710; 2:6-58; 3:541; 4:517; 5:1751)
RNa 2

Which one of the following statements is true? The *pneumotaxic center*

A. is in the midbrain
B. inhibits inspiratory activity
C. contains the major central chemoreceptor area
D. causes long inspiratory gasps when separated from the more superior parts of the brain
E. causes long expiratory gasps when separated from the more superior parts of the brain

Answer and Explanation
A. It is superior to the apneustic center in the pons.
(B.)
C. This is in the medulla.

6

(1:715; 2:6-64; 3:542; 4:518; 5:1751)
RNgb 1

Which one of the following structures is most important in increasing respiratory minute volume in response to a small increase in the P_{CO_2} of the body fluids?

A. pulmonary chemoreceptors
B. venous chemoreceptors
C. lung receptors
D. the hypothalamus
E. medullary chemoreceptors

Answer and Explanation
E. If a resting subject inhales 5% CO_2, her ventilation will increase three- to fourfold. This great sensitivity to changes in P_{CO_2} is due to the central chemoreceptors.

7

(1:717; 2:6-65;
3:542; 4:520;
5:1760)
RNbg 2

The *peripheral chemoreceptors* are most important because they respond to

A. decreases in P_{O_2} in the venous blood
B. decreases in P_{O_2} in the arterial blood
C. decreases in P_{O_2} in the cerebrospinal fluid
D. increases in P_{O_2} in the venous blood
E. increases in P_{O_2} in the arterial blood

Answer and Explanation

B. Changes in venous P_{O_2} are important if they produce changes in the arterial P_{O_2}. The central chemoreceptors are more sensitive to changes in P_{CO_2} than the peripheral chemoreceptors CO_2, unlike H^+, passes readily through the blood-brain barrier.

8

(1:715; 2:6-65;
3:542; 4:520;
5:1762)
RNg 2

Chemoreceptors in the carotid and aortic bodies send impulses via the ninth and tenth cranial nerves to the respiratory centers. Which one of the following best characterizes their function? They

A. send increasing frequencies of impulses up their nerves as the P_{O_2} of arterial blood increases
B. produce a more rapid increase in ventilation in response to an increased arterial P_{CO_2} than do the central chemoreceptors
C. are less sensitive to hypoxia than the central chemoreceptors
D. are least important in the control of respiration during sleep and barbiturate depression
E. only affect respiratory rate

Answer and Explanation

A. They are stimulated by decreases in P_{O_2}.
B. Although peripheral receptors are *less sensitive* to changes in P_{CO_2}, it takes longer for an increased P_{CO_2} from exercise or breath-holding to reach the medulla than the arterial receptors.
C. Small decreases in P_{O_2} that stimulate the peripheral chemoreceptors have no known action on the central receptors. Large decreases in P_{O_2} that stimulate the peripheral receptors decrease ventilation if the peripheral receptors are *decentralized*.
D. During sleep, when the reticular activating center is less active, and during barbiturate depression, the peripheral receptors become increasingly important in the maintenance of respiration.
E. They affect respiratory rate and depth.

9

(1:718; 2:6-65;
3:543; 4:520;
5:1762)
RNga 2

The *peripheral chemoreceptors* produce a more pronounced increase in ventilation in response to

A. a decrease in arterial P_{O_2} from 150 to 90 mm Hg than from 70 to 40 mm Hg under the usual resting conditions.
B. a change in arterial P_{O_2} from 100 to 80 mm Hg at a P_{CO_2} of 48 mm Hg than at a P_{CO_2} of 40 mm Hg
C. both statements A and B are correct
D. a 30% reduction in the O_2 content of the arterial blood, as in anemia, than to a 30% reduction in arterial P_{O_2}
E. a change in pH from 7.4 to 7.3 than cyanide poisoning

Answer and Explanation

A. The sensitivity of the receptors to decreases in P_{O_2} will depend upon the P_{CO_2}. Under resting conditions, decreases in arterial P_{O_2} from 500 to 100 mm Hg produce only small changes in ventilation, whereas decreases from 70 to 40 mm Hg can produce better than a doubling of ventilation.

B.

D. When a decreased O_2 *content* is so severe that there is a decreased P_{O_2} in carotid body blood then there will be an increased ventilation.

E. Cyanide poisoning, by preventing the peripheral chemoreceptors from utilizing O_2, cause them to increase ventilation. The change in pH mentioned above has less effect on these receptors.

10

(1:717; 2:6-65; 3:542; 4:518; 5:1762)
RgNb 2

If the ninth and tenth cranial nerves are blocked in the neck, the subject will no longer respond to (select the one best answer)

A. hypercapnea by causing an increased respiratory minute volume
B. alkalosis by causing an increased respiratory minute volume
C. hypoxia by causing an increased respiratory minute volume
D. hypercapnea or acidity by causing an increased respiratory minute volume
E. acidity or hypoxia by causing an increased respiratory minute volume

Answer and Explanation

C. Hypercapnea acts directly on the respiratory centers as well as on the peripheral receptors to increase ventilation. When the peripheral *chemoreceptors* are decentralized, breathing 8% O_2 has no action on ventilation, but when they are functioning, they will produce a threefold increase in respiratory rate and a sevenfold increase in respiratory depth.

11

(1:717; 2:6-65; 3:542; 4:535; 5:1760)
RNbg 2

Which one of the following conditions usually causes an increased frequency of impulses in afferent neurons from the *carotid bodies*?

A. CO poisoning
B. anemia
C. both of the above
D. cyanide poisoning
E. a 50% reduction in carotid body blood flow

Answer and Explanation

C. Both A and B are examples of *anemic hypoxia*. In both cases, there is a reduction in the *oxygen content* of the blood, but not in the P_{O_2} of the blood passing through the carotid bodies. Arterial P_{O_2} is independent of oxyhemoglobin concentration. The P_{O_2} of the blood leaving the carotid bodies is also relatively independent of oxyhemoglobin concentration, because of the high level of flow through the bodies.

D. Either cyanide or a low arterial P_{O_2} will stimulate the carotid bodies. Cyanide prevents the utilization of oxygen (*histoxic hypoxia*) and the low P_{O_2} (*hypoxic hypoxia*) exposes the chemoreceptors to a lower concentration of oxygen.

E. The high blood flow that passes through the carotid bodies (80 ml/min or 2000 ml/100 g of tissue) makes them insensitive to most reductions in flow, as well as reductions in oxyhemoglobin concentration.

12

(1:718; 2:6-65;
3:542; 4:520;
5:1760)
RNbg 1

The carotid bodies differ from the

A. aortic bodies in that the carotid bodies are sensitive to a decreased arterial P_{O_2} and the aortic bodies are not
B. respiratory center in the medulla in that the carotid bodies have an opposite response to a decreased arterial P_{O_2}
C. aortic bodies and respiratory center as stated above
D. aortic bodies in that the carotid bodies are sensitive to changes in P_{CO_2}
E. respiratory center in that the respiratory center is sensitive to arterial P_{CO_2}

Answer and Explanation

C. If you decentralize the *carotid bodies*, hypoxic hypoxia causes a depression of respiration

D, E. The peripheral and *central chemoreceptors* are similar in that they are stimulated by increases in P_{CO_2} or H^+. On the other hand, the carotid bodies differ from the other chemoreceptors in that they are more sensitive to increases in H^+ concentration.

13

(1:732; 2:534;
3:558; 4:486;
5:1836)
RNbg 2

What are the effects of *hyperventilation* in the healthy subject on the concentration of blood gases and on excitability?

Answer and Explanation

Hyperventilation causes a decreased P_{CO_2} and an increased pH and P_{O_2}. The CO_2 content of the blood is more than halved, but the O_2 content and the hemoglobin saturation with O_2 are increased very little. When the P_{CO_2} approaches 15 mm Hg, there is a decreased threshold for a number of sensory and motor elements, which causes a series of muscle spasms, fasciculations, and increases in tone. Hypocapnea, alkalosis, and the decreases in plasma Ca^{++} that they produce are the causes of these changes.

14

(1:738; 2:6-67;
3:546; 4:518;
5:1818)
RqgN 3

In human beings, *breath-holding* is involuntarily terminated before there is damage. The point at which the breathing involuntarily begins is called the *breaking point*. List at least three techniques that extend the breaking point.

Answer and Explanation

Breathing 100% O_2 before breath-holding, which increases the body's O_2 reserves. Hyperventilating before breath-holding, which increases the body's O_2 reserves and decreases the P_{CO_2} of the alveoli and blood. Holding one's breath at the maximum inspiratory position.

15

(1:739; 2:6-59;
3:546; 4:517;
5:1754)
RqgN 2

It has been noted in *breath-holding* experiments that lung distention increases the tolerance of the body for CO_2 and hypoxia. What is the mechanism of action?

Answer and Explanation

Stimulation of stretch receptors in the bronchi and lung parenchyma

↓

Stimulation of sensory neurons in the vagus nerve

↓

Inhibition of respiratory centers

The stimulation of stretch receptors in the lung has been shown to both facilitate and inhibit inspiration. When lung distention produces an inhibition of inspiration, this is said to be the *Hering-Breuer reflex*.

16

*(1:719; 2:6-59;
3:547; 4:526;
5:1768)*
RqN 2

Check each of the following statements that is true. Further inspiratory activity is inhibited during *inspiration*

1. by slowly adapting stretch receptors in the adult lung when the inspiratory depth exceeds 1 liter
2. by rapidly adapting stretch receptors in the adult lung when the inspiratory depth exceeds 1 liter
3. by the action of inhaled irritants on the respiratory mucosa

Answer and Explanation

(1.) These receptors were, at one time, considered to be important in adults during quiet breathing but are now generally believed to function in the adult only when the tidal volume exceeds 1 liter (in exercise, for example).
2. The rapidly adapting receptors facilitate further inspiration. Their importance in human beings is uncertain.
(3.) Irritants in the upper and lower respiratory passages cause a reflex inhibition of both inspiration and expiration as well as reflex bronchial constriction. Irritants can also bring about an increased inspiratory effort followed by a cough.

17

*(1:732; 2:6-74;
3:552; 4:523;
5:1851)*
RNbg

Under what conditions does a healthy subject develop a low arterial P_{O_2} but a normal or below normal P_{CO_2} and H^+ concentration?

Answer and Explanation

A decrease in the alveolar P_{O_2} not associated with an increased P_{CO_2} occurs when one ascends to a high altitude. A decreased diffusion capacity for O_2, as occurs in pulmonary edema, can also cause these changes.

18

*(1:729; 2:6-68;
3:549; 4:523;
5:1894)*
RNbg 2

A healthy subject responds to running at a moderate rate with an increased ventilation caused by

A. a decreased venous P_{O_2}
B. a decreased arterial P_{O_2}
C. proprioceptive impulses from moving limbs
D. an increased arterial pH
E. an increased arterial P_{CO_2}

Answer and Explanation

C. Human beings characteristically respond to running with a decreased venous P_{O_2}, little or no change in arterial P_{O_2}, an increased facilitation of respiratory activity by sensory impulses from the appendages, a decreased arterial pH due, in part, to the increased lactic acid concentration in the blood, and a decreased arterial P_{CO_2}. The decreased venous P_{O_2} apparently is of little importance, since the venous side seems to lack effective hypoxia-sensing elements. Apparently, the increased ventilation that occurs during running is sufficient to prevent an important decreased arterial P_{O_2} and an important increased arterial P_{CO_2}, but not sufficient to prevent a decreased arterial pH.

19

*(1:733; 2:6-82;
3:557; 4:552;
5:1901)*
Rgba 1

Under what conditions is O_2 *toxic?* What symptoms does O_2 toxicity produce?

Answer and Explanation

When the P_{O_2} of the inspired air is maintained at 300 mm Hg for 50 hours, toxicity symptoms begin to appear. Breathing 80% to 100% O_2 for 8 hours at 1 atmos pressure also causes toxicity symptoms. These symptoms include respiratory tract irritation, nasal con-

(Continued)

gestion, sore throat, coughing, decreased surfactant production, and in the newborn, *retrolental fibroplasia* and blindness. Prolonged exposure to a P_{O_2} in excess of 760 mm Hg can cause muscle twitching, dizziness, convulsions, coma, acidosis, and death.

20

(1:722; 2:5-19; 3:558; 4:524; 5:1915)
Rgba 1

As the CO_2 in the inspired air increases from 0% to 10% at atmospheric pressure, the respiratory minute volume increases in human beings from 8 liters/min to 80 liters/min. What happens if the *hypercapnea* increases still further?

Answer and Explanation
As the CO_2 concentration approaches 20%, there will develop a marked depression of the central nervous system, a decreased ventilation, diminished sensory acuity, confusion, convulsions, coma, and finally, death.

21

(2:6-82; 3:559; 4:552; 5:1924)
RgNa 1

Are there any conditions in which N_2 has a dangerous action on the body? What are they?

Answer and Explanation
Yes, they include the *bends* (nitrogen bubbles sometimes form on rapid decompression) and nitrogen narcosis (depression of the central nervous system at a P_{N_2} greater than 4 atmos). A diver breathing air will, at a depth greater than 120 ft, start to experience a sense of euphoria that, at lower depths, will produce a state resembling intoxication ("rapture of the deep" or *nitrogen narcosis*). At depths greater than 150 ft, divers have been reported to have offered their mask to a passing fish.

22

(2:6-79; 3:559; 4:552; 5:1924)
Rga 2

List the useful functions served by N_2 in the body.

Answer and Explanation
N_2 prevents *atelectasis* and tends to decrease the rate of combustion. Since N_2 is absorbed much less rapidly from alveoli than O_2, it delays alveolar collapse when an alveolar duct is plugged by mucus.

23

(2:6-65; 3:542; 4:526; 5:1828)
Rgba 3

A patient is brought to the emergency room suffering from an overdose of a barbiturate. He exhibits hypoventilation due to *respiratory center depression*. He is given 100% O_2 and his respiratory minute volume decreases markedly, but his mixed venous plasma P_{O_2} rises to 130 mm Hg. The patient probably

A. is now well oxygenated and needs no additional treatment
B. should be switched to 95% O_2 + 5% CO_2
C. should receive a vasoconstrictor agent
D. should be treated for systemic acidosis
E. should be treated for systemic alkalosis

Answer and Explanation
A. Barbiturates, morphine, and other narcotics depress the respiratory centers. This results in a hypoxic hypoxia, which acts through the peripheral *chemoreceptors* to maintain respiration if the depression is not excessive. By giving 100% O_2, you treat the hypoxia but in so doing remove some of the stimulation to the respiratory centers. The further reduction in ventilation that results exaggerates the existing hypercapnea and therefore increases respiratory acidosis. This condition should be treated by the administration of $NaHCO_3$.

B. Although it is true that by switching a subject with a normal P_{CO_2} from breathing 100% O_2 to breathing 95% O_2 + 5% CO_2 you can double or triple his ventilation, this principle does not hold in this case. The subject is hypercapnic because of his hypoventilation. Increasing his hypercapnea will have little effect on ventilation and will increase the associated acidosis. At a P_{CO_2} greater than 76 mm Hg, ventilation is decreased by the addition of CO_2 to the arterial blood.

C. The high venous P_{O_2} is not a serious problem. Oxygen toxicity, if present, is best treated by lowering the alveolar P_{O_2}.

D. The inadequate ventilation is causing respiratory *acidosis*.

24

(1:732; 2:6-75; 3:553; 4:544; 5:1859)
Rgp 2

What useful changes occur when one becomes *acclimitized* to a high altitude?

Answer and Explanation

- *Hyperventilation:* increased respiratory rate, arterial P_{O_2}, and arterial pH; decreased arterial P_{CO_2}. [handwritten: But alveolar P_{O_2} decreased / caused pulmonary vasoconstriction]
- *Polycythemia:* increased hemoglobin concentration in the blood, increased O_2-carrying capacity of the blood, increased blood volume.
- *O_2 dissociation curve shifts to the right:* increased concentration of 2,3-diphosphoglycerate, hemoglobin releases more O_2 at a P_{O_2} of 40 mm Hg.
- *Increased concentration of systemic capillaries.*
- *Increased efficiency during exercise.*

25

(1:698; 2:6-67; 3:542; 4:518; 5:1802)
RNgc 2

A decrease in *plasma pH*

A. causes an increase in ventilation through the stimulation of the carotid bodies
B. is frequently associated with a decrease in arterial P_{CO_2} and an increase in ventilation in metabolic acidosis
C. both of the above statements are true
D. may not cause as great a decrease in the pH of the cerebrospinal fluid because the blood-brain barrier and blood-cerebrospinal fluid barrier are not freely permeable to H^+
E. all of the above statements are true

Answer and Explanation
E.

26

(1:677; 2:6-77; 3:553; 4:544; 5:1851)
RCbq 2

Factors that tend to increase the pulmonary arterial pressure in individuals who have lived at altitudes of 12,000 ft for a week include

A. an increased hematocrit
B. a low alveolar P_{O_2}
C. both of the above
D. acidosis
E. both A and D

Answer and Explanation
C. The increased *hematocrit* causes an increased blood viscosity. The low alveolar P_{O_2} causes pulmonary *vasoconstriction*.
D. The increased ventilation at high altitudes causes *alkalosis*.

27

(1:722; 2:6-74;
3:558; 4:523;
5:1807)
RNbg 2

A moderate hypoxia differs from a moderate asphyxia in that in hypoxia, the direct

A. stimulation of the central chemoreceptors is a more important mechanism for increasing ventilation

B. stimulation of the aortic and carotid sinus is a more important mechanism for increasing ventilation

C. stimulation of the central and peripheral chemoreceptors is an equally important mechanism for increasing ventilation

D. stimulation of the aortic and carotid bodies is a more important mechanism for increasing ventilation

E. inhibition of the peripheral chemoreceptors is a more important mechanism for increasing ventilation

Answer and Explanation

D. In *asphyxia,* there is *hypoxia* plus *hypercapnea.* Hypercapnea stimulates both the central and peripheral (aortic and carotid bodies) chemoreceptors. Hypoxia stimulates the peripheral receptors to send more nerve impulses to the respiratory centers. Its direct action on the centers is depression.

19

Filtration, Reabsorption, and Secretion

1

*(1:875, 2:5-55;
3:576; 4:422;
5:1186)*
KEadfiu 1

Answer questions 1 through 17 by writing in the blanks that precede each question the appropriate structure or structures (f, g, etc.) that answer each question. Structures:

    f. Bowman's capsule
    g. proximal tubule
    h. descending loop of Henle
    i. ascending loop of Henle
    j. distal tubule
    k. collection duct

The number in parentheses at the end of a question, when present, represents the number of appropriate matches for that question.

1. _____ Where does the active reabsorption of Na^+ occur?
2. _____ Where does Tm limited reabsorption of Na^+ occur?
3. _____ Where does over 60% of the reabsorption of water occur? (1)
4. _____ Where does the secondary active reabsorption of glucose and amino acids occur? (1)
5. _____ Where does the active reabsorption of ketone bodies, albumin, creatine, sulfate, uric acid, and ascorbic acid occur? (1)
6. _____ Where does the active reabsorption of K^+ occur? (1)
7. _____ Where does the secretion of K^+ occur? (2)
8. _____ Where does the secretion of H^+ occur? (3)
9. _____ Where does the secretion of NH_3 occur? (3)
10. _____ Where does the passive reabsorption of urea occur? (3)
11. _____ Where is the luminal membrane that is least permeable to water in the presence of ADH? (1)
12. _____ Where does inulin move into the nephron?
13. _____ What structure(s) contain(s) a hypertonic urine in the absence of ADH?
14. _____ What structure(s) contain(s) a hypotonic urine in the absence of ADH?
15. _____ Where are the sites of action of aldosterone? (2)
16. _____ Where are the sites of action of parathormone? (2)
17. _____ Where is the site of action of vasopressin? (1)

Answer and Explanation

    1. g, i, j, k. In *reabsorption* a substance moves from the lumen of the nephron toward the blood.

(Continued)

2. **j.** The *transport maximum* (Tm) is the greatest rate that a transport system can move a solute. It occurs at a concentration of solute that *saturates* the transport system. The *carriers* for Na^+ elsewhere in the nephron do not saturate at the Na^+ concentrations to which they are exposed.

3. **g.** The high level of Na^+ reabsorption and the high *permeability* to water combine in the proximal tubule to move large volumes of water toward the blood.

4. **g. 5. g.** The reabsorption of most nutrients is dependent on (i.e., secondary to) the *active transport* of Na^+. (1) Glucose, for example, is moved across the luminal membrane of the tubule by a carrier *(facilitated diffusion)* which is also moving Na^+ from the lumen. (2) The Na^+ is then actively transported (ATP is catabolized) to the intercellular space of the tubule where (3) it diffuses into the blood. (4) The high intracellular glucose *concentration* in the tubular cell then drives the *diffusion* of the glucose into the blood. If Tm for these nutrients is not exceeded they will be totally reabsorbed in the proximal tubule.

6. **g.**

7. **j, k.** *Secretion* in the nephron is the movement of a particle toward the lumen of the nephron. K^+ secretion into the distal tubule and collecting duct frequently equals K^+ reabsorption in the proximal tubule. On the other hand, (1) decreased K^+ *ingestion*, (2) increased Na^+ entering the distal tubule, (3) systemic *acidosis,* and (4) decreased secretion of *aldosterone* all decrease K^+ secretion.

8. **g, j, k.** (1) H^+ is moved into the tubular lumen by an *exchange carrier* (antiport), (2) which moves Na^+ out of the lumen and into a cell on a 1 H^+ to 1 Na^+ basis. On the basolateral (other) side of the cell (3) Na^+ is being carried out of the cell toward the blood (reabsorption) by a pump mechanism (4) which is bringing K^+ into the cell. (5) The K^+ will compete with H^+ for transport into the tubular lumen. Therefore *K^+ depletion* tends to produce *acidosis* (due to increased H^+ secretion) and acidosis tends to produce *hyperkalemia* (due to decreased K^+ secretion).

9. **g, j, k.** Most of the NH_3 is formed in the cells of the nephron from glutamine (catalyzed by glutaminase) and glutamic acid (catalyzed by glutamic dehydrogenase). The NH_3 diffuses freely into the tubular fluid because it is *lipid soluble.* In the tubular fluid NH_3 combines with H^+ and forms NH_4^+, which is relatively insoluble in lipid and therefore remains in the tubular lumen. When the urine is alkaline no NH_4^+ is formed. In chronic acidosis the production of NH_4^+ increases over a period of about 4 days.

10. **g, j, k.** As water is reabsorbed from the nephron (at the proximal tubule, descending limb, distal tubule, and collecting duct), *urea* becomes more concentrated in the nephron lumen and diffuses toward the blood through most of those parts of the nephron that are permeable to urea. Since urea from the collecting duct tends to accumulate in the interstitial space of the renal medulla (where it contributes to the high *osmolarity* there), it diffuses into the nephron lumen in the descending tubule.

11. **i.** The ascending limb is also poorly permeable to urea, and an area of active Na^+ reabsorption. The Na^+ reabsorption contributes to the *hypertonicity* of the peritubular space in the *renal medulla.* The *ADH* (antidiuretic hormone = vasopressin) increases the *permeability* of the collecting duct to water. In the absence of ADH, the collecting duct is impermeable to water.

12. **f.**

13. **h.** It is surrounded by a *hypertonic* solution and is permeable to water.

14. **i, j, k.** There is an active reabsorption of Na^+ from the tubules in these areas.

15. **j, k.** *Aldosterone* increases (1) Na^+ reabsorption and (2) K^+ and H^+ secretion.

16. **g. j.** *Parathormone* decreases *phosphate* reabsorption in the proximal tubule and increases Ca^{++} reabsorption in the distal tubule. Despite this latter action, we frequently find increased Ca^{++} excretion in *hyperparathyroidism.* This is because in hyperparathyroidism there (1) is an increased plasma Ca^{++}, which (3) presents the distal tubule with a Ca^{++} level that exceeds the Tm for Ca^{++}.

17. **k.** *Vasopressin* (= ADH = antidiuretic hormone) increases the permeability of the collecting duct to water and urea. There results a reabsorption of water and a more concentrated urine.

2

In the kidney, the *glomerular capillary pressure* of a healthy adult (select the one best answer)

*(1:833; 2:5-33;
3:567; 4:405;
5:1172)*
KCpu 2

A. is lower than the pressure in the efferent renal arteriole
B. decreases when the afferent renal arteriole constricts
C. rises by 8% to 10% when the aortic pressure rises by 10%
D. is usually less than the pressure in a patent capillary in the deltoid muscle
E. all of the above statements are true

Answer and Explanation
A. The efferent arteriole carries blood from the glomerular capillary and therefore would be exposed to either a lower or the same pressure as in the capillary.
(B.) The constriction of the afferent arteriole increases the resistance to flow into the capillary.
C. At aortic pressures above 70 mm Hg, the afferent and efferent arterioles tend to keep the pressure in the glomerular capillaries constant.
D. The pressure in the glomerular capillary is about 50 mm Hg and that in the deltoid capillary is about 30 mm Hg.

3

The *glomerulus*

*(1:831; 2:5-29;
3:568; 4:406;
5:1169)*
Kudi 2

A. is impermeable to all molecules with a molecular weight (M) over 5000
B. contains no active transport systems ("pumps") that produce an important effect on the composition of the glomerular filtrate
C. produces a filtrate with a lower concentration of amino acids than found in plasma
D. produces a filtrate with a higher concentration of urea than found in plasma
E. all of the above statements are correct

Answer and Explanation
A. *Inulin* has an M of 5500 and has a concentration in the glomerular filtrate that is 98% of its concentration in plasma. *Hemoglobin* has an M of 68,000 and has a concentration in the filtrate that is 3% of that in the plasma.
(B.) Active transport is important in other parts of the nephron, but the major job of the glomerulus is to produce an *ultrafiltrate* of plasma. The glomeruli of the two kidneys produce approximately 125 ml of this filtrate per minute. *125 ml/min*
C, D. The concentrations of *amino acids, urea, glucose, ketone bodies, fatty acids,* and dissolved *electrolytes* in the filtrate are the same as those in plasma. The main factor that keeps a small particle from filtration is *protein binding.* This occurs in the case of calcium, potassium, and a number of the hormones.

4

The *glomerular filtration rate*

*(1:831; 2:5-29;
3:568; 4:408;
5:1170)*
Kudi 2

A. is greater than 50% of the plasma flow to the glomeruli
B. falls to approximately 25% of normal when the mean arterial pressure changes from 100 to 25 mm Hg
C. is decreased by a decrease in plasma colloid osmotic pressure
D. increases ipsilateral to a ureteral obstruction → ↓ GFR
E. none of the above statements are true

Answer and Explanation
A. The glomerular filtration rate (GFR) is normally about 125 ml/min and the plasma flow to the glomeruli is about 660 ml/min (= ERPF = C_{PAH}; see question 10).
B. It falls to 0 at arterial pressures less than 70 mm Hg.

(Continued)

C. This will cause either no change or an increased GFR.

D. This will decrease the filtration pressure across the glomerulus and, therefore, decrease GFR.

(E.)

5

Glucose and amino acid

A. reabsorption is most marked in the distal convoluted tubule

B. transport is primarily by active secretion into the tubular fluid

C. transport from the lumen of the nephron depends on Na^+ transport

D. transport is controlled by parathormone

E. transport is blocked by aldosterone

(1:845; 2:5-59; 3:380; 4:414; 5:1176)
Kdiu 1

Answer and Explanation

(C.) Glucose and Na^+ bind to a *common carrier* in a membrane lining the nephron. See question 1(5).

D. Parathormone controls Ca^{++} and phosphate reabsorption.

E. Aldosterone facilities Na^+ transport in the distal tubule.

6

In the *distal convoluted tubule,* the cells

A. contain large quantities of carbonic anhydrase which they use in the secretion of H^+ (Acid - Carbonic acid H₂CO₃)

B. can reabsorb Na^+ in exchange for H^+ secretion across the lumenal membrane

C. both A and B are correct

D. reabsorb over 40% of the glomerular filtrate

E. determine the final composition of the urine

(1:883; 2:5-75; 3:580; 4:454; 5:1195)
Kui 2

Answer and Explanation

A. *Carbonic anhydrase* catalyzes the formation of H_2CO_3 from CO_2 and H_2O.

(C.)

D. Approximately 80% of the glomerular filtrate is reabsorbed in the proximal convoluted tubule.

E. The collecting ducts also modify the composition of the urine.

7

A patient is treated with a drug that inhibits the action of *carbonic anhydrase.* Which symptom is he least likely to have?

A. an increased urinary excretion of K^+

B. an increased urinary excretion of Na^+

C. both of the above

D. a decreased plasma pH

E. a decreased urine volume

(1:883; 2:5-75; 3:580; 4:454; 5:1197)
Kuic 3

Answer and Explanation

A, B, C. In the absence of carbonic anhydrase (CA) activity there is less H^+ excreted and less HCO_3^- reabsorbed. Since Na^+ and K^+ reabsorption are paired with H^+ excretion and HCO_3^- reabsorption, Na^+ and K^+ reabsorption will be reduced (i.e., Na^+ and K^+ excretion will be increased) when CA activity is reduced.

D. Less excretion of H^+ causes more plasma H^+ (i.e., a decreased pH).

(E.) The increased excretion of Na^+ and K^+ will increase the urine volume.

8

(1:861; 2:5-63;
3:568; 4:426;
5:1193)
KCdf 2

The renal *clearance* of (select the one best answer)

A. a substance is measured in mg/ml
B. a substance is measured in mg/min
C. sodium is decreased by the injection of aldosterone
D. inulin, at a plasma concentration of 60 mg%, is lower than at a plasma concentration of 120 mg%
E. para-aminohippurate (PAH) at a plasma concentration of 60 mg% is higher than at a plasma concentration of 120 mg%

Answer and Explanation

A, B. Clearance is a measure of the ml of plasma cleared (i.e., completely depleted) of a particular substance, *x*, per minute and can be calculated as follows:

$$C_x = \frac{\text{quantity of } x \text{ excreted (mg/min)}}{\text{concentration of } x \text{ in plasma (mg/ml)}}.$$

C. *Aldosterone* increases the reabsorption of Na^+ from the nephron and therefore decreases sodium clearance.
D. Inulin clearance is the same at these plasma concentrations. It is filtered but not secreted or reabsorbed.
E. PAH clearance is less at the higher plasma concentration. It is filtered and secreted.

9

(1:838; 2:1176;
3:571; 4:418;
5:1176)
Ku 2

If a substance has a *transport maximum (Tm)* for reabsorption, this means

A. reabsorption is only passive
B. only a constant fraction of the substance will be reabsorbed
C. statements A and B are both correct
D. below a *threshold* level, all of the substance will be reabsorbed
E. phlorhizin blocks reabsorption

Answer and Explanation

D.
E. Phlorhizin inactivates the carrier system for (1) glucose, fructose, galactose, and xylose but does not affect the carriers for (2) sulfate and thiosulfate, nor for (3) arginine, lysine, ornithine, and cystine.

10

(1:864; 2:5-43;
3:574; 4:418;
5:1174)
Ku 2

The *renal plasma flow* can be estimated by using a formula for flow that is similar to the formula devised by Fick for the estimation of cardiac output. In this procedure,

A. para-aminohippurate (PAH) is injected at a rate such that its tubular maximum will not be exceeded
B. inulin is injected at a rate such that its tubular maximum will not be exceeded
C. para-aminohippurate is injected at a rate such that its tubular maximum will be exceeded
D. inulin is injected at a rate such that its tubular maximum will be exceeded
E. urea is injected

Answer and Explanation

A. PAH, when injected at a rate of less than 80 mg/min (= Tm_{PAH}), will have a renal vein concentration of almost 0 (zero) and a renal artery concentration equal to that in an arm vein. This makes it easy to estimate the renal arteriovenous concentration difference for PAH: PAH concentration in arm vein − 0.
B, C, D, E. Neither inulin (filtered) nor urea (filtered and reabsorbed) is completely re-

(Continued)

moved from the blood in one circuit. PAH (filtered and secreted), at the proper concentration, comes much closer to this. Approximately 10% of the blood passing through the normal kidney does not pass through functional nephrons (some goes to the renal capsule). Therefore, if one uses PAH to estimate renal plasma flow in a healthy subject, the estimate (ERPF: effective renal plasma flow) will be lower than the true value but will be a good index of the "functional renal mass." In certain diseases of the kidney, the number of nonfunctional units may be increased, and the ERPF may be even more divergent from the true renal plasma flow.

11

(1:864; 2:5-43; 3:566; 4:417; 5:1174)
Kum 2

During the infusion of PAH into a patient, the concentration of PAH in the cephalic vein stabilized at 0.02 mg/ml of plasma (= P_{PAH}). At this time, the two kidneys were producing 1 ml of urine per minute (= $\dot{V}_U$), and the concentration of PAH in the urine was 16 mg/ml (= U_{PAH}). What was the *PAH clearance* (= C_{PAH})? What was the *effective renal plasma flow* (= ERPF)?

Answer and Explanation

$$C_{PAH} = ERPF = \frac{(U_{PAH})}{P_{PAH}} \times (\dot{V}_U)$$

$$= \frac{16 \text{ mg/ml of urine}}{0.02 \text{ mg/ml of plasma}} \times (1 \text{ ml of urine/min})$$

$$= 800 \text{ ml of plasma/min.}$$

12

(1:862; 2:5-34; 3:566; 4:417; 5:1173)
Kfu 2

In questions 8 through 11, we noted that the clearance of PAH is used to indicate the effective renal plasma flow. It is a good index of this aspect of renal function because it is filtered and secreted but not reabsorbed. Therefore, almost all blood passing through the nephron, if it contains low concentrations of PAH, will be depleted of PAH. What characteristics of an agent used to test for *glomerular filtration* would be desirable? What agents meet these requirements (indicate one normally found in the blood and one not normally found in the blood)?

Answer and Explanation
The glomerular capsule should be freely *permeable* to it. It should not be *secreted* or *reabsorbed*. It should not affect body function. It should be easy and inexpensive to use. *Inulin* and *creatinine* come the closest to meeting these criteria. Unfortunately, in human beings, the creatinine clearance is higher than the inulin clearance. This discrepancy occurs because, in humans (unlike dogs), there is some creatinine secretion.

13

(1:863; 2:5-44; 3:566; 4:416; 5:1175)
Kfu 2

In a healthy individual, what percentage of the effective renal plasma flow would you expect to pass into the *glomerular capsule*?

A. less than 5%
B. between 15% and 20%
C. between 40% and 50%
D. between 70% and 80%
E. greater than 90%

Answer and Explanation
B.

14

(1:865; 2:5-44;
3:570; 4:418;
5:1179)
Kfu 2

The *filtration fraction* for the kidney is derived from the clearance of what two agents? What is the formula for filtration fraction?

Answer and Explanation
In a healthy subject:

$$\text{the filtration fraction} = \frac{\text{inulin clearance}}{\text{PAH clearance}} = \frac{125 \text{ ml/min}}{660 \text{ ml/min}} = 0.19.$$

This is the answer to question 13.

15

(1:865; 2:5-44;
3:570; 4:462;
5:1179)
Kfua 3

Acute glomerulonephritis, during its early stages, causes a pronounced decrease in glomerular filtration and a less marked decrease in tubular function. *Pyelonephritis*, on the other hand, causes a pronounced decrease in tubular function and a less marked decrease in glomerular filtration. How can one use the concept of the *filtration fraction* to distinguish these two clinical entities?

Answer and Explanation
In both these diseases, the C_{In} (inulin clearance = GFR = glomerular filtration rate) and the C_{PAH} (para-aminohippurate clearance = filtered and secreted) is reduced. In glomerulonephritis, however, the filtration fraction is decreased and in pyelonephritis it is increased.

16

(1:838; 2:5-37;
3:572; 4:418;
5:1177)
KCua 2

Glucose is noted to be at a plasma concentration of 100 mg% in patient f and 400 mg% in patient g. Which patient has the higher *glucose clearance*? What mechanisms would be responsible for this difference?

Answer and Explanation
Patient f has a glucose clearance of 0. In patient g, the quantity of glucose in the glomerular filtrate is so high that the *active transport* system for glucose reabsorption is overwhelmed. The plasma concentration of glucose at which glucose begins to appear in the urine is called the *renal plasma threshold* for glucose and normally is about 200 mg of glucose per 100 ml of plasma. The maximum quantity of glucose that can be reabsorbed in 1 minute is called the Tm_G (tubular maximum reabsorptive capacity for glucose). Since patient g has exceeded his renal plasma threshold for glucose and has probably also exceeded his Tm_G, he will have a clearance for glucose greater than 0.

17

(1:836; 2:5-40;
3:572; 4:416;
5:1177)
Kfua 2

The C_{In}/Tm_{PAH} ratio is used to determine glomerulotubular preponderance. How would it change (increase, decrease, stay constant) from normal in acute glomerulonephritis?

Answer and Explanation
The ratio decreases.

18

(1:836; 2:5-40;
3:572; 4:416;
5:1177)
KCdf 2

Which one of the following statements is most consistent with a filterable substance *being actively reabsorbed* from the renal tubular lumen?

A. its renal clearance value is lower than that of inulin
B. its renal clearance value is higher than that of inulin
C. the ratio of its rate of urinary excretion/plasma concentration is the same as that for glucose

(Continued)

D. the ratio of its rate of urinary excretion/plasma concentration is greater than that for glucose

E. its concentration in the distal tubule is higher than that in plasma

Answer and Explanation

A. This could be due to either active or passive reabsorption, since inulin is filtered, but not reabsorbed or secreted.

B. This is due to secretion, not reabsorption.

Ⓒ. D. Any substance that is filtered by the glomeruli and has a renal clearance of 0 (as does glucose) must be actively reabsorbed. The ratio mentioned in this question equals clearance:

$$C = \frac{\text{rate of excretion}}{\text{plasma concentration}} = \frac{\text{mg/min}}{\text{mg/ml of plasma}} = \text{ml of plasma/min.}$$

19

(1:874; 2:5-76; 3:578; 4:418; 5:1195)
KWdf 2

Which one of the following substances does not have a *Tm* value: albumin, arginine, beta hydroxybutyrate, glucose, hemoglobin, phosphate, sulfate, urea, uric acid?

Answer and Explanation

The only substance mentioned above that is not actively transported across the cells of the nephron and therefore does not have a transport maximum is urea.

20

(1:831; 2:5-29; 3:574; 4:421; 5:1170)
Kp Wp 2

The ratio of the amount of *inulin* excreted per minute to its arterial plasma concentration in the resting subject will

A. decrease in severe bilateral ureteral obstruction

B. increase in response to a decrease in plasma colloid osmotic pressure

C. change as noted in A and B

D. be dependent on active transport mechanisms

E. be positively correlated with the rate of inulin infusion

Answer and Explanation

A. True. Ureteral obstruction increases the *perivascular pressure* for the glomerular capillaries and therefore decreases glomerular filtration.

B. True. A decrease in glomerular capillary *colloid osmotic* pressure increases glomerular filtration.

Ⓒ. This ratio equals glomerular filtration rate.

21

(1:865; 2:5-41; 3:574; 4:415; 5:1179)
Kfmd 2

A substance that has a renal *clearance* 20 times that of *inulin* is probably

A. only filtered at the glomeruli

B. only secreted by the tubules

C. filtered and secreted

D. synthesized in the tubules and secreted

E. filtered and reabsorbed

Answer and Explanation

A. Inulin is filtered at the glomeruli, but not reabsorbed, secreted, or synthesized. Substances with a clearance greater than that for inulin must be moved into the nephron by a mechanism in addition to filtration.

B, C. Filtration and secretion collectively can clear a greater quantity of blood than either alone. Agents that are both filtered and secreted but not reabsorbed (PAH and *Diodrast,* for example) have a clearance somewhat less than 5 times that for inulin (i.e., their filtration fraction is about 0.2). A clearance 20 times that of inulin clearly indicates a mechanism in addition to filtration, secretion, and reabsorption.

(D) E. Reabsorption would result in a clearance less than that for inulin.

22

(1:874; 2:5-77; 3:578; 4:424; 5:1186)
Ku Wc 2

Select the one best statement. *Urea*

A. is secreted by the distal convoluted tubule
B. has a clearance greater than that for inulin
C. both A and B are correct
D. is formed primarily in the ascending part of the loop of Henle and in the distal convoluted tubule
E. clearance increases as the volume of urine excreted increases

Answer and Explanation

A, B, C. Urea is filtered and passively reabsorbed but not secreted and therefore its clearance is less than that for inulin (filtered but not reabsorbed or secreted).

D. Urea is formed in the liver.

(E.) Less urea is passively reabsorbed when less water is passively reabsorbed. In experiments in humans in which urine flow is modified by changing water intake, the following data are obtained: at a urine flow of 1 ml/min, 0.2 mM of urea were excreted per minute, and at a urine flow of 3 ml/min, 0.3 mM of urea were excreted per minute.

23

(2:1-131; 3:578; 4:874; 5:1648)
Dkcu 1

Which one of the following substances is the major source of urea?

A. dietary purines
B. dietary proteins
C. dietary pyrimidines
D. dietary phospholipids
E. beta hydroxybutyric acid

Answer and Explanation

B.

24

(1:861; 2:5-41; 3:574; 4:415; 5:1179)
Ku Wd 2

The *clearances* of substances f, g, and h are studied at different *concentrations* in the blood. The following data are obtained:

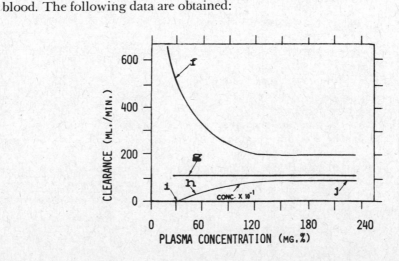

(Continued)

Which one of the following statements best characterizes substance f? Substance f in the nephron is

A. secreted
B. reabsorbed
C. filtered
D. filtered and reabsorbed
E. filtered and secreted

Answer and Explanation

A. When increasing concentrations of a substance in the plasma cause a progressive decrease in the clearance of that substance, then the secretory Tm for that substance has been exceeded. Its large clearance at lower plasma concentrations probably means that it is also filtered.

Ⓔ

25

(1:861; 2:5-36; 3:574; 4:415; 5:1176)
Ku Wd 2

Which one of the following statements best characterizes substance h in question 24? Substance h in the nephron is

A. filtered and actively secreted
B. filtered and passively reabsorbed
C. filtered, passively reabsorbed, and actively reabsorbed
D. filtered and synthesized
E. filtered, synthesized, and secreted

Answer and Explanation

A. If it were actively secreted, there would be a progressive decrease in clearance with an increase in plasma concentration.
B, Ⓒ The progressive increase in *clearance* with increase in plasma concentration suggests active reabsorption.
D, E. A substance that is filtered and secreted, or filtered and synthesized, would have a larger clearance than what is shown for either substance h or g.

26

(1:861; 2:5-37; 3:572; 4:419; 5:1174)
Ku Wd 2

Which one of the following statements concerning curve h in question 24 is true? The plasma *concentration* at point

A. i represents the transport maximum
B. i represents the splay
C. i represents the threshold
D. j represents the splay
E. j represents the threshold

Answer and Explanation

A. The transport maximum is reached somewhere beyond point j.
B. The deviation between the *threshold* concentration and the *transport maximum* concentration is called the *splay*.
Ⓒ The threshold is the lowest plasma concentration at which the ability of the transport system to move all of a particular substance across a barrier is exceeded.

27

(1:836; 2:5-40; 3:573; 4:416; 5:1174)
KuWd

All of the followng substances are filtered into the glomerular capsule. Indicate which ones will produce a *clearance* curve in man similar to f, g, or h in question 24. One of these substances has a Tm value of 375 mg/min and another of 1 mg/min. Therefore, in making your comparisons, concentrate on the shape of the clearance-concentration curve, not absolute values of concentration.

_____ inulin   _____ penicillin   _____ lysine
_____ glucose   _____ albumin   _____ hemoglobin
_____ PAH   _____ arginine   _____ diodrast
_____ sulfate   _____ ketone bodies   _____ phosphate
_____ ascorbic acid

Answer and Explanation

g: inulin   f: penicillin   h: lysine
h: glucose   h: albumin   h: hemoglobin
f: PAH   h: arginine   f: diodrast
h: sulfate   h: ketone bodies   h: phosphate
h: ascorbic acid

You will note, in the above table, that the body conserves nutrients such as arginine and ketone bodies (acetoacetic acid and beta hydroxybutyric acid) and other useful substances such as sulfates and albumin by active reabsorption (curve h). The Tm for curve h may range from 1 mg/min (hemoglobin) to 30 mg/min (albumin) to 375 mg/min (glucose).

28

(1:844, 2:5-37; 3:574; 4:415; 5:1179)
KuWa 2

Toward the end of World War II, Karl Beyer and his associates noted that the injection of PAH decreased the *excretion* of penicillin in the urine. What would you suggest was its mechanism of action? The PAH

A. competes with penicillin for a site on a carrier molecule in one of the reabsorptive mechanisms
B. prevents active reabsorption
C. either of the above could be correct
D. increases filtration
E. competes with penicillin for a site on a carrier molecule in one of the secretory mechanisms

Answer and Explanation

A, B, C. These changes would increase the excretion of penicillin if there were a carrier molecule for the reabsorption of penicillin or if penicillin were actively reabsorbed.
D. If PAH acted to increase filtration it would either produce no change or increase penicillin excretion.
(E.) In *Tm limited active transport* (either secretion or reabsorption) many substances apparently *compete* with one another for sites on a *carrier molecule*. This is presumably the case with (1) PAH, phenol red, penicillin and diodrast, (2) choline, histamine, and thiamine, (3) glucose, fructose, and phlorhizin, (4) the amino acids, arginine, lysine, and cystine, and (5) sulfate and thiosulfate.

Probenin   uricosuric

29

(1:843; 2:5-40; 3:570; 4:410; 5:1178)
KuWd 2

In arthritis, there is the deposition of urate crystals in the joints. If you wished to treat arthritis by decreasing the concentration of circulating *uric acid*, you could (select the one best answer)

A. keep your patient's caloric intake constant and carbohydrate intake low
B. block active reabsorption of uric acid in the kidney   *Probeneuid*
C. both of the above
D. block active secretion of uric acid in the kidney
E. both A and D

(Continued)

Answer and Explanation

A. Uric acid is produced from the catabolism of nucleoproteins. If you wished to decrease the production of uric acid, a diet low in meats (particularly the nonmuscular meats such as liver), meat extracts, and legumes would be far more effective than a diet low in carbohydrate.

(B.) The drug probenecid does this and therefore increases the loss of uric acid from the body.

D. There is evidence that uric acid is both actively secreted and reabsorbed. To block active secretion would increase its concentration in the body.

30

(1:869; 2:5-64; 3:584; 4:417; 5:1194)
Kmbf 2

The following data are obtained from a patient:

24-hour urine sample
Total volume:	1440 ml
Sodium concentration:	120 mEq/liter
Potassium concentration:	100 mEq/liter
Creatinine concentration:	200 mg/100 ml
Urea concentration:	2050 mg/100 ml

Plasma sample taken at the midpoint during the urine collection
Sodium concentration:	140 mEq/liter
Potassium concentration:	5 mEq/liter
Creatinine concentration:	1 mg/100 ml
Urea concentration:	25 mg/100 ml

What is the rate of *potassium excretion*?

A. less than 0.2 mEq/min
B. 0.2 mEq/min
C. 0.3 mEq/min
D. 0.4 mEq/min
E. more than 0.4 mEq/min

Answer and Explanation

A. K excretion rate = (K concentration in urine) × (urine production).

K concentration in urine = (100 mEq/liter) × (1 liter/1000 ml) = 0.1 mEq/ml.

Urine production = (1440 ml/day) × (1 day/24 hr) × (1 hr/60 min) = 1 ml/min.

K excretion rate = (0.1 mEq/ml) × (1 ml/min) = 0.1 mEq/min.

31

(1:861; 2:5-34; 3:578; 4:417; 5:1174)
Kmbf 2

On the basis of the data for the case in question 30, calculate the *urea clearance*.

A. less than 78 ml/min
B. 78 ml/min
C. 80 ml/min
D. 82 ml/min
E. more than 82 ml/min

Answer and Explanation
D.

$$C_{ur} = \left(\frac{U_{ur}}{P_{ur}} \right) \times \dot{V} = \left(\frac{\text{urine conc. of urea}}{\text{plasma conc. of urea}} \right) \times (\text{urine flow})$$

$$= \left(\frac{2050 \text{ mg}/100 \text{ ml of urine}}{25 \text{ mg}/100 \text{ ml of plasma}} \right) \times (1 \text{ ml of urine/min})$$

$$= 82 \text{ ml of plasma/min}.$$

32

(1:845; 2:5-54;
3:571; 4:415;
5:1176)
Kmbf 3

On the basis of the data for the case in question 30, estimate the rate of *sodium reabsorption.*

A. less than 28 mEq/min
B. 29 mEq/min
C. 31 mEq/min
D. 33 mEq/min
E. more than 34 mEq/min

Answer and Explanation
A. Na reabsorbed = (Na filtered) − (Na excreted) + (Na secreted).

Na filtered: creatinine clearance is frequently used clinically to estimate the rate of glomerular filtration. The values obtained using exogenous creatinine are not an accurate nor reliable measure of filtration, but the values obtained in this case using endogenous creatinine yield data almost as good as one would get from an inulin clearance study:

$$C_{Cr} = \left(\frac{U_{Cr}}{P_{Cr}} \right) \times (\dot{V})$$

$$= \left(\frac{200 \text{ mg}/100 \text{ ml of urine}}{1 \text{ mg}/100 \text{ ml of plasma}} \right) \times (1 \text{ ml of urine/min})$$

$$= 200 \text{ ml of plasma/min.}$$

Na filtered = (Na concentration in plasma) × (glomerular filtration rate)
= (0.140 mEq/ml) × (200 ml/min) = 28 mEq/min.

Na excreted = (Na concentration in urine) × (urine production)
= (0.120 mEq/ml) × (1 ml/min) = 0.120 mEq/min.

Na secreted = 0.

Na reabsorption = (28 mEq/min) − (0.12 mEq/min) = 27.88 mEq/min.

33

(1:861; 2:5-76;
3:571; 4:418;
5:1195)
Kmbf 3

On the basis of the data from the case in question 32, calculate the fraction of the filtered *urea* that was *excreted.*

A. 0.04
B. 0.1
C. 0.2
D. 0.4
E. 1.0

Answer and Explanation
D. Fraction of urea excreted = (urea excreted)/(urea filtered).

Urea excreted = (urea concentration in urine) × (urine production)
= (20.5 mg/ml) × (1 ml/min) = 20.5 mg/min.

Urea filtered = (urea concentration in plasma) × (plasma filtered)
− (0.25 mg/ml) × (200 ml/min) = 50 mg/min.

Fraction of urea excreted = (20.5 mg/min)/(50 mg/min) = 0.41.

34

(1:871; 2:5-50; 3:577; 4:422; 5:1180)
KdiW 2

A mechanism in the kidney that produces marked increases in the tonicity of the peritubular liquid that surrounds the loop of Henle is called *countercurrent multiplication*. Which one of the following relationships is not part of this mechanism?

A. the impermeability of the ascending limb prevents the diffusion of Na^+ through the cells that form the limb
B. active transport of Na^+ into the peritubular space from the lumen of the descending limb
C. active transport of Na^+ into the peritubular space from the lumen of the ascending limb
D. diffusion of Na^+ into the descending limb
E. hypertonicity of the fluid in the loop of Henle

Answer and Explanation
A. In countercurrent multiplication (1) the ascending limb by virtue of its impermeability and active transport system creates (2) a hypernatremic environment in the peritubular fluid, which causes (3) a hypernatremia in the lumen of the permeable descending limb, which causes (4) hypernatremia in the ascending limb, which permits (5) an even greater transfer of Na^+ from the ascending limb into the peritubular fluid.
B. This apparently does not occur. The Na^+ current is into the descending limb.

35

(1:877; 2:5-53; 3:189; 4:413; 5:1186)
Kudi 2

The *loop of Henle* differs from the medullary *collecting duct* in that the loop of Henle (select the one best answer)

A. in its descending limb, in the presence of antidiuretic hormone, is permeable to water and the collecting duct is not
B. in its thick ascending limb actively transports ions that serve to make the medullary extracellular substance markedly hypertonic and the collecting duct does not
C. in its thin segment contains, in the presence of antidiuretic hormone, a hypertonic solution and the collecting duct does not
D. statements A, B, and C are all correct
E. statements A, B, and C are all incorrect

Answer and Explanation
A. Both are permeable to water. Antidiuretic hormone increases the permeability of the collecting duct to water.
B. Both are responsible for the hypertonicity of the medullary peritubular compartment.
C. Both contain a hypertonic solution in the presence of antidiuretic hormone (ADH). The ADH increases the osmotic pressure of the fluid in the medullary collecting duct by increasing its permeability to water (i.e., by increasing water reabsorption).
E.

36

(1:829; 2:5-23; 3:565; 4:403; 5:1169)
Kuf 2

Select the one best statement. The *juxtamedullary nephrons*

A. constitute over 40% of the nephrons in the kidney
B. have a short loop of Henle
C. lie solely in the renal medulla
D. lie solely in the renal cortex
E. are not well characterized by any of the above statements

Answer and Explanation
A. They constitute less than 15% of the nephrons. The *cortical nephrons* constitute the rest.
B. The cortical nephrons have a short loop of Henle that extends into the outer zone of the medulla. The juxtamedullary nephrons, like the *collecting ducts*, extend well into

the inner zone of the medulla. Some have loops of Henle that extend to the papillary tip of the medulla.

C, D. Both types of nephrons have their glomerulus in the cortex and their loop of Henle in the medulla.

(E.)

37

(1:845; 2:5-46; 3:575; 4:413; 5:1176)
Kid 2

During an experiment on a man, fluid from the last part of the proximal tubule was sampled after the administration of a drug. This sample had the following characteristics:

Inulin concentration:	30% higher than that found in plasma
Na^+ concentration:	150 mM/liter

What is the most likely action of the drug? It

A. decreased the glomerular filtration rate
B. increased inulin secretion into the proximal tubule
C. did both of the above
D. decreased NaCl transport from the proximal tubule
E. decreased urea transport in the proximal tubule

Answer and Explanation
A. Normally, the *inulin concentration* would be 400% higher than that in plasma, and the Na^+ concentration would be about 150 mM/liter. Decreases in filtration will not cause large changes in inulin concentration.
B. Inulin is neither secreted nor reabsorbed.
(D.) Normally, the active transport of NaCl from the lumen of the *proximal tubule* is responsible for a diffusion of *water* from the lumen that causes a fourfold increase in inulin concentration in the luminal fluid. The net result is that the luminal fluid has an osmotic pressure and Na^+ concentration similar to that in the glomerular filtrate. In other words, *NaCl reabsorption* in the proximal tubule is a mechanism for the concentration of inulin and other poorly reabsorbed substances.
E. The active transport of *urea* has not been demonstrated in humans.

38

(1:945; 2:5-37; 3:572; 4:414; 5:1180)
KuiF. 2

An osmotic diuresis resulting from hyperglycemia differs from water diuresis in that in the osmotic diuresis there is

A. a substantial decrease in the reabsorption of water in the proximal tubule and the proximal part of the loop of Henle
B. a decreased K^+ excretion
C. an increased concentration of urea in the urinary bladder
D. a decreased concentration of circulating antidiuretic hormone
E. none of the above changes

Answer and Explanation
(A.) Approximately 70% of the *water* in the glomerular filtrate is reabsorbed in the *proximal tubule*. This is due to (1) the permeability of the tubule to water and (2) the osmotic gradient between the tubular and peritubular fluid. Abnormally high concentrations in the glomerular filtrate of substances that are not readily *reabsorbed* (glucose in the case of hyperglycemia) prevent the maintenance of this gradient and therefore decrease the reabsorption of water in the proximal tubule. In water diuresis, there is also a reduced reabsorption of water, but this occurs at the *distal tubule* and *collecting duct*, not at the proximal tubule.
B. Potassium excretion will be little affected by diuresis.

(Continued)

C. Urea concentration in the urine is decreased during diuresis.

D. If the *diuresis* causes an increase in the *tonicity* of the blood, there will be an increased release of *antidiuretic hormone* (ADH) if the hypothalamus and pituitary are functioning normally. A decrease in ADH release is one mechanism that produces diuresis in response to a water load.

Questions 39 through 41 consist of two main parts: a statement and a reason for the statement. Choose the best answer for each of these questions from the options A through E below and mark your choice in the space provided.

A. both the statement and the reason are true and are related as cause and effect (i.e., as indicated)

B. both the statement and the reason are true but are not related as cause and effect

C. the statement is true but the reason is false

D. the statement is false but the reason is an accepted fact or principle

E. both the statement and the reason are false.

39

(1:870; 2:5-46; 3:575; 4:410; 5:1188)
Kid 2

Statement

_____ Tubular fluid in the distal third of the proximal tubule has an osmolarity of 200 mOsm because

Reason

there is an active transport of sodium chloride in the proximal two-thirds of the proximal tubule.

Answer and Explanation

D. This fluid has approximately the same tonicity as the *glomerular filtrate* (300 mOsm). The active *reabsorption* of NaCl from the *proximal tubule* does not appreciably affect the *osmolarity* of its contents because the tubule is highly permeable to *water*.

40

(1:870; 2:5-56; 3:575; 4:421; 5:1188)
Kid 2

Statement

_____ The site of highest osmolarity in the nephron of a well hydrated subject is in the loop of Henle because

Reason

Na^+ is actively secreted into the tubular fluid of the loop of Henle.

Answer and Explanation

C. The mechanisms responsible for the high osmolarity in the *loop of Henle* are (1) the high *tonicity* of the fluid surrounding the loop and (2) the high permeability to *water* of its descending limb. The ascending limb actively reabsorbs Na^+.

41

(1:867; 2:7-62; 3:370; 4:433; 5:1588)
CEKi 2

Statement

_____ A hemorrhage that causes a 20% reduction in blood volume will cause an increased reabsorption of Na^+ in the distal tubule because

Reason

hemorrhage causes an increased secretion of renin and aldosterone.

Answer and Explanation

A. *Renin* facilitates the formation of angiotensin II, which stimulates the adrenal cortex to release *aldosterone*, which acts on the *distal tubule* to increase the reabsorption of Na^+.

The following diagram represents the types of cells that might be found between (1) the lumen of a nephron or gastrointestinal tract and (2) a capillary. Answer questions 42 through 48 on this diagram. Write the one best answer (A, B, C, D, E, F) for each question in the blank provided.

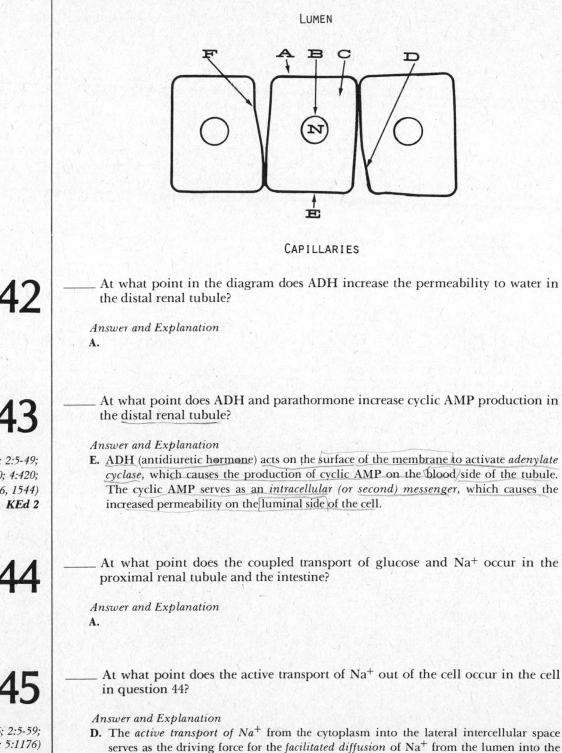

42 _____ At what point in the diagram does ADH increase the permeability to water in the distal renal tubule?

Answer and Explanation
A.

43 _____ At what point does ADH and parathormone increase cyclic AMP production in the distal renal tubule?

(1:877; 2:5-49; 3:190; 4:420; 5:1486, 1544)
KEd 2

Answer and Explanation
E. ADH (antidiuretic hormone) acts on the surface of the membrane to activate *adenylate cyclase*, which causes the production of cyclic AMP on the blood side of the tubule. The cyclic AMP serves as an *intracellular (or second) messenger*, which causes the increased permeability on the luminal side of the cell.

44 _____ At what point does the coupled transport of glucose and Na$^+$ occur in the proximal renal tubule and the intestine?

Answer and Explanation
A.

45 _____ At what point does the active transport of Na$^+$ out of the cell occur in the cell in question 44?

(1:846; 2:5-59; 3:380; 4:51; 5:1176)

Answer and Explanation
D. The *active transport of Na$^+$* from the cytoplasm into the lateral intercellular space serves as the driving force for the *facilitated diffusion* of Na$^+$ from the lumen into the cytoplasm. Since glucose and Na$^+$ share the same carrier in the luminal membrane, the

(Continued)

influx of *glucose* into the cytoplasm (and therefore into the blood) will, within limits, be directly related to the concentration of Na^+ in the lumen.

46

(1:846; 2:5-59; 3:580; 4:51; 5:1199)
Kd 2

_____ At what point is the H^+-Na^+ exchange carrier (antiport) found in the renal tubule?

Answer and Explanation

A. The pattern we see in each of these questions is a specialization of the two sides of the membrane, the *luminal side* (A) and the *capillary side* (D and E). There are (1) specialized *receptors* on the capillary side, (2) an active Na^+-K^+ *pump* extruding Na^+ from the cell on the capillary side (D or E) and sometimes providing a *concentration gradient* that drives the transport system on the luminal side, (3) a mechanism for moving water and nutrients from the lumen (increased permeability, Na^+-coupled transport of glucose) on the luminal side, and (4) a mechanism for moving H^+ into the lumen (H^+-Na^+ exchange carrier) on the luminal side.

47

_____ At what point is the H^+-K^+ ATPase found in the stomach?

Answer and Explanation

F.

48

(1:777; 2:49; 3:395; 4:805; 5:1296)
DKd 2

_____ At what point does HCO_3^- diffusion occur in the cell in question 47.

Answer and Explanation

E. In the stomach, like the kidney, H^+ is actively secreted across the luminal membrane. In the kidney the part of luminal membrane where this occurs forms the intracellular *canaliculi*. This loss of H^+ from the cytoplasm drives the following reaction to the right:

$$CO_2 + HOH \rightleftharpoons H_2CO_3 \rightleftharpoons HCO_3^- + H^+.$$

This increases intracellular $[HCO_3^-]$, and causes HCO_3^- to diffuse out of the cell on the capillary side, and Cl^- diffuses into the cell on the capillary side and out on the luminal side. The net result is HCl moves into the lumen and a *bicarbonate* moves into the blood.

20
Acid-Base Balance

1

*(1:696; 2:5-20;
3:592; 4:592;
5:1102)*

KRabu 1

Write in Column I the best definition from Column II for each condition listed.

Column I (Condition)

1. _____ Uncompensated metabolic acidosis
2. _____ Compensated metabolic acidosis
3. _____ Uncompensated respiratory acidosis
4. _____ Compensated respiratory acidosis

Column II (Definition)

A. Normal pH and high bicarbonate concentration in the blood plasma
B. Near normal pH and a $[HCO_3^-]/[H_2CO_3]$ ratio much lower than 20/1
C. Abnormally low blood pH due to excess fixed acid
D. Abnormally low pH and a high bicarbonate concentration in the plasma
E. Abnormally high blood pH

Answer and Explanation

1. **C.** *Fixed acids* are nonvolatile H^+ donors, such as *lactic acid* and H_2SO_4; and the acid salts, NH_4Cl and NaH_2PO_4.

2. **B.** In *metabolic acidosis*, the respiratory system responds to the increased $[H^+]$ by increasing *ventilation*. The net result is a decreased P_{CO_2} and a marked decrease in $[HCO_3^-]$, because of the decreased P_{CO_2} and because H^+ drives the following equilibrium to the right:

$$H^+ + HCO_3^- \leftrightharpoons H_2CO_3.$$

As a result the $[HCO_3^-]/[H_2CO_3]$ ratio changes from 20 (at a pH of 7.4) to a much lower number (1.1 at a pH of 6.06).

3. **D.** In this condition the retention of the acid-forming gas, CO_2, causes the decreased pH and the increased bicarbonate.

4. **A.** Inadequate ventilation causes an increased plasma P_{CO_2}, $[H^+]$, and $[HCO_3^-]$. The *kidney* is able to return the $[H^+]$ to normal by increasing the excretion of H^+ and a number of H-containing salts. These salts (NaH_2PO_4, $NaHSO_4$, NH_4Cl) in a sense "smuggle" the H^+ out of the body without as markedly affecting urine pH as H^+ plus Cl^- do. The kidney is generally much better, although slower, in compensating for a respiratory acidosis or alkalosis than the respiratory system is for compensating for a

(Continued)

metabolic pH change. This is because the respiratory system uses only one mechanism (ventilation) to control four different parameters (PO_2, PCO_2, $[H^+]$, $[HCO_3^-]$). Therefore, in bringing one parameter (H^+) into balance, it upsets the balance of one or more other parameters (CO_2 and HCO_3^- for example). When we decrease $[H^+]$ by increasing ventilation, we also decrease PCO_2. Since a decreased PCO_2 causes a decreased ventilation, the decreased PCO_2 prevents the compensation for the high $[H^+]$ from being complete:

$$\text{Increased} \rightarrow \text{Increased} \rightarrow \begin{array}{c} \text{Decreased} \rightarrow \\ [H^+] \\ \text{Decreased} \rightarrow \end{array} \text{Decreased}$$

Increased $[H^+]$ → Increased ventilation → Decreased $[H^+]$ / Decreased PCO_2 → Decreased ventilation.

Under healthy, resting conditions, the plasma bicarbonate is about 24 mM/liter, but during *respiratory acidosis* or metabolic alkalosis it may rise to 40 mM/liter.

2

(1:698; 2:5-20; 3:592; 4:455; 5:1102)
KRabu 2

Write in column I each of the appropriate signs form Column II for each of the conditions listed. Each question has either two or three correct matches.

Column I (Condition)
1. ____ Metabolic alkalosis
2. ____ Compensated metabolic alkalosis
3. ____ Respiratory alkalosis
4. ____ Compensated respiratory alkalosis

Column II (Sign)
f. high plasma pH
g. low plasma pH
h. near normal plasma pH
i. normal plasma pH
j. high plasma $[HCO_3^-]$
k. low plasma $[HCO_3^-]$
l. a urine pH between 6.5 and 4.5

Answer and Explanation

1. **f, j, l.** (1) A high alkaline ingestion ($NaHCO_3$ or vegetables) causes an alkaline urine. *Renal failure* can cause an acid urine. (j) Alkalosis drives the following reaction in the plasma to the right:

$$OH^- + H_2CO_3 \rightleftharpoons HOH + HCO_3^-.$$

2. **h, j.** The $[HCO_3^-]$ is increased further by the *compensation* (i.e., *decreased ventilation*).

3. **f, k, l.** The kidney is not able to produce a *urinary pH* below 4.5 or above 8.0. Hyperventilation causes *respiratory alkalosis* by eliminating an excess of the acid-forming gas, CO_2, and by so doing decreases plasma $[HCO_3^-]$ as well as $[H^+]$.

4. **i, k.** *Renal compensation* (increased excretion of bicarbonate, decreased excretion of H^+ and acid salts) exaggerates the decreased bicarbonate concentration, but is usually able to bring the elevated pH back to normal. The kidney, unlike the lung, is usually able to change the plasma pH without upsetting the other parameters it controls (water balance, electrolyte balance, and the elimination of urea, for example). The kidney responds to respiratory alkalosis by excreting more $NaHCO_3$ or H_2CO_3. Which molecule is eliminated in greatest quantity will be determined by the body's need to *conserve Na or K*. In respiratory acidosis the kidney eliminates NH_4Cl, NaH_2PO_4, or KH_2PO_4. Which molecule is eliminated is determined by the body's need to conserve Na, K, or both. The NH_4^+, unlike Na^+ and K^+, is only important as a part of the kidney's mechanism for maintaining acid-base balance.

3

*(1:696; 2:512;
3:592; 4:452;
5:1197)*
KRabu 2

Write in the space provided the cause(s) from Column II (f, g, etc.) for each of the conditions listed in Column I. Each item in column II should be used once.

Column I (Condition)
1. _____ Metabolic acidosis
2. _____ Metabolic alkalosis
3. _____ Respiratory acidosis
4. _____ Respiratory alkalosis

Column II (Cause)
f. asphyxia
g. diabetes mellitus
h. hyperventilation
j. ingestion of $NaHCO_3$
k. regurgitation of gastric contents
l. strenuous running
m. ingestion of vegetables and fruit
n. ingestion of proteins
o. chronic renal disease

Answer and Explanation

1. **g, i, l, n, o.** Metabolic acidosis may be caused by (g and l) an increased production of acid (ketone bodies, lactate, etc.) or (i and l) an increased intake of OH^--removing salt or acid-forming nutrients, or (o) a decreased excretion of fixed acid. The ingestion of NH_4Cl drives the following reaction to the right:

$$OH^- + NH_4Cl \rightleftharpoons NH_4OH + Cl^-.$$

In the healthy subject on an average U.S. *diet*, the kidneys excrete 40 to 80 mEq of *nonvolatile acid* per day and the lungs eliminate about 13,000 mEq of *volatile acid* (CO_2) per day. In chronic renal disease the kidney has a reduced ability to produce NH_3 and therefore a reduced ability (1) to eliminate H^+ (i.e., the urine contains less NH_4Cl) and (2) to prevent a highly acid urine (i.e., the urine is less well buffered and therefore may approach a pH of 4.5).

2. **j, k, m.** Metabolic alkalosis may be caused by (j and m) the ingestion of an alkaline salt or nutrient, or (k) the loss of an acid secretion (*gastric juice*).

3. **f.** *Asphyxia* equals hypoxia + hypercapnea. The hypercapnea causes the acidosis.

4. **h.** *Hyperventilation* decreases plasma PCO_2

4

*(1:693; 2:4-104;
3:592; 4:448;
5:1411)*
Dbc 2

The addition of a fixed acid such as H_2SO_4 to the blood during protein catabolism may produce little change in the pH of the blood because of the *blood buffer* systems. Which buffer system in the healthy subject will usually combine with the most H^+? Name some other important blood buffer systems.

Answer and Explanation

Hemoglobin, under these circumstances, will combine with the most H^+. Other systems that buffer changes in H^+ are plasma proteins, the $NaHCO_3$-H_2CO_3 pair (the principal buffer system in the entire extracellular fluid), and the Na_2HPO_4-NaH_2PO_4 pair.

5

*(1:887; 2:5-72;
3:581; 4:456;
5:1201)*
Kbca 2

In prolonged acidosis, the production of NH_3 by the kidney increases over several days. What functions does this NH_3 serve?

Answer and Explanation

It helps transport H^+ out of the body and, in so doing, spares other transport systems (NaH_2PO_4 and KH_2PO_4) and thus prevents the depletion of Na^+ and K^+ from the body.

6

(1:848; 2:5-68;
3:582; 4:449;
5:1197)
Kbgc 1

Select the one best statement. Renal *bicarbonate reabsorption*

A. reaches a plasma threshold at about the resting plasma bicarbonate concentration
B. has a transport maximum that is elevated by increases in P_{CO_2} and aldosterone
C. both statements A and B are correct
D. has a stable Tm
E. is not active

Answer and Explanation
C.

7

(1:696; 2:5-19;
3:592; 4:458;
5:1201)
KRca 2

A 65-year-old woman complaining of dyspnea enters the hospital. The following data are obtained from her and from a normal healthy subject:

	Patient	Normal
Arterial P_{O_2} (mm Hg):	75	95
Arterial P_{CO_2} (mm Hg):	55	40
Arterial hemoglobin (g/100 ml):	18	15
Arterial pH:	7.35	7.40

The patient is suffering from

A. respiratory alkalosis
B. respiratory acidosis
C. metabolic alkalosis
D. metabolic acidosis
E. anemia

Answer and Explanation
B. When the *pH* is low (= 7.35) and the P_{CO_2}, HCO_3^-, or H_2CO_3 are high (P_{CO_2} = 55 mm Hg in the arteries) the condition is respiratory acidosis.

8

(1:882; 2:5-69;
3:580; 4:453;
5:1198)
Kbci 1

The rate of secretion of H^+ by the nephron is (check each correct statement)

1. greater in the proximal tubule than in the distal tubule ✓
2. increased in response to aldosterone
3. increased when plasma K^+ increases
4. inversely related to the rate of reabsorption of bicarbonate
5. decreased by carbonic anhydrase inhibitors ✓
6. inversely related to the intracellular P_{CO_2}

Answer and Explanation
1. An active transport system apparently moves H^+ into the tubular lumen.
2. Aldosterone facilitates the conservation of Na^+ and the excretion of H^+ and K^+.
3. *Hydrogen ion secretion* is inversely related to plasma K^+ concentration.
4. It is directly related to *bicarbonate reabsorption*.
5.
6. An increase in P_{CO_2} causes an increase in H_2CO_3 and H^+. It therefore increases H^+ secretion.

21
Control of the Kidney

1

*(1:866; 2:5-46;
3:567; 4:429;
5:1193)*
KEano 1

A decreased pressure in the carotid sinus causes

A. an increased renal adrenergic sympathetic tone
B. renal vasoconstriction
C. both of the above
D. an increased secretion of renin by the juxtaglomerular cells
E. all of the above

Answer and Explanation

E. When the pressure in the arterial or cardiopulmonary *stretch receptors* is reduced there is an increased renal, adrenergic, sympathetic tone that causes an increased resistance to renal blood flow and an increased secretion of *renin*. In the decentralized kidney or *transplanted kidney*, most renal functions seem to remain normal, since the *autoregulatory mechanisms* that control renal vasoconstriction and release of renin remain intact.

2

*(1:867; 2:3-170;
3:510; 4:433;
5:1588)*
KCEa 2

In a series of experiments, one group of animals has both kidneys removed and a second group has a sham operation. Each group is then bled until the *arterial pressure* decreases to 10 mm Hg. The animals are then permitted 40 minutes to recover. At the end of this period, it is found that the sham *nephrectomized* animals have an arterial pressure 30 mm Hg higher than the nephrectomized animals. Which one of the following hypotheses do these observations support? In response to hemorrhage the kidney

A. (by conserving water) can increase the arterial pressure
B. (by producing an agent that either directly or indirectly causes an increased sysmetic resistance) increases the arterial pressure
C. does both A and B
D. releases a vasodilator agent
E. does both A and D

Answer and Explanation

A. The kidney conserves water by decreasing urine volume. In the nephrectomized animals there would be no urine production, so, on this basis, the nephrectomized animals would not have a lower pressure after hemorrhage.

(Continued)

B. This observation is consistent with the data presented.

D. A vasodilator would decrease arterial pressure.

3

*(1:867; 2:3-170;
3:372; 4:433;
5:1588)*
KCEa 1

What is the agent released by the kidney in question 2 that tends to counter the hypotensive action of hemorrhage? Where in the kidney is it produced?

Answer and Explanation

It is *renin*, and it is produced by juxtaglomerular cells surrounding the renal afferent arterioles.

4

*(1:867; 2:5-63;
3:370; 4:433;
5:1588)*
KCEa 1

What is the mechanism whereby *renin* produces an increased arterial pressure?

Answer and Explanation

Renin catalyzes the formation of the potent vasoconstrictor, angiotensin II:

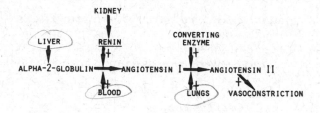

5

*(1:867, 2:5-63;
3:371; 4:433;
5:1593)*
KEui 3

What actions does *angiotensin II* have that are not related to its ability to produce *vasoconstriction* (select the one best answer)? Angiotensin II

A. increases extracellular fluid volume

B. increases K^+ and H^+ excretion

C. statements A and B are both correct

D. decreases the reabsorption of Na^+ by the kidney

E. statements A, B, and D are correct

Answer and Explanation

A. Angiotensin II facilitates the release of aldosterone from the adrenal cortex. Aldosterone, by causing the kidney to reabsorb more *osmotically active material* (Na^+ and Cl^-) than it excretes, increases extracellular fluid volume.

C.

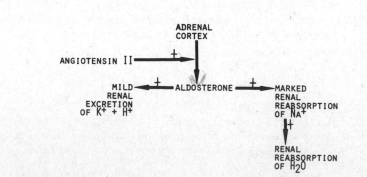

6

(1:1007; 2:5-47;
3:189; 4:423;
5:1486)
KECq 2

Which one of the following statements is incorrect? The *antidiuretic hormone* (ADH)

A. is released in response to a hypertonic solution infused into the arteries serving the anterior hypothalamus
B. decreases blood volume
C. is released during water deprivation
D. is released in the well-hydrated subject
E. is released in response to stimulation of the supraoptic nuclei of the hypothalamus

Answer and Explanation

B.

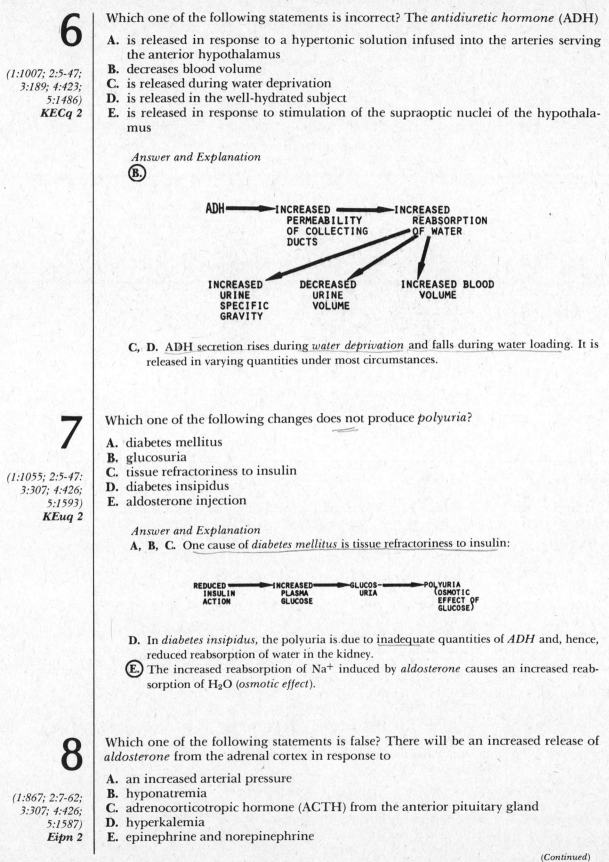

C, D. ADH secretion rises during *water deprivation* and falls during water loading. It is released in varying quantities under most circumstances.

7

(1:1055; 2:5-47:
3:307; 4:426;
5:1593)
KEuq 2

Which one of the following changes does not produce *polyuria*?

A. diabetes mellitus
B. glucosuria
C. tissue refractoriness to insulin
D. diabetes insipidus
E. aldosterone injection

Answer and Explanation

A, B, C. One cause of *diabetes mellitus* is tissue refractoriness to insulin:

REDUCED ➡ INCREASED ➡ GLUCOS- ➡ POLYURIA
INSULIN   PLASMA   URIA   (OSMOTIC
ACTION   GLUCOSE   EFFECT OF
                                       GLUCOSE)

D. In *diabetes insipidus,* the polyuria is due to inadequate quantities of *ADH* and, hence, reduced reabsorption of water in the kidney.
E. The increased reabsorption of Na^+ induced by *aldosterone* causes an increased reabsorption of H_2O *(osmotic effect)*.

8

(1:867; 2:7-62;
3:307; 4:426;
5:1587)
Eipn 2

Which one of the following statements is false? There will be an increased release of *aldosterone* from the adrenal cortex in response to

A. an increased arterial pressure
B. hyponatremia
C. adrenocorticotropic hormone (ACTH) from the anterior pituitary gland
D. hyperkalemia
E. epinephrine and norepinephrine

(Continued)

Answer and Explanation

A. Increases in arterial pressure decrease the secretion of aldosterone:

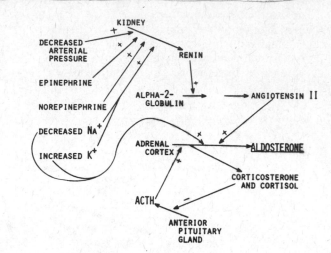

9

Which one of the following statements is <u>incorrect</u>? *Parathyroid hormone*

A. can cause an increased muscle tone and hyperreflexia
B. decreases the renal reabsorption of phosphate
C. increases the renal reabsorption of Ca^{++}
D. can cause a demineralization of bone
E. can cause an increase in urinary Ca^{++}

(1:842; 2:7-48;
3:320; 4:581,
5:1541)
KENi 2

Answer and Explanation

A. Parathormone (PTH) produces an increased plasma Ca^{++}. It is a decreased plasma Ca^{++} that causes tetany (increased muscle tone and hyperreflexia).

B, C, D, E.

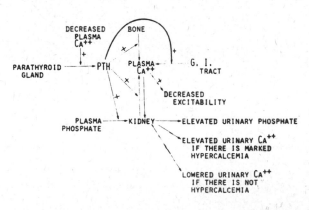

10

A patient is noted to have a markedly reduced plasma colloid osmotic pressure and Na^+ excretion. After the intravenous infusion of a concentrated albumin solution, the patient has a rapid weight loss. The signs, prior to treatment, are consistent with a diagnosis of

(1:867; 2:5-63;
3:307; 4:426;
5:1593)
DEqu 3

A. edema
B. an elevated aldosterone secretion
C. both A and B are correct
D. a low secretion of antidiuretic hormone
E. both A and D are correct

Answer and Explanation
A, B, (C.)

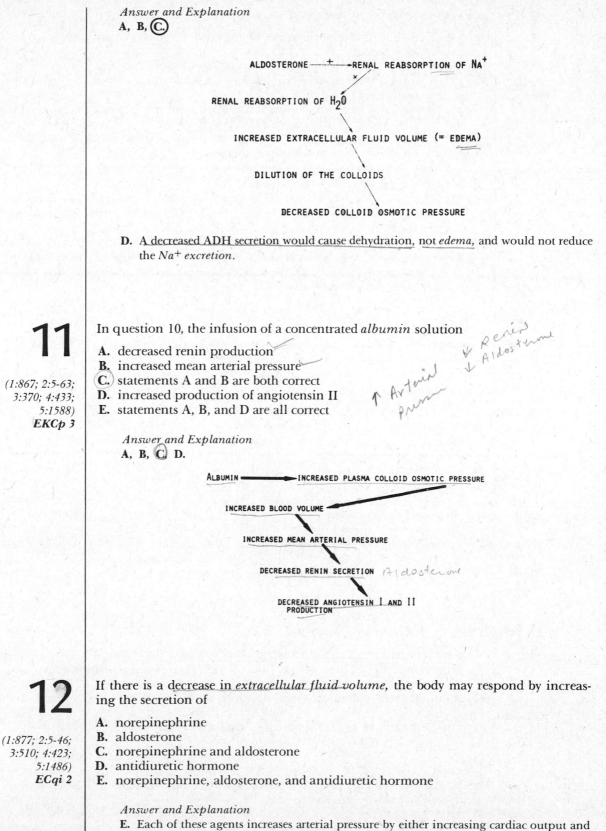

ALDOSTERONE ————+———→ RENAL REABSORPTION OF Na⁺

RENAL REABSORPTION OF H₂O

INCREASED EXTRACELLULAR FLUID VOLUME (= EDEMA)

DILUTION OF THE COLLOIDS

DECREASED COLLOID OSMOTIC PRESSURE

D. A decreased ADH secretion would cause dehydration, not *edema,* and would not reduce the Na^+ excretion.

11

(1:867; 2:5-63; 3:370; 4:433; 5:1588)
EKCp 3

In question 10, the infusion of a concentrated *albumin* solution

A. decreased renin production
B. increased mean arterial pressure
(C.) statements A and B are both correct
D. increased production of angiotensin II
E. statements A, B, and D are all correct

Answer and Explanation
A, B, (C) D.

ALBUMIN ————————→ INCREASED PLASMA COLLOID OSMOTIC PRESSURE

INCREASED BLOOD VOLUME

INCREASED MEAN ARTERIAL PRESSURE

DECREASED RENIN SECRETION

DECREASED ANGIOTENSIN I AND II PRODUCTION

12

(1:877; 2:5-46; 3:510; 4:423; 5:1486)
ECqi 2

If there is a decrease in *extracellular fluid volume,* the body may respond by increasing the secretion of

A. norepinephrine
B. aldosterone
C. norepinephrine and aldosterone
D. antidiuretic hormone
E. norepinephrine, aldosterone, and antidiuretic hormone

Answer and Explanation
E. Each of these agents increases arterial pressure by either increasing cardiac output and peripheral resistance or by increasing blood volume.

13

(1:891; 2:5-101; 3:587; 4:472; 5:912)
KMNu 2

Select the one incorrect statement.

A. the distention produced by urine flowing into the minor renal calyces triggers a calyceal contraction
B. the renal pelvis undergoes rhythmic dilation and contraction
C. after closure of the pelviureteric junction, a peristaltic wave in the ureter carries urine toward the bladder
D. as the bladder fills, there is a progressive increase in intravesicular pressure similar to what one finds when a balloon is filled with air
E. the stimulation of autonomic neurons originating from S-2, S-3, and S-4 will initiate the contraction of the urinary bladder

Answer and Explanation

D. The *urinary bladder,* unlike a balloon, *accommodates* for increases in volume by adjusting its tone. The injection of 100 ml of liquid into the bladder causes an initial increase in pressure of about 3 cm H_2O, but this is followed by a decrease in pressure (accommodation). It is not until the volume of the adult bladder exceeds 250 to 450 ml that intravesicular pressure exceeds 5 cm H_2O. During micturition, the pressure in the bladder exceeds 30 cm H_2O. Pressure in the upper *ureter* averages 15 cm H_2O and in the pelvic ureter averages about 30 cm H_2O during ureteral systole.

14

(1:891; 2:5-100; 3:587; 4:472; 5:912)
NMua 2

The *residual volume* for an organ is the volume of fluid remaining in its lumen after a maximal contraction. The approximate residual volume in the adult for the left ventricle is 25 ml and for the lungs is 1400 ml. What is the residual volume for the urinary bladder? It is

A. between 10 and 20 ml in the healthy subject
B. lowered after a transection of the cord at L-3
C. lowered in the late stages of tabes dorsalis where there is sclerosis of the posterior columns and spinal roots
D. all of the above are false
E. all of the above are true

Answer and Explanation

D. Under normal conditions, micturition leaves a residual volume of 0.09 to 2.34 ml. Since urine is a medium that supports the growth of microorganisms, the higher urine residual volumes seen after certain central nervous system disorders may lead to infection of the urinary system which, at its worst, can be fatal.

15

(1:890; 2:5-97; 3:588; 4:472; 5:912)
NMua 2

Select the one incorrect statement.

A. contraction of the detrusor muscle widens the posterior urethra
B. urine remaining in the urethra of the male at the end of micturition is expelled by the contraction of the bulbocavernosus muscle
C. voiding can be voluntarily terminated by the contraction of the external urethra sphincter
D. urine flow can be accelerated by the Valsalva maneuver
E. the decentralized bladder is unable to contract

Answer and Explanation

A. A number of physiologists persist in referring to the neck of the bladder as the *internal urethral sphincter.* There is little anatomic or physiologic data to support this thesis. Apparently, the contraction of the *detrusor muscle* reduces the resistance to outflow by both shortening and widening the posterior urethra (= prostatic urethra in the male).

In short, there is little to indicate that in micturition there is an "inhibition of the internal urethral sphincter."

D. In the *Valsalva maneuver,* an increased intraabdominal pressure is produced during a forced expiration with the glottis closed.

E. The *decentralized bladder* is initially distended and flaccid but eventually attains a small volume, contracts frequently in response to distention, but has contractions of short duration. This results in the frequent involuntary expulsion of small volumes of urine and a potentially dangerous large residual volume for the bladder.

22
Mouth, Esophagus, and Stomach

1

(1:773; 2:2-13;
3:393; 4:802;
5:1292)
DNEf 2

Which one of the following statements is most correct?

A. the secretion of saliva is primarily under hormonal control
B. the secretion of saliva is increased by parasympathetic stimulation
C. the ratio of salivary/plasma iodine concentration lies between 0.8 and 2.0
D. the concentration of Na^+ in the saliva is increased by aldosterone
E. the increased concentration of ketone bodies in the plasma in diabetes mellitus does not affect their concentration in the saliva

Answer and Explanation

A, **B.** Although the content of the *saliva* may vary with changes in hormonal concentrations, the major way the body initiates the secretion of saliva is through the stimulation of parasympathetic neurons to the salivary glands.
C. The ratio sometimes reaches 60.
D. The reverse is true.
E. Increased plasma concentrations of ketone bodies, urea, and mercury can also cause increases of these substances in the saliva.

2

(1:771; 2:2-36;
3:392; 4:804;
5:1265)
Dc 2

Which one of the following statements is incorrect? Saliva

A. is essential for the complete digestion of starch
B. prevents dental caries
C. prevents decalcification of the teeth
D. is a well-buffered solution that tends to maintain a pH of about 7.0 in the mouth
E. produced by the serous gland cells of the mouth contains a higher concentration of salivary amylase and a lower concentration of mucin than the saliva from the mucous gland cells of the mouth

Answer and Explanation

A. Pancreatic amylase can digest starch in the absence of salivary *amylase*.
B. Patients with a deficient secretion of saliva have a higher incidence of *caries* than normal subjects.
C, D. The saliva, by keeping the pH of the mouth from becoming acid, prevents the decalcification of the *teeth*.

3

1:753; 2:2-104;
3:392; 4:789;
5:1322)
MNRa 2

Which one of the following statements is incorrect?

A. removal of the epiglottis frequently causes aspiration pneumonitis
B. elevation of the larynx is an important part of volitional swallowing
C. liquids swallowed by an erect individual may reach the esophagogastric junction while it is still closed
D. the esophagogastric junction usually prevents regurgitation of the gastric contents into the esophagus, and in this way tends to prevent heartburn
E. food lodged in the esophagus initiates an involuntary reflex that carries the food to the stomach, but does not include elevation of the larynx

Answer and Explanation

A. It is the elevation of the larynx that prevents the *aspiration* of solids and liquids.
B. Voluntary *swallowing* becomes very difficult if the *larynx* is immobilized or if one attempts to swallow while the mouth remains open. Each of these events is part of a series that eventually leads to the delivery of food or water to the stomach.
C. Normally, the junction will open before the peristaltic wave reaches it, but since liquids move downhill more rapidly than peristalsis, they sometimes reach the junction while it is still closed.
E. This phenomenon is sometimes called *secondary peristalsis*.

4

(1:759; 3:383;
4:791; 5:1259)
DMc 2

Which one of the following statements is most correct?

A. some of the fat in ingested food may remain in the stomach for as long as 5 hours
B. water in the stomach is more slowly absorbed than water in the intestine
C. pancreatic juice has a higher concentration of bicarbonate than gastric juice
D. the pyloric sphincter remains open most of the time
E. all of the above are correct

Answer and Explanation

A. *Fat* slows gastric emptying.
B. *Water* placed in the intestine is absorbed in about 10 minutes. Water in the stomach is absorbed in about 60 minutes.
C. Gastric juice has a lower *bicarbonate* concentration than any of the alkaline secretions of the alimentary tract (i.e., saliva, pancreatic juice, bile, and the succus entericus).
E.

5

(1:760; 2:2-116;
3:180; 4:832;
5:1366)
DNEM 2

Which one of the following statements is incorrect? The *stomach*

A. decreases its motility in response to H$^+$ in the duodenum
B. decreases its motility in response to fatty acids in the duodenum
C. decreases its motility in response to the hormone, enterogastrone
D. is essential for vomiting in the adult *(in infant)*
E. decreases its motility in response to an enterogastric reflex *inhibits vagal parasympathetic tone to the stomach.*

Answer and Explanation

A. H$^+$ and the products of protein digestion in the duodenum decrease gastric motility by means of the enterogastric reflex.
B, C. Fatty acids, triglycerides, and phospholipids act on the mucosa of the duodenum to cause the release of *enterogastrone* which is carried in the blood to the stomach where it inhibits gastric motility and thus delays the emptying of the stomach.
D. In the adult, unlike the infant, *vomiting* is more frequent in the absence of a stomach than in its presence. In the adult, vomiting is caused primarily by a buildup in pressure in the abdomen in response to a closure of the glottis and a contraction of the diaphragm and other skeletal muscles.
E. This reflex inhibits vagal parasympathetic tone to the stomach.

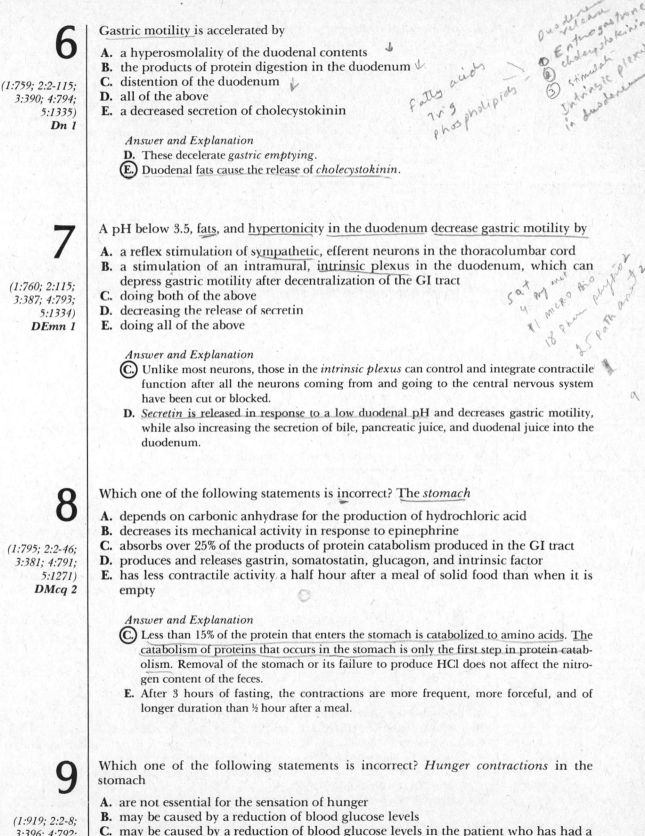

6

(1:759; 2:2-115;
3:390; 4:794;
5:1335)
Dn 1

Gastric motility is accelerated by

A. a hyperosmolality of the duodenal contents
B. the products of protein digestion in the duodenum
C. distention of the duodenum
D. all of the above
E. a decreased secretion of cholecystokinin

Answer and Explanation
D. These decelerate *gastric emptying*.
E. Duodenal fats cause the release of *cholecystokinin*.

7

(1:760; 2:115;
3:387; 4:793;
5:1334)
DEmn 1

A pH below 3.5, fats, and hypertonicity in the duodenum decrease gastric motility by

A. a reflex stimulation of sympathetic, efferent neurons in the thoracolumbar cord
B. a stimulation of an intramural, intrinsic plexus in the duodenum, which can depress gastric motility after decentralization of the GI tract
C. doing both of the above
D. decreasing the release of secretin
E. doing all of the above

Answer and Explanation
C. Unlike most neurons, those in the *intrinsic plexus* can control and integrate contractile function after all the neurons coming from and going to the central nervous system have been cut or blocked.
D. *Secretin* is released in response to a low duodenal pH and decreases gastric motility, while also increasing the secretion of bile, pancreatic juice, and duodenal juice into the duodenum.

8

(1:795; 2:2-46;
3:381; 4:791;
5:1271)
DMcq 2

Which one of the following statements is incorrect? The *stomach*

A. depends on carbonic anhydrase for the production of hydrochloric acid
B. decreases its mechanical activity in response to epinephrine
C. absorbs over 25% of the products of protein catabolism produced in the GI tract
D. produces and releases gastrin, somatostatin, glucagon, and intrinsic factor
E. has less contractile activity a half hour after a meal of solid food than when it is empty

Answer and Explanation
C. Less than 15% of the protein that enters the stomach is catabolized to amino acids. The catabolism of proteins that occurs in the stomach is only the first step in protein catabolism. Removal of the stomach or its failure to produce HCl does not affect the nitrogen content of the feces.
E. After 3 hours of fasting, the contractions are more frequent, more forceful, and of longer duration than ½ hour after a meal.

9

(1:919; 2:2-8;
3:396; 4:792;
5:1380)
MNbD 2

Which one of the following statements is incorrect? *Hunger contractions* in the stomach

A. are not essential for the sensation of hunger
B. may be caused by a reduction of blood glucose levels
C. may be caused by a reduction of blood glucose levels in the patient who has had a bilateral, supradiaphragmatic vagotomy

(Continued)

D. are inhibited by hyperglycemia
E. occur in diabetes mellitus

Answer and Explanation

A, B, (C), D. It is hypothesized that there are a number of cells within the satiety center of the *hypothalamus* that decrease the activity of the center as the quantity of available glucose diminishes. These cells collectively are called *glucostats*. In hypoglycemia, they initiate the sensation of hunger and, through *vagal parasympathetic neurons*, hunger contractions. In other words, hunger contractions are not thought of as the cause of the hunger sensation but are merely associated with hunger. On the other hand, they may be so strong as to cause a painful sensation (*hunger pangs*) in addition to the *sensation of hunger*.

E. In *diabetes mellitus* there is *hyperglycemia*, but the blood sugar is not readily available for cell metabolism. Therefore, the glucostat responds to this condition as it would to hypoglycemia. There are many situations such as this in the body. We have noted, for example, that in CO poisoning the important factor in O_2 transport is not the concentration of hemoglobin, but rather the concentration of AVAILABLE hemoglobin. This perspective is also important in considering the action of hormones bound to plasma proteins (thyroxine, for example).

10

(1:795; 2:2-44; 3:377; 4:805; 5:1293)
DMf 2

Which one of the following statements is incorrect? The *stomach*

A. responds to an increase in its contents from 600 to 1600 ml with less than a 5 mm Hg increase in pressure
B. empties more rapidly in response to a liquid meal than a solid meal
C. during emptying, produces a more forceful contraction in its pyloric portion than in its body
D. contains a pacemaker area near its cardiac portion
E. secretes the following enzymes: pepsin, trypsin, lipase, amylase

Answer and Explanation

A. The stomach, like the urinary bladder, shows receptive relaxation. The increase in gastric volume mentioned above might very well produce no change in intragastric pressure.

C. The pyloric part of the stomach is more muscular than the more cephalad portions. The contractions in the body serve to mix the gastric contents with enzymes and acid. The contractions in the pyloric antrum serve to move liquid chyme into the intestine and solid material into the body.

(E.) The stomach does not secrete trypsin or amylase. The digestion of starch that occurs in the stomach is due to salivary amylase.

11

(1:760; 2:2-116; 3:180; 4:832; 5:1336)
NMfa 2

Which one of the following statements is incorrect? *Vomiting* in the adult

A. begins with salivation and a sense of nausea
B. is controlled by a center in the reticular formation of the medulla
C. in response to apomorphine, a number of other emetic drugs, radiation, and uremia can be abolished by lesions in the area postrema
D. in response to irritation of the mucosa in the upper GI tract can be abolished by lesions in the area postrema
E. is caused by visceral afferent neurons in sympathetic nerves and the vagi

Answer and Explanation

B, C, (D). The area postrema has been hypothesized to contain a *chemoreceptor trigger*

zone that is sensitive to certain agents, such as apomorphine, and that sends stimuli to the vomiting center. Sensory impulses from the GI tract do not pass through the trigger zone.

12

(1:760; 2:2-110;
3:180; 4:832;
5:1336)
GDMf 2

In the healthy *newborn infant*, as opposed to the adult (select the one *incorrect* statement):

A. skeletal muscle contraction is more important in causing vomiting
B. the esophagogastric junction is less effective in preventing regurgitation
C. the gastric juice is less acid
D. heartburn (painful irritation of the esophagus) is less frequent
E. regurgitated gastric contents bubble out the mouth more often

Answer and Explanation

Ⓐ Vomiting or regurgitation in the infant is not prevented by skeletal muscle blockade, whereas in the adult it is. In the adult the skeletal muscle contraction may be so marked during vomiting that there is pain. In the infant, vomiting is almost totally a gastroesophageal phenomenon and is seldom if ever painful.

B, C, E. The higher *pH* of the infant's gastric contents protects him from *heartburn* during regurgitation, since heartburn occurs when the esophagus is exposed to a pH below 4.0.

13

(1:799; 2:2-38;
3:394; 4:805;
5:1277)
GDf 1

Which one of the following statements is incorrect? In the

A. stomach, pepsinogen is activated to pepsin by HCl
B. stomach, pepsin is formed and, in the duodenum, is inactivated by the higher pH
C. stomach, in the pyloric gland area there is a complete replacement of cells every 2 days
D. stomach, mucus forms a lining greater than 1 mm thick over the membranes and keeps their pH near 7.0
E. newborn infant, the stomach is less permeable than the stomach of the adult

Answer and Explanation

E. The infant's digestive tract is generally more permeable than that of the adult. This is probably one of the reasons young people are more likely to show allergic reactions. In some animals, this lack of a GI barrier permits the antibodies in the mother's milk to enter the plasma of the young. There is little evidence to indicate that this is the case in human infants, however.

14

(1:811; 2:4-38;
3:384; 4:59; 5:1282)
Dba 2

Which one of the following statements is most correct? After the *stomach is removed*, the patient who receives no further treatment will usually develop

A. symptoms due to a decreased absorption of iron
B. symptoms due to a decreased absorption of vitamin B_{12}
C. an anemia due to the decreased absorption of iron and vitamin B_{12}
D. a carbohydrate loss in the feces that is in excess of 25% of the normally digestible carbohydrate intake (i.e., excluding cellulose and bran)
E. all of the above statements are correct

Answer and Explanation

A. The acid released by the stomach promotes iron absorption.
B. The parietal cells of the stomach produce a glycoprotein that is needed to transport

(Continued)

vitamin B_{12} into the circulation. This so-called *intrinsic factor* is not needed if the B_{12} is injected intravenously.

C. The iron deficiency causes a microcytic hypochromic anemia, and the B_{12} deficiency causes a megaloblastic anemia.

D. Removal of the stomach only slightly affects carbohydrate metabolism and absorption in the alimentary tract.

15

(1:756; 2:2-44; 3:394; 4:791; 5:1259)
Dba 2

Which of the following statements is most correct? After the *stomach is removed*, the patient who receives no further treatment will usually develop

A. exaggerated fluctuations in plasma glucose concentration after a meal
B. exaggerated fluctuations in blood volume
C. marked increases in the lipid content of the feces
D. all of the above
E. increased secretion of pancreatic juice

Answer and Explanation

A, B. The stomach, by storing its contents and slowly releasing them to the intestine, restricts the quantity of hypertonic solution and glucose that enters the more permeable intestine per minute immediately after a meal. In the absence of the stomach, *hyperglycemia* followed by hypoglycemia and potentially serious decreases in *blood volume* may occur after a meal. In addition, the amount of food eaten during a single meal is decreased, and the individual must eat more frequently in order to meet his caloric needs.

C. In a normal subject, the absorption of fat from the alimentary tract exceeds 90%. After gastrectomy, it may be less than 30%. This is probably due to the loss of gastric digestion and HCl production and, as a result, a decreased stimulation of pancreatic secretion.

D.

E. There is a decreased secretion of pancreatic juice after gastrectomy.

16

(1:779; 2:2-12; 3:396; 4:786; 5:1299)
DNEf 2

Which one of the following statements is most correct? Supradiaphragmatic, bilateral *vagal section* prevents *gastric secretion* in response to

A. feelings of hostility
B. distention of the stomach
C. feelings of hostility and distention of the stomach
D. the products of protein catabolism in the duodenum
E. all of the above statements are correct

Answer and Explanation

A. It is through the parasympathetic fibers in the vagi that the brain exerts its major control over the stomach.

17

(1:780; 2:2-21; 3:389; 4:807; 5:1299)
DNEf 2

In 1964, Gregory and Tracey purified *gastrin*. It is best characterized as a linear peptide that

A. is released by the stomach in response to the stimulation of preganglionic parasympathetic neurons to the stomach
B. is released by the stomach in response to the distention of the stomach
C. is released in response to the products of protein digestion in the duodenum
D. increases gastric secretion
E. all of the above statements are correct

Answer and Explanation

E.

18

(1:779; 2:2-49;
3:397; 4:808;
5:1299)
DNEf 2

The *control* of the secretions of the *stomach* has been divided into three phases. What are they?

Answer and Explanation
The three phases are (1) the cephalic, (2) the gastric, and (3) the intestinal phase.

19

(1:779; 2:2-49;
3:397; 4:808;
5:1299)
DNEf 2

List the mechanisms important in each of the three phases of gastric *control*. Include in your list specific neural and hormonal systems.

Answer and Explanation
- *Cephalic phase:* (1) facilitation of secretion by *parasympathetic* neurons and (2) inhibition of secretion by *sympathetic* neurons.
- *Gastric phase:* (1) facilitation of secretion by *reflexes*, (2) facilitation of secretion by the *myenteric plexus*, (3) facilitation of secretion by the release of *gastrin*, and (4) direct action on the stomach by its contents.
- *Intestinal phase:* (1) hormonal facilitation (possibly by intestinal gastrin) and (2) hormonal inhibition (possibly by *enterogastrone* or *secretin*).

A hypothesis on how these mechanisms function is presented in the following illustration.

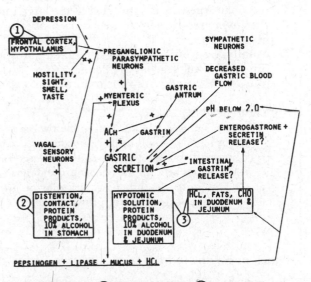

GASTRIC DIGESTION: ① CEPHALIC PHASE,   ② GASTRIC PHASE,
③ INTESTINAL PHASE OF CONTROL

A HYPOTHESIS ON THE CONTROL OF GASTRIC SECRETION

20

(1:753; 2:2-52;
3:396; 5:1298)
Dn 1

Acid secretion in the *stomach* is inhibited by

A. an H_1 *histamine* blocker (benadryl)
B. an H_2 histamine blocker (cimetidine)
C. *atropine*
D. both A and C
E. both B and C

Answer and Explanation
E.

23
Small and Large Intestine

1

(1:765; 2:2-122; 3:406; 4:796; 5:1336)
Ddc 1

Which one of the following statements is incorrect?

A. the small intestine is about 3 meters long in the healthy subject
B. the small intestine is connected to the large intestine at the jejunocecal sphincter
C. the chyme received from the stomach usually remains in the small intestine for 2 hours or longer
D. usually, over 50% of the digestion and absorption of nutrients is complete by the time the chyme reaches the ileum
E. glucose absorption from the intestine can occur against a concentration gradient

Answer and Explanation
A. These data were obtained on human volunteers by intubation. When the small intestine is removed from the body, it is longer as a result of loss of smooth muscle tone.
B. The small intestine empties into the large intestine at the *ileocecal sphincter*. The *jejunum* lies between the duodenum and the ileum.
E. Glucose in the intestine is absorbed by an *active transport* system similar to that for glucose reabsorption in the proximal tubule of the renal nephron.

2

(1:812; 2:2-91; 3:383; 4:822; 5:1259)
Did 2

Which one of the following statements is incorrect?

A. nicotinic acid, nicotinamide, pantothenic acid, biotin, biocytin, thiamine, thiamine phosphate, folic acid, and flavine mononucleotide are absorbed primarily by active transport
B. K^+ is actively secreted into the lower half of the small intestine and into the colon
C. parathormone increases the absorption of Ca^{++} from the small intestine
D. Fe^{++} enters the mucosa cell of the upper small intestine, where it either combines with apoferritin or moves into the blood
E. aldosterone exerts a strong influence in facilitating Na^+ absorption in the colon but has a minor influence on the small intestine

Answer and Explanation
A. Vitamin B_{12} (molecular weight = 1357) is one of the few vitamins that requires an *active transport* system.
B. Feces contain an average of 75 mEq of *potassium* per liter, whereas the chyme delivered to the colon contains about 5 mEq of potassium per liter.
C. *Calcium* absorption can occur against a concentration gradient.

E. Feces contain an average of 32 mEq of *sodium* per liter, whereas the chyme delivered to the colon contains 121 mEq of sodium per liter. A decreased plasma *aldosterone* concentration in the plasma causes an increased fecal sodium concentration.

3

(1:814; 2:2-100; 3:384; 4:828; 5:1284)
Da 2

Vitamin B_{12} deficiency may result from

A. total gastrectomy
B. ileal resection
C. total gastrectomy or ileal resection
D. inadequate consumption of fresh fruits and leafy green vegetables
E. high dietary intake of phosphates

Answer and Explanation

A, B, C. The absorption of B_{12} is greatly facilitated by its combination with a muco-protein produced by the stomach *(intrinsic factor)*. Since absorption of the B_{12}–mucoprotein occurs in the terminal ileum, either *gastrectomy* or ileal resection can cause a B_{12} deficiency.
D. Although fresh fruits and leafy green vegetables are excellent sources of vitamin C, they are inadequate sources of B_{12}. Liver, meat, eggs, and milk are good sources of B_{12}.

4

(1:818; 2:2-74; 3:379; 4:825; 5:1257)
DWdc 2

The upper third of the *small intestine* has the capacity to *actively absorb* large quantities of (select the one *incorrect* statement)

A. bile salts
B. pyrimidines
C. triglycerides
D. Na^+
E. neutral amino acids and glucose

Bile salts + vit B_{12} absorbed in ileum

Answer and Explanation

A. The upper third of the tract includes the duodenum and jejunum. If the *bile salts* were absorbed in this area, it would interfere with fat digestion.

5

(1:816; 2:2-58; 3:380; 4:819; 5:1302)
Dc 1

adult → .5–.7 gm protein
child → 4. gm

Which one of the following statements is incorrect?

A. a 2-year-old requires four to eight times the protein intake of a 35-year-old man
B. after the removal of 60% of the pancreas, over 20% of the ingested protein is lost in the feces
C. after the removal of the total pancreas, over 40% of the ingested protein is lost in the feces
D. proteolytic enzymes are released by the chief cells of the stomach, the exocrine cells of the pancreas, and the exocrine cells of the intestine
E. nucleases released by the cells of the intestine catabolize nucleic acids to pentoses, purine, and pyrimidine bases

Answer and Explanation

A. An adult requires 0.5 to 0.7 g of *protein* per day. A child (1 to 3 years) requires 4 g per day.
B. There is usually sufficient pancreatic reserve so that removal of 60% of the *pancreas* does not affect either protein or fat catabolism.

6

(1:815; 2:2-59;
3:381; 4:817;
5:1278)
Dc 1

Pancreatic lipase in the intestine catalyzes the conversion of dietary triglycerides to what *end products*?

Answer and Explanation

Lipid catabolism can be characterized as follows:

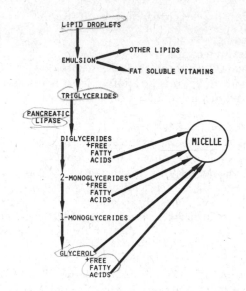

7

(1:786; 2:2-68;
3:382; 4:818;
5:1278)
Dc 1

In the scheme presented in question 6, what is responsible for the *emulsification* of the lipid droplets?

Answer and Explanation

The bile salts or bile acids (conjugates of cholic and chenodeoxycholic salts with glycine and taurine). They represent 0.7% of human bile.

8

(1:816; 2:2-70;
3:382; 4:824;
5:1278)
Dc 2

In the scheme presented for *lipid catabolism* in question 6, what is a micelle? A *micelle* is (select the one *incorrect* statement)

A. a small polymolecular aggregate containing bile salts, monoglycerides, and fatty acids as its major constituents
B. one-millionth the volume of a fat droplet
C. frequently seen inside the cells that line the intestine
D. a structure that speeds lipid absorption
E. a structure that helps maintain the saturation of the chyme with fatty acids and monoglycerides

Answer and Explanation
A. It may also contain fat soluble vitamins and cholesterol.
B. It has a diameter of 4 to 6 μm.
C, D, E. The micelle is an intraluminal structure that speeds absorption of fatty acids and monoglycerides by delivering its contents to the outer surface of the intestinal epithelial cell. Fatty acids and monoglycerides are absorbed chiefly by the epithelial cells of the duodenum and jejunum. Most of the conjugated bile salts, on the other hand, are absorbed in the terminal ileum.

9

(1:818; 2:2-97; 3:282; 4:824; 5:1278)
Dcb 2

Which one of the following statements is incorrect?

A. most monoglycerides and long-chain (C_{14} or longer) fatty acids that enter the mucosal cells are reesterified to triglycerides

B. most short- and medium-chain fatty acids and some glycerol move from the mucosal cell to the hepatic portal blood

C. chylomicrons are aggregates of protein, cholesterol, saturated triglycerides, phospholipids, and free fatty acids and are 0.1 to 3.5 μm in diameter

D. chylomicrons move from the mucosal cell into the lymph capillaries and ducts

E. chylomicrons seldom remain in the lymph ducts longer than a half hour

Answer and Explanation

C. The *chylomicron* is 89% to 93% *triglyceride*, 5% to 9% phospholipid, 1% to 7% free *fatty acid*, and 0.7% to 1.5% *cholesterol* by weight. Protein covers about 10% of the particle's surface.

D. Chylomicrons are unable to enter *blood* capillaries. They are able to enter the more permeable lymph capillary.

E. Lymph flow is slow, and therefore the nutrients carried by the lymph will be added to the blood over a period of hours. Absorbed fat may remain in the lymph channels for periods up to 12 hours.

10

(1:788; 2:2-69; 3:402; 4:871; 5:1278)
Dbc 1

Which one of the following statements is incorrect?

A. the bile acids, cholic and chenodeoxycholic acid, are synthesized in liver cells from cholesterol

B. bile acids are changed in the liver to deoxycholic and lithocholic acids

C. many microorganisms in the intestine deconjugate taurocholic and glycocholic acid

D. bile acids are much less effective detergents than their conjugated salts

E. bacterial overgrowth in the small intestine can cause a decreased fat absorption

Answer and Explanation

A. They are sometimes called the *primary bile acids*.

B. *Deoxycholic and lithocholic acids* are produced in the small or large intestine from the primary bile acids or bile salts by bacterial action. They are sometimes called the *secondary bile acids*. *Cholic acid* and *chenodeoxycholic acid* are conjugated in the liver with *taurine* and *glycine*.

E. This results from the deconjugation of the bile salts by the microorganisms.

11

(1:791; 2:2-70; 3:402; 4:825; 5:1278)
DCbc 2

Which one of the following statements is incorrect?

A. over 90% of the bile salts are absorbed into the blood

B. under 10% of the bile salt pool is lost each day

C. removal of the terminal ileum causes an increased synthesis of bile salts by the liver

D. if over 50% of the bile salts are lost each day, the bile salt pool will decrease

E. loss to the feces of all or most of the bile salts each day causes a watery diarrhea

Answer and Explanation

B. Fifteen percent to 25% of the *bile salt pool* is lost each day. This is because the pool may circulate twice during the digestion of a single meal.

(Continued)

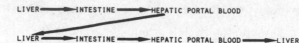

C. The terminal ileum is where most of the bile salts are absorbed.

D. If over 33% of the pool is lost, the liver is unable to produce enough bile salts to replenish the loss. Normally 12 to 30 g of bile salts are sent into the duodenum per day. In the absence of bile salt absorption, only 3 to 5 g enter the duodenum per day.

E. Bile salts act as an *osmotic cathartic*.

12

(1:789; 2:2-71; 3:402; 4:871; 5:1278)
DCbc 2

Which one of the following statements is incorrect?

A. an increased excretion of conjugated bilirubin in the urine is caused by obstruction of the common bile duct by a gallstone or an increased destruction of erythrocytes

B. most of the urobilinogen is formed in the intestine

C. an increased excretion of urobilinogen in the urine can be caused by liver damage

D. the bile pigments markedly facilitate the absorption of fat

E. urobilinogen is excreted in the bile

Answer and Explanation

A. Obstruction of the *common bile duct* decreases the loss of *bile pigments* in the feces. An increased destruction of *erythrocytes* (in hemolytic anemia, for example) will cause an increased production of *bilirubin*.

B, C. *Urobilinogen* is produced in the intestine by bacteria. Most of it is excreted in the feces, but some also diffuses into the blood and is excreted in the urine and bile. After liver damage, the urinary excretion of urobilinogen may increase from 0.5 to 2.0 mg/day to 5 mg/day.

(D.) The bile pigments, unlike the bile salts, are waste products of hemoglobin catabolism and play no role in fat absorption.

13

(1:790; 2:2-79; 3:403; 4:871; 5:1309)
DMc 2

Which one of the following statements is incorrect?

A. the gallbladder normally has a capacity of less than 60 ml

B. bile pigments, bile salts, and cholesterol may be 5 to 20 times more concentrated in the healthy gallbladder than in the hepatic duct

C. bile in the gallbladder has a higher concentration of sodium than in the hepatic duct

D. in the normal subject, all bile in the hepatic duct passes to the gallbladder

E. the gallbladder will contract in response to emotional stimuli such as hostility, as well as in response to food in the mouth

Answer and Explanation

A. In the adult it varies between 14 and 60 ml.

B, C. The gallbladder, by means of the active transport of Na^+, Cl^-, and HCO_3^-, removes water from the bile and modifies the pH of the bile. Its high concentration of sodium results from the sodium being held in association with bile salts in *osmotically inactive* micelles. The concentration of sodium in gallbladder bile is approximately twice that in hepatic duct bile.

(D.) *Bile* may move from the *hepatic ducts* to the common hepatic duct past the *cystic duct* and into the bile duct and duodenum.

E. The gallbladder, like the stomach, responds to cephalic stimuli.

14

(1:790; 2:2-80;
3:404; 4:872;
5:1309)
Dba 2

Which one of the following statements is incorrect?

A. cholecystectomy usually causes jaundice
B. cholecystectomy decreases the bile salt pool by more than 50%
C. cholecystectomy is followed by a progressive dilation of the bile duct
D. obstruction of the common bile duct usually increases the prothrombin time
E. obstruction of the common bile duct usually causes a clay-colored stool

Answer and Explanation

(A) B, C. After *cholecystectomy*, there is a dilation of the bile duct, but the major problem remains a reduced storage capacity for bile salts. The maintenance of a low *plasma bilirubin* concentration by the liver is not a problem because, as bile is produced by the liver, the sphincter of the ampulla is forced open even in the absence of nervous or humoral stimuli. In other words, cholecystectomy impairs the digestive function of the biliary system but not its excretory function.

D, E. Obstruction of the *common bile duct* interferes with both the digestive and excretory functions of the biliary system. By preventing the release of *bile salts* into the duodenum, more fat is lost in the stool and with it the fat-soluble *vitamins A, D, E,* and K. Vitamin K is necessary for the production of adequate quantities of *prothrombin* and other procoagulants. With inadequate vitamin K absorption, the concentration of these procoagulants in the plasma decreases and the prothrombin time increases. The clay-colored stool results in part from a decreased concentration of the brown pigment, *urobilinogen*, and the increased concentration of lipids in the stool (steatorrhea).

15

(1:793; 2:2-83;
3:403; 4:872;
5:1313)
Dcia 2

Which one of the following statements is incorrect?

A. *Escherichia coli*, in the biliary system, will conjugate bilirubin and glucuronic acid and, in so doing, cause the production of gallstones
B. the calcium salt of bilirubin is less soluble in water than the glucuronic acid conjugate of bilirubin
C. when the calcium salt of bilirubin precipitates, cholesterol crystals aggregate around it
D. if the concentration of cholesterol in bile rises above 10%, cholesterol crystals will usually precipitate
E. if the concentration of bile salts falls below 40%, cholesterol crystals will usually precipitate

Answer and Explanation

(A) B, C. *Escherichia coli* have beta glucuronidase activity and tend to hydrolyze the *bilirubin conjugate* to free bilirubin. When the free bilirubin forms a calcium salt, this salt comes out of solution and forms the nucleus for a *gallstone*. The formation of conjugates prevents this process. The destruction of the bilirubin conjugate by a microorganism initiates it.

D, E. *Cholesterol* is insoluble in water. The bile salts and lecithin form with cholesterol the small, water soluble aggregates called *micelles*.

16

(1:791; 2:2-77;
3:389; 4:812;
5:1310)
DNME 2

Which one of the following statements is incorrect?

A. cholecystokinin is released from the duodenum in response to fat, egg yolk, and meat
B. cholecystokinin is carried in the blood to the gallbladder where it facilitates the contraction of the bladder

(Continued)

C. chewing food can reflexly stimulate parasympathetic neurons in the vagus to initiate the contraction of the gallbladder

D. hypoglycemia reflexly stimulates vagal parasympathetic neurons to increase bile secretion

E. vagal stimulation causes the release of an agent, gastrin, which decreases bile secretion

Answer and Explanation

A, B, C. Some important aspects in the control of the gallbladder can be summarized as follows:

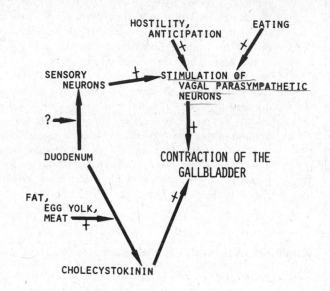

D, E. The vagus increases the secretion of bile through the action of acetylcholine on liver cells and by stimulating the release of gastrin, which also acts on liver cells.

17

(1:789; 2:2-77; 3:390; 4:812; 5:1310)

DENc 2

Which one of the following statements is incorrect?

A. bile salts absorbed from the ileum may decrease the synthesis of additional bile salts but increase the volume of bile produced

B. secretin causes the production of a bile that has a lower concentration of bile salts and HCO_3^- and a higher pH than bile produced in response to bile salts alone

C. gastrin increases the volume of bile produced

D. sympathetic stimulation decreases the production of bile

E. cholecystokinin increases the production of bile

Answer and Explanation

B. Secretin stimulates the production of *bile* with a lower concentration of *bile salts*, a higher concentration of HCO_3^-, and a higher pH than that produced by bile salts alone. The following diagram represents a hypothesis on some of the ways bile secretion is controlled. In this diagram, the stronger influences have been underlined.

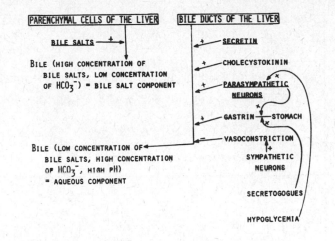

18

(1:810; 2:2-91; 3:384; 4:822; 5:1283)
D 1

Which one of the following statements is correct?

A. most of the water we drink is absorbed in the stomach
B. most of the vitamin B_{12} in our diet is absorbed in the stomach and first part of the small intestine
C. both of the above are true
D. most of the taurocholate in the alimentary tract is absorbed in the jejunum
E. most of the calcium is absorbed in the duodenum

Answer and Explanation
E.

19

(1:785; 2:2-63; 3:390; 4:811; 5:1304)
DEq 1

Name two *hormones* that are produced by *duodenal* cells and that increase either the volume or the enzyme content of the exocrine secretion of the *pancreas*.

Answer and Explanation
Secretin and cholecystokinin (pancreozymin).

20

(1:784; 2:2-59; 3:377; 4:819; 5:1302)
Dc 1

Chymotrypsinogen

A. is secreted primarily by the stomach
B. is converted to chymotrypsin by the action of trypsin
C. is converted to chymotrypsin by the action of acid
D. uses starch as its primary substrate
E. uses triglycerides as its primary substrate

Answer and Explanation
A. It is secreted by the *pancreatic acinar cells.*
B. The *precursor* to *trypsin* (trypsinogen) is also secreted by the pancreas.
D, E. Trypsin and *chymotrypsin* reduce *polypeptides* to smaller peptides.

21

(1:781; 2:2-61;
3:377; 4:816;
5:1302)
Dc 2

Define the term *enzyme*. Contrast the types of enzymes contained in the *bile* and *pancreatic* juice and give examples of each.

Answer and Explanation

An enzyme is a protein that acts as a catalyst. It is produced in the body.

- Bile contains no enzymes. The bile salts produce physical changes (emulsification), not chemical changes.
- Pancreatic juice: (1) *proteolytic* enzymes (trypsin, chymotrypsin); (2) *nucleic acid-splitting* enzymes (nuclease, ribonuclease, deoxyribonuclease); (3) pancreatic *lipase*; (4) pancreatic *amylase*.

22

(1:781; 2:2-63;
3:390; 4:811;
5:1304)
DNc 1

The pH of the duodenum is reduced to 4.0. There follows the secretion of a large volume of *pancreatic juice* with a high concentration of *bicarbonate* and a low concentration of *enzymes*. In addition, *gastric secretion* decreases and *bile secretion* increases. Apparently the *acidity in the duodenum* has produced these responses by causing the

A. release of gastrin
B. release of pancreozymin
C. release of cholecystokinin
D. release of secretin
E. stimulation of adrenergic sympathetic neurons

Answer and Explanation

A. The procedure that will produce the most pronounced increase in pancreatic secretion is the release of secretin. Gastrin facilitates a pancreatic secretion with a high concentration of enzymes and facilitates gastric secretion.

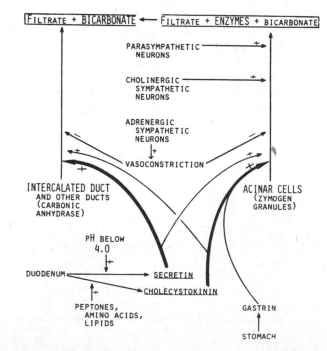

B, C. Pancreozymin and cholecystokinin are the same hormone. It facilitates the secretion of a pancreatic juice with a high concentration of enzymes and a low concentration of bicarbonate.

Ⓓ **E.** Stimulation of these neurons decreases pancreatic blood flow and tends to decrease secretion. Stimulation of cholinergic neurons to the pancreas, on the other hand, may increase pancreatic secretion.

23

(1:773; 2:2-44; 3:392; 4:802; 5:1304)
DEc 1

What are *zymogen granules?*

Answer and Explanation
Zymogen granules are discrete masses in secretory cells that contain products synthesized by the ribosomes and/or other cellular elements. Since they accumulate in cells in their presecretory state and disappear during secretion, they are believed to be storage forms of enzymes or, in the case of certain endocrine glands, proteinaceous hormones.

24

(1:795; 2:2-91; 3:377; 4:813; 5:1313)
Dc 1

What *enzymes* produced by the *intestinal mucosa* are most important in intraluminal digestion?

Answer and Explanation
Enterokinase (enteropeptidase) activates trypsinogen. The oligosaccharidases catabolize polysaccharides. There are also peptidases and nucleases.

25

(1:784; 2:2-58; 3:377; 4:816; 5:1297)
Dc 1

What are the specific terms used to designate the precursors of *trypsin, chymotrypsin, carboxypeptidase A, elastase,* and *phospholipase A?*

Answer and Explanation
Their precursors are trypsinogen, chymotrypsinogen, procarboxypeptidase A, proelastase, and prophospholipase A, respectively.

26

(1:795; 2:2-94; 3:377; 4:817; 5:1265)
Dc 1

What are the products that result from the action of sucrase on its *substrate* and *nuclease* on its substrate?

Answer and Explanation

Sucrose $\xrightarrow{\text{sucrase}}$ glucose + fructose.

Nucleic acid $\xrightarrow{\text{nuclease}}$ pentoses + purine bases + pyrimidine + pyrimidine bases.

27

(1:799; 2:2-86; 3:410; 4:825; 5:1313)
DcG 1

What is the role of *desquamation* in intestinal digestion?

Answer and Explanation
With the exception of enterokinase, it is believed that all the intestinal *enzymes* are derived from cells sloughed from the tips of the intestinal villi. It has been estimated that in a healthy subject, there is a complete replacement of all the cells lining the small intestine every 3 days. In 1961, Crosby estimated that about 250 g of epithelial cells are shed into the lumen of the small intestine each day.

28

(1:811; 2:2-91; 3:385; 4:61; 5:1282)
Dia 1

Which one of the following statements is incorrect?

A. excess iron in the body causes hemochromatosis (i.e., pigmentation of the skin, pancreatic damage and a resultant diabetes, cirrhosis of the liver, gonadal atrophy)
B. in hemochromatosis, iron absorption decreases but does not cease
C. in hemochromatosis, most of the excess iron is excreted in the urine
D. most iron is absorbed in the duodenum and first part of the jejunum
E. iron is held in the body in the iron-containing proteins ferritin and hemosiderin and is transported in the plasma in combination with the globulin transferrin (siderophilin).

Answer and Explanation

C. Most of the excess *iron* is eliminated in the *feces*. This is due to (1) the decreased *absorption* of iron and (2) the *sloughing* of mucosal cells (desquamation) that have served as a major *reservoir* of iron.

29

(1:811; 2:2-91; 3:385; 4:61; 5:1282)
Dib 1

Which one of the following statements is incorrect? In a normal healthy subject,

A. most of the body's iron is in the hemoglobin molecule
B. an increase in dietary iron causes little or no increase in iron absorption
C. ferrous iron is absorbed more readily than ferric iron
D. an excess of iron in the plasma will be followed by an increased migration of plasma iron into mucosal cells
E. hemorrhage is followed after an interval of 3 to 4 days by increased iron absorption

Answer and Explanation

A. There are approximately 4 g of *iron* in the body. Two to 2.5 g are in *hemoglobin*, 0.5 to 1 g is stored, 150 mg are in myoglobin, 5 mg are in plasma, and the rest is scattered throughout the body in cytochrome, peroxidase, etc.
B. When the diet contains 0.01 mg of iron, approximately 0.001 mg is absorbed. When it contains 10 mg about 1 mg is absorbed.
C. Fe^{++} is absorbed 2 to 15 times more readily than Fe^{+++}.
E. It has been suggested that this long interval is, in part, due to the slow migration in the upper intestine of mucosal cells from the area of formation to the tips of the villi.

30

(1:764; 2:2-120; 3:387; 4:788; 5:1339)
DNM 2

Which one of the following statements is incorrect?

A. decentralization of the intestine causes a disappearance of peristalsis
B. decentralization and atropinization of the intestine does not cause the disappearance of segmenting movements
C. distention of the intestine causes the release of a substance, serotonin (5-hydroxytryptamine), which increases the frequency and force of the peristaltic contractions
D. the stimulation of parasympathetic neurons to the intestine increases intestinal secretion and motility
E. peristaltic waves characteristically travel less than 15 cm and then disappear

Answer and Explanation

A. The *intrinsic nerve plexus* is able to continue to coordinate *peristalsis* in the absence of any connection with the central nervous system.
B. *Segmenting movements* do not require the presence of any neurons.
E. In man, unlike the rabbit, for example, peristaltic waves passing through long segments of the small intestine (*peristaltic rush*) are abnormal.

31

(1:763; 2:2-119; 3:407; 4:787; 5:1336)
DMc 1

Which one of the following statements is incorrect?

A. antiperistaltic waves occur at almost as great a frequency as peristaltic waves
B. the stimulation of sympathetic neurons to the intestine causes vasoconstriction, the contraction of the muscularis mucosa, and an increase in the motility of the villi
C. in the decentralized jejunum, distention and HCl increase the force of peristaltic waves
D. the absorption of glucose and galactose is inhibited by phlorhizin and other nutrients
E. the rate of D-glucose absorption is more rapid than that for pentose or D-ribose

Answer and Explanation

A. *Reverse peristalsis* in human beings is either rare or absent. This does not mean that once chyme has entered the ileum it does not move back into the jejunum. Segmentation, changes in tone, and peristalsis (i.e., waves of contraction moving toward the anus) may push some of the chyme toward the stomach as well as in the opposite direction.

B, C. The following is a hypothesis on the control of intestinal motility:

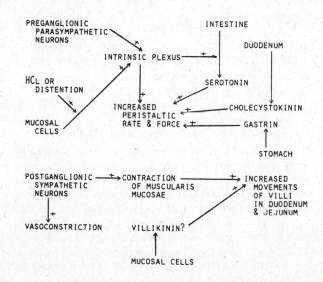

D, E. *Glucose* and galactose are actively absorbed from the intestine by a system that has a Tm (transport maximum) and is blocked by phlorhizin. It is because of this active transport that glucose absorption is faster than for many smaller molecules.

32

(1:765; 2:2-122; 3:409; 4:796; 5:1340)
DMNa 1

Which one of the following statements is incorrect? In human beings, the

A. ileocecal sphincter, unlike the cardiac and pyloric sphincters, is usually closed
B. surgical removal of the ileocecal sphincter is followed by the regurgitation of large volumes from the large intestine into the small intestine
C. ileocecal sphincter is controlled, in part, by the myenteric plexus in the cecum
D. ileocecal sphincter relaxes when a peristaltic wave in the ileum comes within 2 cm of it
E. ileocecal sphincter increases its frequency of opening in response to food distending the stomach

Answer and Explanation

B. It has been suggested that the major importance of the *ileocecal sphincter* is (1) to retard emptying of the ileum into the colon and (2) to retard migration of the *microorganisms* in the colon to the ileum.

C, D, E. The following is a hypothesis on the control of the ileocecal sphincter:

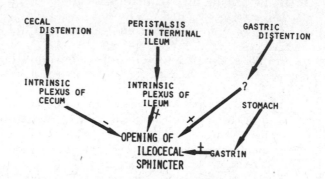

33

(1:765; 2:2-119; 3:409; 4:797; 5:1320)
DMNa 2

Which one of the following statements is incorrect?

A. the first part of a test meal reaches the cecum in about 4 hours, the splenic flexure in 6 hours, the hepatic flexure in 9 hours, and the sigmoid colon in 12 hours

B. in defecation, the contractions are frequently sufficient to empty the distal colon as far as the splenic flexure

C. diarrhea may result from volumes of ileal fluid in excess of 2 liters per day entering the colon

D. diarrhea may result from magnesium or sulfate salts entering the colon

E. absence of the myenteric plexus in the descending colon can cause megacolon

Answer and Explanation

CHSS

A. The first part of a test meal reaches the *cecum* in about 4 hours, the *hepatic flexure* in 6 hours, the *splenic flexure* in 9 hours, and the *sigmoid colon* in 12 hours.

B. The descending colon and sigmoid colon characteristically contain solid material, whereas that in the transverse colon is semisolid and that in the cecum is fluid. II·Ac

C. In cholera, the delivery of large volumes of isotonic fluid to the lumen of the jejunum and ileum by mucosal cells can cause a loss in the feces of 12 liters of water per day.

D. The salts of magnesium, sulfate, citrate, and tartrate are poorly absorbed in the stomach and intestines and can therefore serve as *osmotic cathartics.*

E. Megacolon, like achalasia of the esophagus (enlargement due to food lodging there), is caused by the absence of a *myenteric plexus.*

34

(1:768; 2:2-119; 3:411; 4:798; 5:1230)
DNqa 2

Which one of the following statements is most correct? The *defecation reflex*

A. is facilitated by food entering the stomach
B. is eliminated by destruction of the lumbar cord
C. statements A and B are both correct
D. is eliminated by the paralysis of skeletal muscle
E. statements B and D are both correct

Answer and Explanation

A.

B. C. The *sacral cord* contains the parasympathetic neurons that cause the contraction of the rectum and the relaxation of the internal anal sphincter.

D. The defecation reflex can occur in the absence of skeletal muscle contraction.

35

*(1:768; 2:2-125;
3:411; 4:798;
5:1343)*
DMqa 2

List, in chronologic order, the events that initiate *defecation* in an individual whose skeletal muscles are paralyzed.

Answer and Explanation
Mass movement of feces into rectum

- Rectal distention
- Stimulation of stretch receptors in rectum
- Sensory impulses to cord, medulla, cerebral cortex
- Stimulation of parasympathetic neurons in the sacral cord
- Contraction of the rectum and relaxation of the internal anal sphincter

36

*(1:768; 2:2-125;
3:411; 4:798;
5:1343)*
DMq 2

List, in chronologic order, the events that initiate a *voluntary defecation*.

Answer and Explanation
Strong, maintained contraction of the diaphragm

- Closure of the glottis
- Contraction of the chest muscles
- Contraction of the abdominal muscles
- Movement of feces into the rectum and initiation of the defecation reflex

37

*(1:765; 2:2-89;
3:409; 4:825;
5:1264)*
Dig 2

Which one of the following statements is incorrect? In the colon

A. Na^+ and water are removed from the chyme
B. the concentration of K^+ and HCO_3^- in the chyme is increased
C. any glucose or amino acids in the chyme are actively absorbed
D. microorganisms produce vitamins (riboflavin, nicotinic acid, biotin, folic acid, and vitamin K) that will be absorbed into the blood
E. the amount of flatus produced will more than double if one switches from an average diet to one in which 25% of the caloric intake is pork and beans

Answer and Explanation
C. The colon does not actively absorb glucose or amino acids. The value of nutritive enemas is thought to be negligible.
E. Staying for 2 weeks on a diet high in beans has been shown to increase the volume of flatus from 17 to 203 ml/hour.

38

*(1:765; 2:2-126;
3:412; 4:831;
5:1264)*
DCi 2

Which one of the following statements is incorrect?

A. more protein in the feces comes from bacteria than from the diet
B. an increase in the cellulose content in the diet tends to cause an increase in the frequency of bowel movements
C. constipation is associated with a restlessness, dull headache, loss of appetite, and occasionally nausea, all of which are due to the absorption of toxins from the feces
D. an individual who defecates fewer than three times a week should not necessarily be considered constipated or unhealthy
E. ammonium ions are produced in the colon

Answer and Explanation
A. Approximately 10% of the dry weight of the feces is *bacteria*.
B. In a series of healthy men, it was shown that by changing the *fiber content of the diet*
(Continued)

from 30 to 100 mg/kg/day, the interval between defecations changed from 30 hours to 17 hours. This is not to deny that one's attitudes do not also affect the frequency of *defecation*.

C. D. Apparently, these symptoms are due to distention of the *rectum* rather than *toxins*. What few toxins do enter the blood (NH_4^+, for example) pass via the portal circulation to the liver, where they are removed before they can enter the general circulation. An individual who defecates only once a week, if he does not have the symptoms listed above, should be considered healthy and not *constipated*. Unfortunately, many confuse what is healthy with what is average.

E. Stool water contains an average of 14 mEq of NH_4^+ per liter. NH_4^+ is produced in the colon by the oxidative deamination of amino acids and by the hydrolysis of urea.

24
Energy Balance

1

(1:915; 2:1-95; 3:225; 4:838; 5:107)
Wh 1

Which one of the following substances yields less than 4 kcal/mol when its *phosphate bond* is hydrolyzed?

A. creatine phosphate
B. adenosine diphosphate
C. adenosine triphosphate
D. glucose-6-phosphate (G-6-P)
E. guanosine triphosphate

Answer and Explanation

A, B, C, E. These substances yield 10 to 12 kcal/mol when their terminal phosphate bond is hydrolyzed.

D. G-6-P yields 2 to 3 kcal/mol when its phosphate bond is hydrolyzed. This bond is not considered a *high-energy phosphate bond*.

2

(1:915; 2:1-95; 3:230; 4:844; 5:105)
WDc 1

Which one of the following statements is incorrect?

A. the anaerobic conversion of 1 mol of glucose to 2 mol of pyruvate results in a net production of 19 mol of ATP from ADP and inorganic phosphate   =2 mol of ATP
B. the aerobic conversion of 1 mol of glucose to CO_2 and H_2O via the Embden-Myerhof pathway and citric acid cycle results in the net production of 38 mol of ATP from ADP and inorganic phosphate
C. ATP is a precursor to cyclic AMP (cyclic adenosine-3′,5′-monophosphate)
D. catecholamines ("the first messenger") cause the intracellular release of cyclic AMP ("the second messenger") in the heart, adipose tissue, and liver
E. the pentoses generated by the hexosemonophosphate shunt are used in the production of nucleotides

Answer and Explanation

A. In glycolysis, there is a net production of 2 mol of ATP.   anaerobically
D. The *cyclic AMP* released in the heart increases the force of contraction, in adipose tissue causes lipolysis, and in the liver causes glycogenolysis.

3

(1:933; 2:1-95; 3:228; 4:839; 5:1646)
Wlbc 2

Which one of the following statements is <u>most correct</u>? The *blood sugar* concentration during hypoglycemia is elevated by

A. the liver, because it has a high concentration of the enzyme glucose-6-phosphatase

B. skeletal muscle, because it has a high concentration of the enzyme glucose-6-phosphatase

C. the liver and skeletal muscle, because they have a high concentration of the enzyme glucose-6-phosphatase

D. the liver and skeletal muscle, because they have a low concentration of the enzyme glucose-6-phosphatase

E. the liver, because it has a low concentration of the enzyme glucose-6-phosphatase

Answer and Explanation

(A.) In the liver and kidney, the following reactions occur:

$$GLYCOGEN \rightleftharpoons \rightleftharpoons G\text{-}6\text{-}P \rightleftharpoons \rightleftharpoons PYRUVIC\ ACID \rightleftharpoons \rightharpoonup$$

with LACTIC ACID $\updownarrow$ and $CO_2 + H_2O$ above pyruvic acid,
HEXOKINASE and GLUCOSE-6-PHOSPHATASE acting on GLUCOSE $\leftrightarrow$ G-6-P.

B. Skeletal muscle can release large concentrations of *lactic acid* into the blood but no glucose. This is because it lacks the enzyme (glucose-6-phosphatase) for <u>converting glucose-6-phosphate (G-6-P) to glucose.</u> It can, however, perform the following:

$$GLYCOGEN \rightleftharpoons \rightleftharpoons G\text{-}6\text{-}P \rightleftharpoons \rightleftharpoons PYRUVIC\ ACID$$

with GLUCOSE $\uparrow$ G-6-P, LACTIC ACID $\updownarrow$ above pyruvic acid, and pyruvic acid $\downarrow\uparrow$ leading to $CO_2 + H_2O$.

4

(1:933; 2:7-81; 3:278; 4:840; 5:1510)
Ecl 1

Gluconeogenesis in the liver is decreased by

A. epinephrine

B. glucagon

C. thyroxine

D. all of the above

E. insulin

Answer and Explanation

E. Insulin (1) inhibits hepatic enzymes that facilitate the catabolism of proteins and fats, (2) facilitates the production of glycogen and fatty acids from glucose-6-phosphate, and (3) lowers the blood sugar. Epinephrine and glucagon antagonize these actions.

5

(1:939; 2:7-130;
3:233; 4:841;
5:1667)
Wbc 1

After a *fast* of 1 week, the major source(s) of *blood glucose* is(are)

A. glycogenolysis
B. gluconeogenesis from fatty acids   *Heart*
C. glycogenolysis and gluconeogenesis from fatty acids
D. glycogenolysis and gluconeogenesis from glycerol
E. glycogenolysis and gluconeogenesis from amino acids and glycerol

Answer and Explanation
A. After 48 hours of fasting, most of the body's glycogen has been depleted.
B, C. Fatty acids do not form glucose in important quantities.
D, (E.) *Glycerol* and the gluconeogenic *amino acids* (cystine, cysteine, glycine, serine, valine, threonine, for example) form glucose. The so-called ketogenic amino acids include leucine, isoleucine, phenylalanine, and tyrosine.

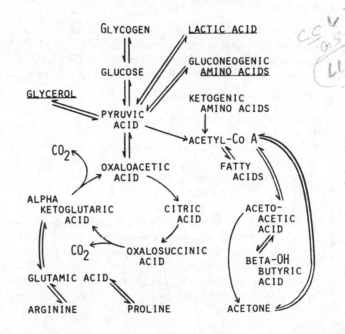

6

(1:939; 2:7-130;
3:241; 4:853;
5:1617)
Wclb 1

What is a ketone body? List three.

Answer and Explanation
A ketone body is a molecule that contains a C=O group (i.e., a ketone group) and is produced in human beings and other vertebrates in the liver from acetyl-CoA. The ketone bodies are (1) acetoacetic acid, (2) beta-OH butyric acid, and (3) acetone.

7

(1:859; 2:1-134;
3:236; 4:866;
5:1648)
Wlc 1

Which one of the following statements is incorrect? In a healthy person

A. the liver produces most of the body's urea + *Brain also*
B. most of the ammonia formed from the deamination of amino acids is changed to uric acid   *(urea) + to some uric acid*
C. skeletal muscle degeneration is one cause of an increased concentration of creatinine in the urine
D. the purines released by the catabolism of nucleotides may be further changed to uric acid
E. proline is not an essential amino acid because it can be readily produced from glutamic acid as well as other amino acids

(Continued)

Answer and Explanation

A. The liver and brain are probably the only two organs in the body that produce *urea*.

B. Most of the *ammonia* produced in human beings is changed to urea. Small amounts are used to produce *uric acid*. Glutamine serves to transport nitrogen released during *deamination* to the kidney, where it may be excreted as NH_4^+, or to the liver, where it may be converted to urea.

C. *Creatinine* is formed by the removal of phosphate from creatine-phosphate.

D. The urine contains about 42 mEq of uric acid and 1820 mEq of urea per liter.

E. The *essential amino acids* include (1) valine, (2) leucine, (3) isoleucine, (4) threonine, (5) methionine, (6) phenylalanine, (7) tryptophan, and (8) lysine.

8

(2:1-151; 3:239; 4:865; 5:105)
Wcu 2

Patient p is on a protein-rich, calorically adequate diet. Patient l is on a protein poor, calorically adequate diet. Patient p will have a daily excretion of (check each correct answer):

1. urea that is greater than four times that for patient l
2. ammonia that is greater than four times that for patient l
3. inorganic sulfate that is greater than four times that for patient l
4. creatinine that is greater than four times that for patient l

Summarize your conclusion.

A. statements 1 and 2 are correct
B. statements 1 and 3 are correct
C. statements 1, 2, and 3 are correct
D. statements 1, 2, 3, and 4 are correct
E. statement 1 is incorrect

Answer and Explanation

B. Excess *protein intake* causes increased *gluconeogenesis* and the increased excretion of the *nitrogenous waste* product, (1) *urea*, and of (3) *inorganic sulfate*. On the other hand, the quantity of (2) *ammonia* and (4) *creatinine* excreted is determined not by protein ingestion, but by (2) the pH of the plasma and (4) the "wear and tear" on the tissues, respectively. The amount of *uric acid* excretion is related to both protein ingestion and "wear and tear."

9

(1:935; 2:7-136; 3:247; 4:899; 5:1656)
Wc 2

Certain dietary *fatty acids* have been shown to be *essential* to animals in laboratory experiments and are probably also essential to human beings. Some of the nutrients studied include:

- beta-OH butyric acid: $CH_3-CHOH-CH_2-COOH$
- palmitic acid: $CH_3-(CH_2)_{14}-COOH$
- stearic acid: $CH_3-(CH_2)_{16}-COOH$
- oleic acid: $CH_3-(CH_2)_7-CH=CH-(CH_2)_7-COOH$
- linoleic acid: $CH_3-(CH_2)_4-CH=CH-CH_2-CH=CH-(CH_2)_7-COOH$
- linolenic acid: $CH_3-CH_2-CH=CH-CH_2-CH=CH-CH_2-CH=CH-(CH_2)_7-COOH$

The *essential fatty acids* include?

A. all of the above
B. all of the unsaturated fatty acids listed above
C. all of the fatty acids listed above with more than 14 carbon atoms
D. linolenic and linoleic acid
E. palmitic and stearic acid

Answer and Explanation

A. Beta-OH butyric acid is a ketone body produced in quantity by the liver from ketogenic amino acids, carbohydrates, and fatty acids and cannot therefore be considered an essential element (see question 5).

B. Oleic acid is an unsaturated fatty acid (contains a double bond) that can be produced in the body from other fatty acids, amino acids, or carbohydrates.

C. Palmitic, stearic, and oleic acids are not essential dietary constituents.

D. The essential fatty acids are all polyunsaturated and, in addition to the two listed, include arachidonic acid.

10

(1:935; 2:7-136; 3:245; 4:850; 5:1127)
Wbc

Which one of the following statements is incorrect?

A. free fatty acids in plasma are bound to albumin

B. unbound free fatty acids in plasma constitute more than 20% of the total free fatty acids.

C. free fatty acids in plasma can be catabolized to CO_2 and water by most tissues

D. free fatty acids in plasma are in a higher concentration than the ketone bodies in plasma

E. most of the fatty acids in plasma are in phospholipids and triglycerides and are esterified to cholesterol

Answer and Explanation

B. Practically all *free fatty acids* in plasma are bound to albumin.

C. The major exception to this rule is the brain, which depends on carbohydrate for nutrition. The heart and skeletal muscle, on the other hand, usually receive most of their energy from the catabolism of fatty acids.

D. *Ketone bodies*, in the healthy resting subject, have a plasma concentration of 1 mg/100 ml, whereas that for free fatty acids is 12 mg/100 ml.

E. Approximately 50% of the total fatty acids in plasma are in phospholipids, 25% in triglycerides, 20% esterified to cholesterol, and 5% are free (i.e., loosely bound to plasma proteins).

11

(1:930; 2:7-82; 3:278; 4:867; 5:1656)
EWlb 1

Which one of the following statements is incorrect? *Insulin*

A. increases the uptake of glucose by cardiac, skeletal, and smooth muscle, adipose tissue, and the liver

B. has little or no effect on the uptake of glucose by kidney tubules and most of the brain

C. affects glucose uptake primarily by its action on the cell membrane

D. has a half-life of about 40 min

E. slows normal growth and wound healing

Answer and Explanation

E. Insulin increases the uptake of *amino acids* by the cell and increases protein synthesis by the cell.

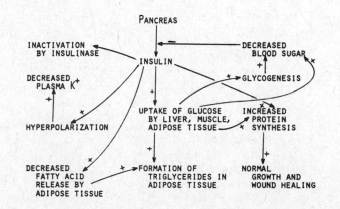

12

*(1:939; 2:7-84;
3:239; 4:969;
5:1638)*
WEca 2

Prolonged *fasting* and *diabetes mellitus* lead to (select the one best answer)

A. increased gluconeogenesis
B. ketonemia
C. a negative nitrogen balance
D. all of the above
E. hyperglycemia

Answer and Explanation

B. The *ketonemia* in both conditions is due to an increased production of ketone bodies by the liver.
C. The negative nitrogen balance is due to an increased catabolism of proteins by the cells and a resultant increased production of *urea* and other nitrogen-containing waste products.
(D.) E. *Hyperglycemia* is characteristic of diabetes mellitus, but in starvation there is a hypoglycemia.

13

*(1:934; 2:1-133;
3:239; 4:901;
5:1195)*
DuWa 2

A positive *nitrogen balance* in the adult occurs (select the one best answer)

A. when the dietary intake of protein is increased in the healthy subject
B. during a debilitating illness
C. statements A and B are both correct
D. during recovery from a debilitating illness
E. during recovery from a debilitating illness or during protein starvation

Answer and Explanation

A. An increased intake of *dietary protein* does not change the nitrogen balance in the healthy subject. The excess protein in the diet is converted to other nutrients (fat and carbohydrate), and the nitrogen in the protein molecules is transferred to nitrogen-containing waste products such as urea.
B, C. During a debilitating illness, the nitrogen excretion usually exceeds its intake and there is a negative nitrogen balance.
(D.) Growth and repair of body tissues causes a positive nitrogen balance.
E. During protein starvation, the excretion of nitrogen in the urine exceeds the intake in the food.

14

*(1:1068; 2:7-103;
3:239; 4:867;
5:1631)*
WEG 2

Which one of the following produces a markedly more positive *nitrogen balance*?

A. testosterone
B. prolactin
C. estrogen
D. chorionic gonadotropin
E. none of the above

Answer and Explanation

A. *Testosterone* increases protein synthesis.

15

(1:996; 2:7-142; 3:330; 4:867; 5:1476)
WGEb 2

Which one of the following statements is most correct? *Growth hormone* (GH)

A. decreases the concentration of blood urea nitrogen *As ↑ Protein synth*
B. increases the transport of neutral and basic amino acids into the cell by a mechanism different than that for insulin
C. has a half-life of 25 min
D. decreases the body's RQ.
E. exerts all of the above actions

Answer and Explanation

A, B. Growth hormone (GH or STH: somatotrophic hormone) facilitates *protein synthesis.*
C. A longer half-life would be inconsistent with GH's role of the minute-to-minute control of body metabolism.
D. GH decreases carbohydrate catabolism and increases fat catabolism. See questions 23 through 25 for a discussion of RQ.
(E.)

16

(1:932; 2:7-54; 3:327; 4:867; 5:1476)
WEGl 2

Name each of the hormones described below.

A. facilitates protein anabolism and closure of the cartilaginous plates of long bones
B. facilitates lipid catabolism and cell proliferation in the cartilaginous plates of the long bones of children
C. facilitates carbohydrate catabolism and prevents lipid catabolism
D. facilitates protein catabolism and hepatic glycogenesis

Answer and Explanation
A. Testosterone.
B. Growth hormone.
C. Insulin.
D. Glucocorticoids, such as cortisol, corticosterone, and cortisone.

17

(1:407; 2:4-98; 3:430; 4:864; 5:1649)
IbcE 1

Which one of the following statements is incorrect? The cells of the *liver* normally

A. produce most of the body's urea and ketone bodies
B. are the only important site for the production of albumin, prothrombin, and fibrinogen
C. are the most important site for the production of gamma globulins
D. inactivate steroid and polypeptide hormones
E. produce the precursor of angiotensin

Answer and Explanation
(C.) Although 60% to 80% of the plasma globulins are produced in the liver, the plasma cells in the lymphatic system are probably the most important source of *gamma globulin.*
D. Cortisol, aldosterone, testosterone, estrogens, insulin, progesterone, growth hormone, prolactin, antidiuretic hormone, and oxytocin are all changed in the liver to less active substances.
E. The liver produces *alpha-2-globulin.*

18

(1:960; 2:7-33;
3:320; 4:931;
5:1506)
WEi 2

List three hormones produced by the *thyroid glands*.

Answer and Explanation

Thyroxine (T_4), triiodothyronine (T_3), and calcitonin.

[handwritten: T_3 or T_4 cause protein catabolism]

19

(1:1025; 2:7-42;
3:260; 4:867;
5:1506)
Webg 1

[handwritten: Protein catabolism]

Which one of the following is not a characteristic human response to *thyroxine or triiodothyronine*?

A. a decreased O_2 consumption by the adult
B. little or no change in the O_2 consumption of the brain, testes, lymph nodes, and spleen as compared to the whole adult
C. an increase in 2,3-diphosphoglycerate in the erythrocyte
D. an increased nitrogen and calcium excretion in the adult
E. an increased cholesterol synthesis associated with a movement of cholesterol from the plasma and into the liver

Answer and Explanation

A, B. Thyroxine and triiodothyronine can increase the O_2 *consumption* of human beings by more than 50% while producing less than a 5% change in the O_2 consumption of the brain.
D. The *protein catabolism* may be so marked in skeletal muscle as to cause *weakness* and *creatinuria*. The action of T_4 and T_3 on bone may cause a mild *osteoporosis* and hypercalciuria.

20

(1:813; 2:4-37;
3:251; 4:910;
5:1130)
Wics 1

Which one of the following statements is incorrect?

A. a dietary deficiency of vitamin A and its precursors can cause a dry skin and night blindness
B. a dietary excess of vitamin A can cause a scaly dermatitis and bone pain
C. a dietary excess of pyridoxine (vitamin B_6) causes hypoirritability and lethargy
D. a dietary deficiency of copper or iron can cause abnormalities in the blood
E. dietary magnesium, manganese, cobalt, bromine, and zinc are necessary in trace amounts

Answer and Explanation

B. The fat-soluble vitamins A and D, unlike the water-soluble vitamins, cause symptoms of *hypervitaminosis*.
C. *Hypovitaminosis* B_6 causes hyperirritability and convulsions. Hypervitaminosis symptoms for this water-soluble vitamin apparently do not occur.
D. *Copper* deficiency causes neutropenia and *iron* deficiency causes anemia.
E. Although essential in trace amounts, these minerals are seldom, if ever, deficient in diets that are otherwise normal.

21

(2:2-99; 3:246;
4:857; 5:1510)
Wlbc 1

An individual on a diet that is practically devoid of *cholesterol* will in many cases have (select the one best answer)

A. a low blood cholesterol level because the body cannot synthesize cholesterol
B. a near normal blood cholesterol level because the liver synthesizes cholesterol and adds it to the blood when the blood level is low
C. a near normal blood cholesterol level because the intestinal flora supply cholesterol to the blood

D. a near normal blood cholesterol because of an increased ability of the intestine to absorb cholesterol

E. a near normal blood cholesterol because of a decreased excretion of cholesterol in the bile

Answer and Explanation

A. Although a lower blood cholesterol may result from dietary restriction, it is not because the body cannot synthesize it.

B. Most human tissues can synthesize cholesterol. The liver is the most important organ for the maintenance of blood cholesterol levels.

C. In human beings, the microorganisms in the intestine probably serve only as a source of certain vitamins (possibly only folic acid in the adult).

D. A change in the absorptive power of the intestine for Ca^{++} may serve to increase the level of blood Ca^{++}. Changes in the absorptive power for cholesterol, on the other hand, are not an important response to hypocholesterolemia.

E. The cholesterol in the bile is reabsorbed into the blood in the intestinal ileum. In short, cholesterol in the bile is not an important route for loss of cholesterol from the body. The bile pigments, on the other hand, are bile elements lost from the body in the feces of the normal subject.

22

(1:915; 3:223; 4:900; 5:1352)
Rhgm 2

A subject walking 5 miles/hour at a 5% slope on a treadmill had an *oxygen consumption* of 2 liters/min (STPD) and a *respiratory quotient* (RQ) of 0.8. If his caloric equivalent for oxygen was 4.8 kcal/liter, approximately how much *heat* must he eliminate from his body in order to maintain a constant body temperature?

A. less than 2 kcal/min
B. between 2 and 4 kcal/min
C. between 4 and 6 kcal/min
D. between 6 and 8 kcal/min
E. over 8 kcal/min

Answer and Explanation

E. Energy liberated = (2 liters of O_2/min) × (4.8 kcal/liter of O_2) = 9.6 kcal/min. See question 32 (page 194) for the distinction between RQ and respiratory exchange ratio (R).

23

(1:917; 2:1-50; 3:221; 4:901; 5:1357)
RWcg 3

If, in the exercise study in question 22, the subject had a *respiratory quotient* (RQ) of 0.75, this lower RQ would

A. cause a decrease in the caloric equivalent for 1 liter of O_2
B. indicate that lipid catabolism is a less important source of energy than protein or carbohydrate catabolism
C. indicate that both of the above statements are true
D. cause a decrease in alveolar minute ventilation
E. cause all of the above

Answer and Explanation

A. RQ = CO_2 produced/O_2 utilized. Therefore, a decrease in RQ means it takes more O_2 to produce the same quantity of CO_2 (i.e., approximately the same number of calories).

B, C. Compared to carbohydrate and protein, lipid is oxygen poor in terms of content. A decrease in lipid catabolism would increase RQ (i.e., decrease the O_2 utilized).

D. Subjects with a low RQ have a greater O_2 consumption. A greater O_2 consumption is usually associated with an increased alveolar minute volume.

24

(1:917; 3:223;
4:901; 5:1357)
WRg 2

Which one of the following statements is correct? In a resting individual in the post-absorptive state, the *RQ* (respiratory quotient) of

A. practically all major organs (heart, brain, skeletal muscle, etc.) is similar
B. the brain is greater than 0.95
C. skeletal muscle is greater than 0.95
D. the heart is greater than 0.95
E. the skin is greater than 0.95

Answer and Explanation
A, (B,) C, D, E. The RQ averages 0.83 for an adult, 0.98 for the brain, and 0.85 for skeletal muscle, heart, and skin.

25

(1:604; 2:1-50;
3:245; 4:901;
5:1357)
RgW 2

Why is the *RQ* of the heart so much lower than that of the *brain*?

Answer and Explanation
The nutrient from which the brain receives its energy is primarily glucose:

$$C_6H_{12}O_6 + 6(O_2) \rightarrow 6(CO_2) + 6(H_2O);$$

$$RQ \text{ for glucose} = \frac{6(CO_2)}{6(O_2)} + 1.00. \quad \checkmark$$

The major class of nutrient from which the heart gets its energy is the fatty acid (supplies approximately 70% of the calories during the postabsorptive state). In the case of stearic acid, for example, the reactions are as follows:

$$C_{18}H_{36}O_2 + 26(O_2) \rightarrow 18(CO_2) + 18(H_2O);$$

$$RQ \text{ for stearic acid} = \frac{18(CO_2)}{26(O_2)} + 0.69. \quad \checkmark$$

26

(1:915; 2:1-131;
3:223)
WhDl 1

The *specific dynamic action* of (select the one *incorrect* statement)

A. food causes an increase in a subject's O_2 consumption after eating
B. food persists for about 6 hours after eating
C. protein is due in large part to the oxidative deamination of amino acids in the liver
D. protein is in excess of 20% of its total caloric value
E. fat is greater than that for protein

Answer and Explanation
E. The specific dynamic action of protein is 30% of its caloric value, whereas that for carbohydrate and fat is about 6% and 4% respectively.

27

(1:915; 3:221;
4:899; 5:1352)
Whg 2

Which one of the following statements is incorrect?

A. the kcal (kilocalorie) is the amount of heat required to raise the temperature of 1 g of water 1°C (from 15 to 16°C)
B. the basal metabolic rate of an average 40-year-old man is about 1600 kcal per 24 hours
C. the total metabolic rate of an average 40-year-old man is about 3000 kcal per 24 hours

D. the total metabolic rate is increased following eating
E. when one changes the room temperature from 30 to 0°C or from 30 to 50°C, the total metabolic rate increases

Answer and Explanation
(A.) 1 dyne-cm = 1 erg = 10^{-7} joules = 2.39×10^{-8} cal = 2.39×10^{-11} *kcal*. The kcal is the amount of heat required to raise the temperature of 1 kg of water from 15 to 16°C.
B. For a woman the value is about 1300, and for an 18-year-old boy about 1800 kcal.
D. This is due to the specific dynamic action of food.
E. Increases in total *metabolic rate* occur in response to body cooling (may cause shivering) or body warming (may cause sweating).

28

(3:245; 4:857; 5:1510)
WhEa 1

Which one of the following statements is incorrect?

A. leafy green vegetables are good sources for folic acid, vitamin C, and vitamin K
B. liver is a good source for thiamine, riboflavin, niacin, pantothenic acid, biotin, and vitamin B_{12}
C. 1 g of fat produces more than twice the number of calories as 1 g of carbohydrate
D. foods, such as margarine, that are derived from vegetables have a higher cholesterol concentration than most foods derived from animals, such as milk, cheese, eggs, and meat
E. estrogens cause a lowering of plasma cholesterol levels

Answer and Explanation
C. One gram of animal fat yields about 9.5 *kcal*, whereas 1 g of starch yields 4.2 kcal, 1 g of cane sugar yields 4.0 kcal, and 1 g of protein yields 4.4 kcal.
(D.) *Cholesterol* is found only in animals and their products.
E. *Atherosclerosis* is characterized as a condition in which cholesterol infiltrates arterial walls. It has been suggested that *estrogens*, in that they lower plasma cholesterol levels, protect premenopausal women from atherosclerosis.

29

(1:932; 2:2-6; 3:273; 4:902; 5:1380)
WEbD 2

Which of the following statements is most accurate and complete?

A. the major feeding and satiety centers lie in the brain stem
B. the high blood sugar that occurs during diabetes mellitus depresses the hunger center
C. insulin facilitates the utilization of sugar by the "glucostats" in the brain
D. appetite is unaffected if the stomach is distended with inert material
E. all of the above statements are true

Answer and Explanation
A. The *feeding center* is in the nuclei of the lateral part of the hypothalamus and the *satiety center* is in the medial hypothalamic nuclei. The amygdaloid nuclei of the limbic system also play a role in controlling appetite.
B, (C.) Appetite is controlled, in part, by the utilization of blood sugar (i.e., arteriovenous glucose difference) in cells that are said to constitute the *glucostat*:

(Continued)

D. Distention of the stomach decreases appetite.

30

(1:880; 2:5-7; 3:188; 4:441; 5:1163)
qpNa 1

Which one of the following statements is incorrect? *Thirst* is

A. not quenched when water is drunk but when the water is absorbed into the blood
B. diminished by lesions in the hypothalamus that do not affect food intake
C. increased by infusing hypertonic solutions into the arteries going to the hypothalamus
D. produced by a fall in blood volume in which there is no change in blood tonicity
E. increased in diabetes mellitus

Answer and Explanation

A. Water in the mouth and stomach temporarily inhibits thirst by the stimulation of receptors and their sensory neurons.

31

(1:947; 2:5-5; 3:249; 4:851; 5:1149)
qh 2

Which one of the following statements is most correct? An increase in the body *fat*

A. causes a decrease in the percentage of body weight due to water
B. decreases survival time in cold (0°C) water
C. raises the specific gravity of the body
D. statements A, B, and C are all correct
E. causes a reduction in the lean body mass

Answer and Explanation

A. Adipose tissue is less than 25% *water,* whereas skeletal muscle is about 75% water, and the total body is about 63% water.
B. Fat improves the *insulating* properties of the body and therefore increases survival time in cold water.
C. Fat lowers the specific gravity of the body.
E. *Lean body mass* stays fairly constant when body fat is increased or decreased.
Lean body mass = (total body weight) − (weight of fat reserves).

32

(1:1048; 2:7-136; 3:241; 4:851; 5:1644)
WCc 1

Which of the following statements is most correct? *Fat depots*

A. convert glucose to fatty acids
B. produce over 20% of the individual's ketone bodies
C. usually constitute about 30% of the total body weight
D. usually constitute a greater percentage of the body weight in men than women
E. usually have a blood flow that is comparable to or greater than that found in other tissues

Answer and Explanation

A. Both the liver and adipose tissue produce fatty acids from *glucose.*
B. Adipose tissue is not an important source for *ketone bodies.* They are produced in the liver.
C. Fat usually constitutes about 15% of the *body weight* in men and 20% in women.
D. The reverse is true.

33

*(3:243; 4:892;
5:1435)*
WhG 1

Which of the following statements is most correct? *Brown fat*

A. is most commonly seen after puberty
B. is most commonly found in the abdomen
C. has a lower concentration of mitochondria than ordinary (i.e., white) adipose tissue
D. statements A, B, and C are all correct
E. releases significant quantities of heat to the body in response to the stimulation of its sympathetic neurons

Answer and Explanation

A, B, C. Brown fat is usually seen in infants as pads in the neck and around the scapulas. It is richer in mitochondria and has a higher metabolic rate than white fat. It is not normally found in adult humans.

(E.) Infants do not have a well-developed shivering mechanism and depend to a great degree on the reflex increase in brown fat metabolism as a mechanism for increased heat production in response to cooling.

25
Body Temperature

1

(1:605; 2:9-138;
3:196; 4:887;
5:1425)
hWCf 1

One mechanism for the elimination of heat from the body is vaporization. List three other mechanisms and define each.

Answer and Explanation
- *Vaporization:* the change of a solid or liquid (water in this case) into a gas (water vapor).
- *Conduction:* transmission of energy (such as heat) between adjacent structures without the movement of either structure.
- *Convection:* the transmission of heat in liquids or gases by fluid moving over and away from a structure (i.e., air moving over the skin).
- *Radiation:* the transmission of energy in all directions to distant objects.

2

(2:9-138; 3:196;
4:888; 5:1425)
HwCf 1

A 25-year-old man clothed only in shoes and shorts is in a room with a temperature of 70°F (21°C) and a humidity of 20%. As he stands quietly, what percentage of the total heat that he loses from his body will be dissipated by the mechanisms listed in question 1 (select the one best statement)?

A. 65% by insensible perspiration
B. 65% by radiation
C. 40% by vaporization and 40% by radiation
D. approximately equal amounts by vaporization, radiation, and convection
E. approximately equal amounts by vaporization, radiation, convection, and conduction

Answer and Explanation
B. Radiation accounts for about 65% of the *heat dissipated*, vaporization 20%, and convection about 15%.

3

(2:9-138; 3:196;
4:888; 5:1431)
hWCf 2

The subject in question 2 would, during a strenuous tennis match, lose most of his body *heat* by

A. conduction
B. convection
C. radiation

D. vaporization
E. no single mechanism

Answer and Explanation
D. During strenuous exercise the amount of heat lost by (1) radiation may increase by 10%, (2) by convection may triple, and (3) by vaporization may increase 25-fold.

4

(2:9-138; 3:196; 4:888; 5:1431)
hWCf 2

If the subject in question 2 is placed in a room with a *temperature* of 37°C and a *humidity* of 20%, (select the one best answer)

A. over 90% of the heat lost will be by vaporization
B. most of the heat lost will be by vaporization
C. most of the heat lost will be by radiation
D. most of the heat lost will be by convection
E. most of the heat lost will be by conduction

Answer and Explanation
A. Radiation, convection, and conduction of heat from the body require a room temperature lower than the temperature of the skin. At a room temperature of 36°C or higher, all of the heat lost from the body is usually by vaporization. This requires, however, a humidity below 100%.

5

(1:605; 2:9-137; 3:195; 4:886; 5:1435)
hWCf 1

In discussing temperature regulation, we can divide the human body into a *core* and a *shell*. What structures make up the shell?

Answer and Explanation
The shell consists of the structures that receive heat from or send heat to the external environment. It is less insulated from that environment than the core and consists of the skin, subcutaneous tissue, hands, feet, and most of the arm and leg.

6

(2:7-113; 3:195; 4:886; 5:1418)
hWf 1

Check each of the following statements that applies to a healthy, naked, 24-year-old woman under resting conditions.

1. her rectal temperature averages 37°C (98.6°F)
2. at room temperatures between 23 and 33°C, her skin temperature will remain from 0 to 2°C below rectal temperature
3. the rectal temperature has a diurnal fluctuation of about 0.6°C, being lowest in most individuals at about 6 A.M.
4. during the menstrual cycle, the most distinct rise in rectal and oral temperature occurs about 2 days before menstruation

Answer and Explanation
①.
2. At a *room temperature* of 23°C, the average *skin temperature* will be 5°C below rectal temperature and the foot will be over 11°C below *rectal temperature*.
③.
④. The most distinct rise occurs after *ovulation*. The rectal temperature returns to the preovulation value several days prior to menstruation.

7

(2:9-137; 3:198;
4:886; 5:1419)
h Wfa 1

Check each of the following statements that is true.

1. during strenuous exercise, the rectal temperature may rise to 40°C
2. during exercise, the skin temperature may either rise (frequently to 39°C) or fall
3. for each rise of 1°C in fever there is a 13% increase in O_2 consumption
4. rectal temperatures above 43°C may be fatal

Answer and Explanation

①.

②. During most exercises, there is an increased blood supply to the skin. This is an important mechanism for *shunting heat* from the core to the shell. During a maximal effort, however, cutaneous blood flow may go to 20% of the preexercise value. Under these circumstances the rectal temperature rises rapidly and the cutaneous temperature falls. The temperature of the skin will also depend on the quantity and type of clothing, the amount of sweating, the humidity, the room temperature, and the amount of air movement.

③.

④. If this temperature is maintained for an hour it is usually *fatal*. There is, however, individual variability.

8

(2:9-137; 3:199;
4:897; 5:1875)
h Wbf 1

Check each of the following statements that is true.

1. human subjects with rectal temperatures between 21 and 24°C for 10 min can be revived if warmed rapidly
2. at a rectal temperature of 24°C, the subject is conscious and shivering
3. at a rectal temperature of 24°C, the heart rate and respiratory rate are slightly elevated
4. at a rectal temperature of 24°C, hemorrhagic problems are exaggerated
5. an individual usually tolerates well a rectal temperature of 32°C maintained for 2 days under hospital conditions

Answer and Explanation

①. Prolonged cooling at these levels causes cardiac standstill or fibrillation.
2. At rectal temperatures below 28°C, the subject is unconscious, not shivering, and unable to recover from cooling by his own mechanisms.
3, 4. At these *low temperatures*, the heart rate, arterial pressure, and respiratory rate are reduced and bleeding is a reduced problem.
⑤. This level of hypothermia slows the heart and decreases body metabolism.

9

(1:607; 2:5-49;
3:197; 4:421;
5:1184)
h Cbf 2

The *countercurrent* mechanism of heat transfer during exposure to the cold explains which one of the following sets of data?

A. heat in arteries of the arm is 37°C, in arteries of the hand is 22°C, in veins of the arm is 36°C
B. heat in arteries of the arm is 37°C, in arteries of the hand is 36°C, in veins of the arm is 35°C
C. heat in arteries of the arm is 35°C, in arteries of the hand is 36°C, in veins of the arm is 37°C
D. heat in arteries of the arm is 36°C, in arteries of the hand is 22°C, in veins of the arm is 37°C
E. heat in arteries of the arm is 33°C, in arteries of the hand is 25°C, in veins of the arm is 22°C

Answer and Explanation

A. During exposure to the cold, blood flow to the hand is reduced, and much of the heat in the arterial blood in the arteries of the arm and forearm is conducted to the cooler veins that lie adjacent to these arteries (the *venae comites*). Thus, the veins and arteries of the arm have blood with a relatively high temperature, and the blood in the hand has a low temperature. This countercurrent heat mechanism is similar to that responsible for Na^+ transfer in the ascending and descending loops of the nephron and is an important mechanism for heat conservation.

10

(2:9-138; 5:1425)
hNbf 1

Which one of the following statements is incorrect? Under resting conditions at a temperature of 75°C,

A. a naked black man absorbs 20% more heat from the sun than a naked white man of similar height and weight
B. approximately one-third of the heat lost due to insensible perspiration is lost in the expired air
C. 580 kcal of heat are lost from the body for each kg of water vaporized on the skin or mucous membranes
D. sweat glands are innervated by postganglionic, cholinergic sympathetic neurons
E. an increased cutaneous blood flow causes an increased thermal conductivity of the skin

Answer and Explanation

Ⓐ Human skin, as far as heat is concerned, can be characterized as a 97% *black body*. In other words, the skins of blacks and whites, although different in the visual ranges, are similar in their heat absorption characteristics. The skins of blacks, on the other hand, let less ultraviolet light (wavelengths 290 to 32 mμ) penetrate and are probably therefore less susceptible to hypervitaminosis D and skin cancer.

C. Of this 580 kcal, 539 kcal is the heat of vaporization and 41 kcal is the heat carried from the body by the vapor.

11

(2:9-139; 3:197;
4:889; 5:1161)
hCiq 2

An athlete ran a 42-km race in 158 minutes. During the race, he lost 3 liters of *sweat*. After the race, he drank 3 liters of water. Several hours later, the subject was studied. Which one of the following conclusions would be most likely? Since sweat contains

A. a higher concentration of Na than the extracellular fluid, there was hyponatremia
B. a lower concentration of Na than the extracellular fluid, there was hypernatremia
C. approximately the same concentration of Na as the extracellular fluid and there was water replacement but not Na replacement, there was an increase in blood volume
D. approximately the same concentration of Na as the extracellular fluid and there was water replacement but not Na replacement, there was a decrease in blood volume
E. a lower concentration of Na than the extracellular fluid and there was water replacement but not Na replacement, there was a decrease in blood volume

Answer and Explanation

A, B, C, D. Sweat losses as great as 10 liters per day have been reported. Sweat is hypotonic and will contain from 9 to 80 mEq of Na per liter. Perfuse sweating without water or Na replacement will tend to produce *hypernatremia* but with water replacement may cause *hyponatremia*. The adrenal cortex responds to hypernatremia by decreasing its output of *aldosterone*. There results an increased concentration of Na in the urine and sweat.

(Continued)

E. Loss of Na + loss of water + gain of water
↓
hypotonicity of extracellular compartment
↓
diffusion of water into the intracellular environment
↓
decreased blood volume

12

(1:327; 2:9-140; 3:197; 4:890; 5:1446)
hNbf 3

A patient receives a *spinal transection* at T-2, and after a period of time, his withdrawal reflexes return below the lesion. If his legs and thighs are now *cooled*, the initial response to this cooling would be (select the one best answer)

A. shivering in the legs because of a direct action of cooling on the legs
B. shivering in the legs because of a spinal reflex
C. shivering in the arms because of a spinal reflex
D. shivering in the legs because of a cooling of the hypothalamus
E. shivering in the arms because of a cooling of the hypothalamus

Answer and Explanation
D. The transection has eliminated functional connections between the *hypothalamus* and the legs.
E. By cooling the lower extremities, the blood has been cooled and therefore the hypothalamus.

13

(1:327; 2:9-140; 3:197; 4:890; 5:1446)
hNbf 3

A patient with a spinal transection at T-8 differs from one with a transection at T-1 in that the former responds to (select the one best answer)

A. cooling of the lower extremities with a more intense vasoconstriction in the hand
B. warming of the lower extremities with a more profuse sweating in the hand
C. both A and B
D. cooling of the lower extremities with a more intense vasodilation in the hand
E. both B and D

Answer and Explanation
A, B, C. A transection at T-8, unlike the one at T-1, does not separate the *hypothalamus* from the *adrenergic sympathetic neurons* to the (1) arterioles of the hand nor the (2) *cholinergic sympathetic neurons* to the sweat glands of the hand.

14

(1:1027; 2:9-143; 3:196; 4:892; 5:1511)
hE 1

Contrast the actions of *epinephrine* and *thyroxine* on heat production.

Answer and Explanation
Epinephrine acts on the heart, skeletal muscle, and many other structures to increase their heat production. Thyroxine also acts on these structures to increase their heat production, and potentiates the action of epinephrine on them as well. Epinephrine is released and produces its *calorigenic action* on cells within seconds after exposure to the cold. Thyroxine has a latency period of about 4 hours and, in part because of protein binding in the plasma, a longer duration of action. The calorigenic action of thyroxine is probably of greater importance than that of epinephrine in the individual's response to a prolonged exposure to the cold. Rats, for example, exposed to low temperatures (7 to 12°C) for periods of 3 weeks, exhibit increases in metabolic rate as high as 16% and a hyperplasia of the thyroid.

15

(2:9-137; 3:197;
4:896; 5:1452)
hW 2

A group of young men living in a cold environment began exercising for 5 hours at 36°C. After 12 consecutive days of this regime, would they show any signs of *acclimatization* to *exercising* at this elevated *temperature*? What would these signs be?

Answer and Explanation

Yes, some of the signs of acclimatization would be the following:

- *Sweating:* the volume of sweat produced during exercise would approximately double (i.e., change from 600 ml/hr during the first hour of exercise on the first day to 1200 ml/hr during the first hour on the twelfth day). The concentration of Na in the sweat would decrease.
- *O$_2$ consumption:* the O$_2$ used during exercise would decrease.
- *Rectal temperature:* with a more effective means of eliminating excess heat (sweating and circulatory changes) and an increased efficiency, the subject would have less marked increases in rectal temperature during exercise.

16

(1:327; 2:9-137;
3:196; 4:891;
5:1445)
Whn 1

An individual exposed to a hot environment will have an increased *body core temperature* which will cause

A. a reflex stimulation of cholinergic sympathetic neurons to the sweat glands
B. a reflex inhibition of adrenergic sympathetic neurons to the blood vessels of the skin
C. both of the above
D. a decreased skin temperature
E. all of the above

Answer and Explanation

C, D. The decreased adrenergic sympathetic tone to the cutaneous vessels shunts more blood to the skin. This causes an increased *skin temperature* and an increased heat loss from the body if the temperature of the external environment is lower than body temperature. *Vaporization* on the skin surface cools the skin and therefore the body, even when the external temperature exceeds body temperature.

17

(1:327; 2:9-143;
3:198; 4:893;
5:1449)
WNabh 1

A fever induced by a bacterial toxin is caused by

A. the production of an endogenous pyrogen by the leukocytes and macrophages
B. a direct action of the toxin on skeletal muscle and the cutaneous blood vessels
C. a direct action of the toxin on the hypothalamus
D. an elevation of the set point in the heat regulatory center of the substantia nigra
E. an inhibition of prostaglandin synthesis and release in the forebrain

Answer and Explanation

A. Polymorphonuclear *leukocytes, monocytes,* splenic and alveolar *macrophages,* and Kupffer cells produce the endogenous *pyrogen* in response to the toxin.
B, C, D, E. The endogenous pyrogen elevates the set point of the *heat regulatory centers* in the preoptic area of the *hypothalamus* (possibly by stimulating the local release of *prostaglandins*).

PART EIGHT
ENDOCRINE CONTROL

26
Hormones

Note the following formulae. Answer questions 1 through 10 on these formulae:

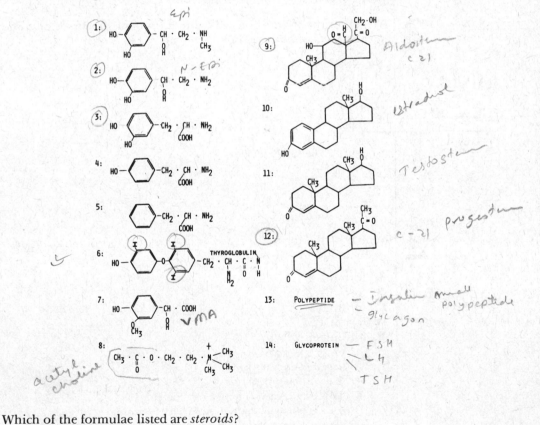

1

Which of the formulae listed are *steroids*?

A. formula 1
B. formula 9
C. formulae 1 and 9
D. formula 7
E. formulae 1, 7, and 9

(1:1035; 2:7-55; 3:294; 4:954; 5:1559)
ECc 2

(Continued)

Answer and Explanation
A. This is a catecholamine.
B. This is aldosterone.
D. This is 3-methoxymandelic acid.

2

(1:1061; 2:7-70; 3:292; 4:712; 5:209)
ECcn 2

Which of the formulae listed are *catecholamines*?

A. formula 2
B. formula 4
C. formulae 2 and 4
D. formula 7
E. formulae 2, 4, and 7

Answer and Explanation
A. Formulae 1, 2, and 3 are all catecholamines, since they contain the catechol group (i.e., the dihydroxybenzene ring) plus the amine group (i.e., $-NH_2$, $-NHCH_3$, etc.).

3

(1:1061; 2:7-70; 3:292; 4:713; 5:904)
ECcn 2

Which of the formulae listed is the immediate precursor of epinephrine and the major secretion of *postganglionic sympathetic neurons*?

A. formula 1
B. formula 2
C. formula 3
D. formula 4
E. formula 6

Answer and Explanation
B.

4

(1:1062; 2:7-70; 3:75; 4:713; 5:904)
ECcu 2

Which of the formulae listed is the product of *epinephrine destruction* and is normally found in the urine?

A. formula 2
B. formula 4
C. formula 5
D. formula 7
E. formula 6

Answer and Explanation
A. This is norepinephrine.
B. This is tyrosine, a precursor of epinephrine.
C. This is phenylalanine, a precursor of tyrosine.
D. This is 3-methoxy-4-hydroxymandelic acid (commonly called vanillylmandelic acid or VMA). About 35% of the secreted catecholamines appear in the urine as VMA.
E. A molecule this large would not be found in the urine of a normal subject.

5

(1:52; 2:9-61; 3:71;
4:712; 5:210)
nECc 2

Which of the formulae listed is the *neurotransmitter* secreted by preganglionic sympathetic neurons?

A. formula 1
B. formula 2
C. formula 6
D. formula 7
E. formula 8

Answer and Explanation
A. This is epinephrine. It is released by the adrenal medulla.
B. This is norepinephrine. It is released by postganglionic, adrenergic sympathetic neurons.
(E.) This is acetylcholine.

6

(1:1035; 2:7-55;
3:294; 4:954;
5:1559)
Ec 2

Which of the formulae listed are C-21 *steroids?*

A. formula 9
B. formula 10
C. formula 11
D. formula 12
E. formulae 9 and 12

Answer and Explanation
E. The C-21 steroids have a two-carbon side chain attached to carbon 17 of the cyclopentanoperhydrophenanthrene ring.

7

(1:1017; 2:7-34;
3:255; 4:933;
5:1496)
Ec 2

Which of the formulae listed will cause the disappearance of edema, an increased oxygen consumption, and the increased catabolism of protein and fat in the patient with *myxedema?*

A. formula 5
B. formula 6
C. formula 8
D. formula 11
E. formula 12

Answer and Explanation
B. Formula 6 is triiodothyronine thyroglobulin.

8

(1:924; 2:7-74;
3:269; 4:960;
5:1640)
Ec 1

Which of the formulae listed are most representative of the hormone *insulin?*

A. formula 5
B. formula 6
C. formula 12
D. formula 13
E. formula 14

Answer and Explanation
D. Insulin is a small polypeptide with a molecular weight of 5734. It contains two chains of amino acids linked by disulfide bridges.

9

(1:988; 2:7-14; 3:323; 4:1000; 5:1473)
Ec 1

Which of the formulae listed best characterizes *follicle-stimulating hormone?*

A. formula 7
B. formula 8
C. formula 12
D. formula 13
E. formula 14

Answer and Explanation
    E. FSH, LH, and TSH are all glycoproteins.

10

(1:936; 2:7-76; 3:282; 4:966; 5:1660)
Ec 1

Which of the formulae listed best characterizes *glucagon?*

A. formula 7
B. formula 8
C. formula 12
D. formula 13
E. formula 14

Answer and Explanation
    D. Both insulin and glucagon are polypeptides produced by the pancreas. Glucagon has a molecular weight of 3485.

11

(1:1036; 2:7-55; 3:294; 4:954; 5:1560)
ESc 2

What hormones are *C-21 steroids?*

A. 11-desoxycortisol, cortisol, and cortisone
B. progesterone and aldosterone
C. estradiol and testosterone
D. progesterone, aldosterone, 11-desoxycortisol, cortisol, and cortisone
E. all of the above

Answer and Explanation
    D. Estradiol has 18 carbons in its molecule, and testosterone is a C-19 steroid (i.e., it has only 19 carbons in its molecule). Formula 10 is estradiol, and formula 11 is testosterone. Formula 9 is aldosterone, and formula 12 is progesterone. They both contain 21 carbons in their molecules.

12

(1:1041; 2:7-55; 3:294; 4:953; 5:1562)
ESc 2

The *17-hydroxycorticoids* include

A. 11-desoxycortisol, cortisol, and cortisone → 17- hydroxy cortioids
B. progesterone and aldosterone
C. estradiol and testosterone
D. progesterone, aldosterone, 11-desoxycortisol, cortisol, and cortisone
E. all of the above

Answer and Explanation
    A. The C-21 steroids with a hydroxyl group in the C-17 position are called 17-hydroxycorticoids. The C-19 steroids with an oxygen in the C-17 position are called 17-ketosteroids and include dehydroepiandrosterone, androsterone, and androstenedione.

13

(1:1042; 2:7-100;
3:295; 4:953;
5:1564)
ESc 2

The *17-ketosteroids* in the urine include

A. androgens and the metabolic products of testosterone inactivation
B. metabolic products of cortisone and cortisol inactivation
C. both A and B
D. metabolic products of corticosterone inactivation
E. all of the above

Answer and Explanation
C. The androgens produced by the adrenal cortex and testis are 17-ketosteroids. Testosterone, on the other hand, is a 17-hydroxycorticoid but is changed in the body to a 17-ketosteroid. In addition about 10% of the cortisol secreted (this includes that which is later changed to cortisone) is converted in the liver to a 17-ketosteroid. In total, the normal adult man eliminates in the urine about 15 mg of 17-ketosteroids per day. About two-thirds of these come from the adrenal cortex and one-third from the testis. The normal adult woman, on the other hand, eliminates about 10 mg/day in her urine.

14

(2:7-10; 3:248;
4:918; 5:1059)
Ec 1

The following molecule has been characterized as a modulator of hormone action. What is its name?

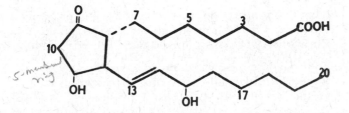

Answer and Explanation
The molecule is *prostaglandin* E_1 (11 alpha, 15-dihydroxy-9-keto-prost-13-enoic acid).

15

(2:7-11; 3:247)
Ec 1

Sixteen *prostaglandins* have been identified and characterized as belonging to one of four series. What are the four series of prostaglandins?

Answer and Explanation
PGE (O at C-9), PGF (OH at C-9), and dehydration products of PGE: PGA and PGB.

16

(2:7-11; 3:247)
Ec 1

The various *prostaglandins* are differentiated by a subscript. The subscript indicates

A. its order of discovery (i.e., PGE_1 was the first PGE discovered)
B. the number of ketone groups in the molecule
C. the number of double bonds in the molecule
D. the number of acid groups in the molecule
E. its order of potency (i.e., PGE_2 is the second most potent PGE)

Answer and Explanation
C.

17

(2:7-11; 3:247; 4:918; 5:1059)
Ec 1

Which one of the following statements is correct? *Prostaglandins*

A. are produced only in the prostate gland
B. have, as their major source, the prostate gland
C. are produced by most of the organs in the body
D. produce biologic changes only at molar concentrations ten times those necessary for most well-established hormones
E. have little or no known action on noncontractile structures

Answer and Explanation

A, B, C. In 1930, Kurzok and Lieb noted that fresh human semen caused uterine strips to either contract or relax. U.S. von Euler, in Sweden, called the active principle prostaglandin. It is now believed that most of the prostaglandin in the semen comes from the seminiferous tubules and not the prostate. The kidneys, GI tract, hypothalamus, and many other structures may also represent important sources.

D. Prostaglandins are active at concentrations of 10^{-9} molar. This means that they are probably more potent than any of the well-established hormones.

E. PGE (1) produces vasodilation, bronchodilation, and contraction of the uterus, (2) produces secretion of the adrenal cortex, corpus luteum, and thyroid, (3) produces natriuresis, and (4) inhibits lipolysis, platelet aggregation, and the secretion of acid by the stomach.

18

(1:905; 2:7-7; 3:26; 4:918; 5:1462)
Ec 2

What one property do each of the following probably have in common: ACTH, FSH, TSH, TRH, LH, chorionic gonadotropin, parathormone, calcitonin, catecholamines, glucagon, ADH, gastrin, prostaglandin E, histamine?

A. they stimulate beta adrenergic receptors
B. they stimulate alpha adrenergic receptors
C. they readily penetrate the cell membrane
D. they cause an increase in cyclic AMP activity in their target cells, but not all cells
E. they cause an increase in cyclic AMP activity in all cells

Answer and Explanation

A. The stimulation of *beta adrenergic receptors* in the liver by epinephrine causes the intracellular activation of c-AMP. Glucagon also stimulates the activation of c-AMP in the liver but not by the stimulation of beta receptors.

C. None of these substances readily penetrate the cell membrane.

D. Hormones are considered the "first messenger" and c-AMP the "second messenger."

19

(1:1065; 2:7-7; 3:231; 4:960; 5:1463)
Ec 1

Which one of the following statements is incorrect?

A. adenylate cyclase is in the cell membrane of certain liver cells
B. adenylate cyclase is activated in the liver cell in response to epinephrine
C. adenylate cyclase catalyzes the conversion of ATP to c-AMP
D. c-AMP inhibits phosphorylase in the liver
E. c-AMP increases the mobilization of Ca^{++} from bone (i.e., increases blood Ca^{++})

Answer and Explanation

D. *Cyclic AMP* activates phosphorylase in the liver.

20

*(1:1047; 2:7-11;
3:28; 4:954; 5:1465)*
Ec 1

Check each of the following statements that is true. Most or possibly all *steroid hormones* are believed to control cell function, for the most part, by

1. causing the cell membrane to release a second messenger other than cyclic AMP
2. penetrating the cell membrane
3. penetrating the nuclear membrane
4. causing the production of messenger RNA

Answer and Explanation

1, (2), (3), (4). Second messengers other than c-AMP have been proposed, but the steroids seem to function usually through (1) their penetration of the *cell membrane*, (2) their formation of a *protein complex* in the cytoplasm, (3) the entry of the complex into the karyoplasm, (4) its facilitation of the formation of *messenger RNA* by *DNA*, (5) the facilitation of protein synthesis in the cytoplasm by m-RNA. Thyroid hormone also penetrates the cell membrane and increases the production of m-RNA.

21

*(1:900; 2:7-118;
3:26; 4:917; 5:1466)*
EW 1

In some of the previous questions, we have noted that a number of hormones affect cell function by (1) causing the release of a messenger from the cell membrane or (2) penetrating the cell membrane. List two other mechanisms that have been suggested for hormone function.

Answer and Explanation
Hormones function by:

1. causing the cell membrane to release an intracellular *messenger*
2. entering the cytoplasm and/or karyoplasm
3. controlling membrane permeability and/or *transport* (insulin, for example, increases the entry of glucose into the cells)
4. a *permissive action* on the membrane [for example, some hormones may modify the activity of *adenylate cyclase* (= adenyl cyclase = guanylate cyclase) in the cell membrane]

22

*(1:938; 2:7-83;
3:283; 4:967;
5:1658)*
Ecbl 3

Contrast the action of glucagon, epinephrine, and insulin on the control of *blood sugar* levels by the liver. Include in your discussion the mechanisms whereby each performs its function.

Answer and Explanation

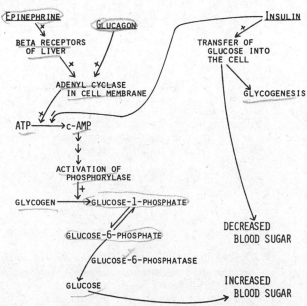

23

(1:1047; 2:7-60; 3:301; 4:948; 5:1569)
ECWb 2

Which one of the following statements is incorrect? *Corticosteroids*

A. in excess can produce a condition similar to diabetes mellitus
B. decrease the vasoconstrictor action of catecholamines
C. play a permissive role in the response of adipose tissue to growth hormone
D. facilitate the release of free fatty acids into the blood by adipose tissue
E. increase blood sugar levels

Answer and Explanation
A. They decrease the sensitivity of cells to insulin.
B. They play a *permissive role* in the response of a number of cells to glucagon, epinephrine, and norepinephrine. They inhibit the enzyme catecholamine-O-methyl-transferase.   COMT
C, D, E. They facilitate gluconeogenesis by causing lipid, protein, and amino acid catabolism and inhibiting the action of insulin (see question 4, chapter 24).

24

(1:908; 2:7-56; 3:297; 4:933; 5:1563)
Ebca 2

Which one of the following statements is incorrect? The *concentration* of *cortisol* in the plasma increases

A. during pregnancy
B. when there is an increased release of CBG by the liver
C. when there is an increased release of ACTH by the pituitary gland
D. when there is an increased release of estrogen into the blood
E. during nephrosis

Answer and Explanation
A, B, C, D. Most of the cortisol in the plasma is bound to corticosteroid-binding globulin (CBG = transcortin) and albumin. The bound material adds to the plasma concentration but is not active. In other words, it does not depress the release of ACTH nor bring about gluconeogenesis. It acts as a reservoir. Increases in the release of CBG by the liver are brought about by estrogen and are noted during pregnancy. Increased cortisol binding causes increased ACTH release and increased cortisol production. This can be summarized as follows:

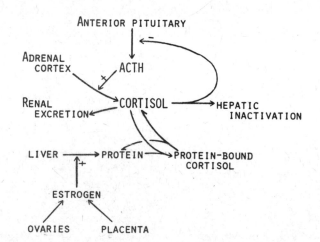

E. During nephrosis there may be a loss of cortisol and CBG in the urine. This will decrease the concentration of cortisol in the plasma.

25

*(1:1022; 2:7-38;
3:258; 4:933;
5:1503)*
Ebca 2

Under what circumstances does a decrease in *PBI* occur that is not associated with a decreased *metabolic rate*? In response to

A. androgens, because they decrease the concentration of TBG
B. dilantin, because it ties up receptor sites on TBG
C. a decreased production of TBPA and albumin
D. all of the above conditions
E. none of the above conditions

Answer and Explanation

D. When there is a decrease in bound thyroid hormones but no change in free thyroid hormones, there will be no change in metabolic rate if all other factors remain unchanged. The metabolic rate is controlled by the level of unbound thyroxine (T_4) and triiodothyronine (T_3). Protein-bound T_4 and T_3 do not affect metabolic rate or the release of TSH (thyroid-stimulating hormone) from the pituitary:

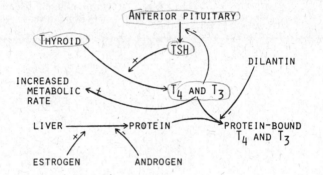

In the average, healthy individual approximately 50% of the T_4 and 60% of the T_3 is carried by TBG (thyroxine-binding globulin). T_4 is also bound to TBPA (thyroxine-binding prealbumin) and albumin, approximately 0.1% being free in the plasama.

26

*(1:1031; 2:7-44;
3:264; 4:939;
5:1502)*
EWa 3

Which one of the following statements is incorrect? In Graves' disease there is usually

A. an enlargement of the thyroid gland
B. a release of LATS, which is an immunoglobulin
C. a release of LATS, which causes an increased release of TSH
D. a release of LATS, which causes an increased release of thyroid hormone
E. a release of LATS, which, unlike TSH, crosses the placental barrier

Hyper thyroidism

Answer and Explanation

A, B, C, D. Graves' disease is a form of *hyperthyroidism* caused by a group of antibodies against TSH (thyroid-stimulating hormone) receptors in the thyroid. These antibodies (thyroid-stimulating immunoglobulins: TSI) include a long-acting thyroid stimulator (LATS). They are produced by lymphocytes and have an effect on the thyroid similar to that produced by *TSH* but, unlike TSH, their production and secretion are not inhibited by *thyroid hormones*. Since LATS causes an increased plasma concentration of thyroid hormones, it will cause a decreased release of TSH:

(Continued)

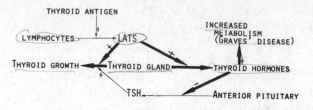

Comment: The concept of autoimmune disease is an important development of modern medicine. It was generally believed that one's own proteins cannot be antigenic. In Burnet's words, the body has a "knowledge" of "self" and "not self." It is now suggested, however, that proteins sequestered or segregated from the antibody-producing system during the early years, when this system is defining "self" and "not self," may leak out later and cause an autoimmune reaction. In 1912, Hashimoto described a disease (lymphocytic thyroiditis) in which antibodies were produced against certain thyroid molecules. In addition, certain collagen diseases (systemic lupus erythematosus and rheumatoid arthritis) and acquired hemolytic anemia involve autoimmunity in their etiology.

27
Thyroid and Parathyroid Glands

1

(1:1022; 2:7-33;
3:256; 4:931;
5:1507)
Ebc 1

3,5,3′—triiodothyronine (T_3) is different from thyroxine (T_4) in that

A. T_3 is less potent than T_4
B. over 70% of T_3 is produced outside the thryroid gland from T_4
C. has a half-life of 7 days, whereas T_4 has a half-life of 1 day
D. most of the T_3 in the plasma is bound to protein
E. the total T_3 concentration in the plasma is more than 60 times that for T_4

Answer and Explanation
A. T_3 is about three times more potent than T_4.
B. T_3 is also produced inside the *thyroid gland* from monoiodotyrosine (MIT).
C. T_4 has a half-life of 7 days and T_3 one of 1 day.
D. E. T_4 serves as a large reservoir of *prohormone* for T_3, as well as having some biologic activity of its own. There is more than 60 times more total T_4 in plasma than T_3. Eighty percent of it is *bound to protein*, whereas only 30% of T_3 is bound to protein.

2

(1:1030; 2:7-36;
3:265; 4:934;
5:1503)
EWa 2

Which one of the following statements is most correct? *Propylthiouracil* prevents the formation of thyroxine (T_4) and triiodothyronine (T_3) from tyrosine and I ; it therefore

A. causes a decreased production of TSH (thyroid-stimulating hormone)
B. tends to produce an increased metabolic rate
C. tends to prevent goiter
D. tends to do all of the above
E. tends to produce edema in the adult

Answer and Explanation
A, B, C, D, E. This goitrogen, by inhibiting the production of thyroid hormones (T_3 and T_4), increases the production of TSH:

(Continued)

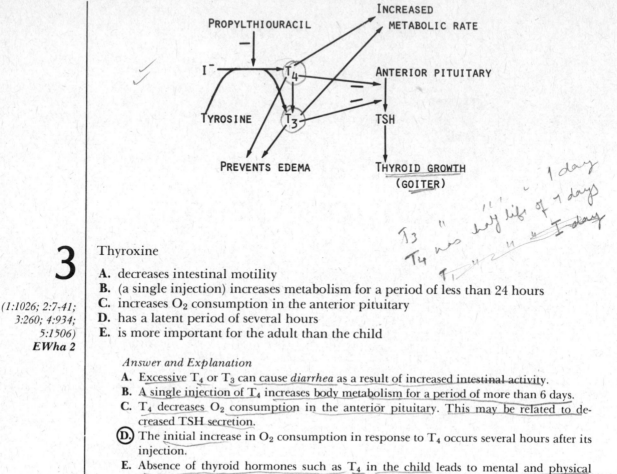

T3 "" body life of 7 days   1 day
T4 " " " " 1 day
T1

3

(1:1026; 2:7-41;
3:260; 4:934;
5:1506)
EWha 2

Thyroxine

A. decreases intestinal motility
B. (a single injection) increases metabolism for a period of less than 24 hours
C. increases O_2 consumption in the anterior pituitary
D. has a latent period of several hours
E. is more important for the adult than the child

Answer and Explanation

A. Excessive T_4 or T_3 can cause *diarrhea* as a result of increased intestinal activity.
B. A single injection of T_4 increases body metabolism for a period of more than 6 days.
C. T_4 decreases O_2 consumption in the anterior pituitary. This may be related to decreased TSH secretion.
D. The initial increase in O_2 consumption in response to T_4 occurs several hours after its injection.
E. Absence of thyroid hormones such as T_4 in the child leads to mental and physical retardation (*cretinism*). Their absence in the adult leads to *myxedema*.

4

(1:1026; 2:7-74;
3:261; 4:934;
5:1463)
EHc 2

Triiodothyronine (T_3)

A. decreases the synthesis of m-RNA and the activity of myocardial adenylate cyclase
B. decreases the sensitivity of adipose tissue to the fat-mobilizing action of epinephrine
C. has a negative chronotropic action on the heart
D. elevates circulating cholesterol levels
E. increases the activity of Na^+ pumps

Answer and Explanation

A, B, C. T_3, by increasing *adenylate cyclase* activity, exerts a *permissive action* on the response of many membranes (adipose tissue, heart, etc.) to epinephrine, as well as other agents.
D. T_3 increases *cholesterol* levels.
E.

5

(1:1020; 2:7-39;
3:257; 4:937;
5:1502)
Eh 2

Thyroid-stimulating hormone (TSH)

A. release is greater in warm-adapted than cold-adapted subjects
B. release is increased during physical trauma and emotional stress
C. does not appreciably affect iodine trapping by the thyroid gland
D. increases the production of thyroglobulin by the thyroid gland
E. has all of the above characteristics

Answer and Explanation
A. Release of TSH is less in warm-adapted subjects.
B. It is less under these conditions.
C, (D.) TSH apparently directly increases the removal of I⁻ from the plasma and the production of thyroglobulin by the thyroid.

6

(1:1020; 2:7-39;
3:181; 4:920;
5:1480)
ENh 2

The *hypophyseal portal circulation*

A. is a series of vessels connecting the capillaries of the hypophysis with the capillaries of the pituitary
B. carries the tripeptide TRF to the anterior pituitary
C. if destroyed, causes a cessation of the secretion of TSH
D. if destroyed, prevents the release of TSH in response to a cold stress
E. statements C and D are both correct

Answer and Explanation
A, B. This circulation carries thyrotropin-releasing factor *(TRF)* to the anterior pituitary.
(C.) D. *TSH* is released even from a pituitary transplanted onto the renal capsule. The transplanted pituitary does not, however, increase in secretion of TSH in response to cold but will decrease its secretion in response to T_4 and T_3.

7

(1:1030; 2:7-45;
3:263; 4:941;
5:1506)
EWca 2

Which one of the following symptoms is not characteristic of *hypothyroidism* in the adult?

A. a below normal mouth temperature
B. a feeling of drowsiness
C. increased body hair
D. a deposition of mucoprotein in the subcutaneous and extracellular spaces that causes edema
E. a negative nitrogen balance

Answer and Explanation
(C.) The *hair* becomes coarse in about 76% of the cases, and there is a loss of hair in about 57% of the cases of hypothyroidism.
D, E. In hypothyroidism, both the synthesis and the degradation of *protein* is reduced. The net result is that more nitrogen is lost from the body than is taken in. On the other hand, there are accumulations of extracellular proteins in a number of areas, which cause increases in *colloid osmotic pressure* in those areas. The mechanisms responsible for these changes are poorly understood.

8

(1:1030; 2:7-45;
3:263; 4:941;
5:1473)
EMha 1

Which one of the following symptoms is not characteristic of *hypothyroidism* in the adult?

A. dry skin
B. muscle weakness
C. exophthalmos
D. a deterioration of the thermoregulatory responses to cold
E. a decreased ability to sweat in response to warming

Answer and Explanation
    C. *Exophthalmos* produced by a factor in the blood of individuals with hyperthyroidism. It may be a derivative of TSH.

9

(1:1030; 2:7-45;
3:263; 4:942;
5:1511)
EGSa 2

Which one of the following statements is most incorrect? *Cretinism*

A. is caused by prolonged, untreated hypothyroidism in the infant
B. is associated with premature sexual development
C. is frequently associated with symptoms that are not totally reversible
D. is associated with mental retardation, potbellies, and enlarged protruding tongues
E. can be caused by an iodine-deficient diet

Answer and Explanation
    A. Untreated hypothyroidism that occurs in the child, as opposed to the infant, produces a less severe series of changes and is called juvenile hypothyroidism.
    B. Cretinism is usually associated with *amenorrhea and sterility.*
    C. The degree of success in treating cretinism with thyroid hormones is inversely related to its duration. If permanent mental retardation is to be prevented, treatment must be begun before the symptoms become marked.
    E. Dietary iodine is essential for the production of T_3 and T_4.

10

(1:1029; 2:7-41;
3:264; 4:939;
5:1493)
ENa 2

Untreated *goiter* is associated with

A. hyperthyroidism
B. hypothyroidism
C. euthyroidism
D. hyperthyroidism and hypothyroidism
E. hyperthyroidism, hypothyroidism, and euthyroidism

Answer and Explanation
    E. Goiter is an enlargement of the *thyroid*. It may occur in response to (1) a prolonged elevation in the *TSH* concentration of the plasma or (2) the production of *LATS*, an immunoglobulin. Some common causes of elevations in TSH are an *iodine*-deficient diet or the presence of *goitrogens* in the diet. In both cases, the goiter is associated with either *hypothyroidism* or euthyroidism. In *euthyroid* goiter due to iodine deficiency, the enlargement of the thyroid makes it possible for increased quantities of iodine to be removed from the blood, and thus it is a useful response to a stress. The production of LATS, unlike the production of TSH, is not decreased in response to thyroid hormones. Therefore, LATS produces a *hyperthyroid* goiter. See questions 25 and 26 in Chapter 26.

11

*(1:1029; 2:7-44;
3:264; 4:939;
5:1511)*
EWba 2

Hyperthyroidism causes

A. an increase in concentration of TBG, TBPA, and albumin
B. a positive nitrogen balance
C. a potentiation of the action of catecholamines
D. a decrease in urinary excretion of Ca
E. lethargy

Answer and Explanation

A. The increase in *PBI* in hyperthyroidism is due to an increase in the concentration of thyroid hormones, not to an increase in the thyroid-binding proteins.

Ⓒ

B, D. In both hypothyroidism and hyperthyroidism, there is a negative *nitrogen balance*. The catabolism of bone protein may in part contribute to the demineralization of *bone* and the resultant increase in urinary Ca seen in hyperthyroidism.

E. Hyperthyroidism causes *nervousness* and hyperreflexia. There is also a muscle *weakness* and tendency to fatigue easily that might be confused with lethargy.

12

*(1:1015; 2:7-33;
3:255; 4:932;
5:1501)*
Ewc 1

Most of the *thyroid hormone* is stored in the thyroid gland as

A. diiodotyrosine (DIT)
B. monoiodotyrosine (MIT)
C. thyroglobulin
D. parathormone
E. rT_3

Answer and Explanation

A, B. Both DIT and MIT are in higher concentration in the thyroglobulin molecule than T_3 and T_4 combined.

Ⓒ Normally, about 30% of the weight of the thyroid is *thyroglobulin*. In the absence of further synthesis, this would meet the needs of the body for 2 or 3 months.

E. Reverse T_3 (rT_3: 3,3',5'-triiodothyronine) is an inert substance that usually represents less than 1% of the iodothyronine secretion of the thyroid gland.

13

*(1:957; 2:7-48;
3:318; 4:986;
5:1544)*
EiMa 2

Which one of the following statements is **most correct?** A patient who, during surgery, has accidentally had all her *parathyroid tissue* removed will probably

A. develop skeletal muscle spasms
B. show improvement in response to vitamin D
C. show improvement in response to calcium gluconate injections
D. show improvement in response to vitamin D or calcium gluconate
E. have all of the above responses

Answer and Explanation

E. Hypoparathyroidism causes a reduction of plasma Ca^{++} and as a result a *hyperreflexia* that causes an increase in muscle tone. Vitamin D will cause an increase in absorption of calcium from the GI tract and a decrease in calcium in the feces.

14

(1:957; 2:7-48;
3:320; 4:981;
5:1545)
Eib 2

Parathormone (PTH) *polypeptid*

A. is a steroid
B. secretion is decreased in response to a decreased level of plasma Ca^{++}
C. inhibits the loss of phosphate by bone
D. is well characterized by all of the above statements
E. increases the formation of 1,25-dihydroxycholecalciferol (vitamin D_3)

Answer and Explanation

A. PTH is a linear polypeptide with a molecular weight of 9500.
B. An increase in plasma Ca^{++} causes a decrease in secretion of PTH. Calcium appears to be the primary determinant for the secretion of PTH.
C. PTH facilitates the loss of calcium and phosphate by bone.

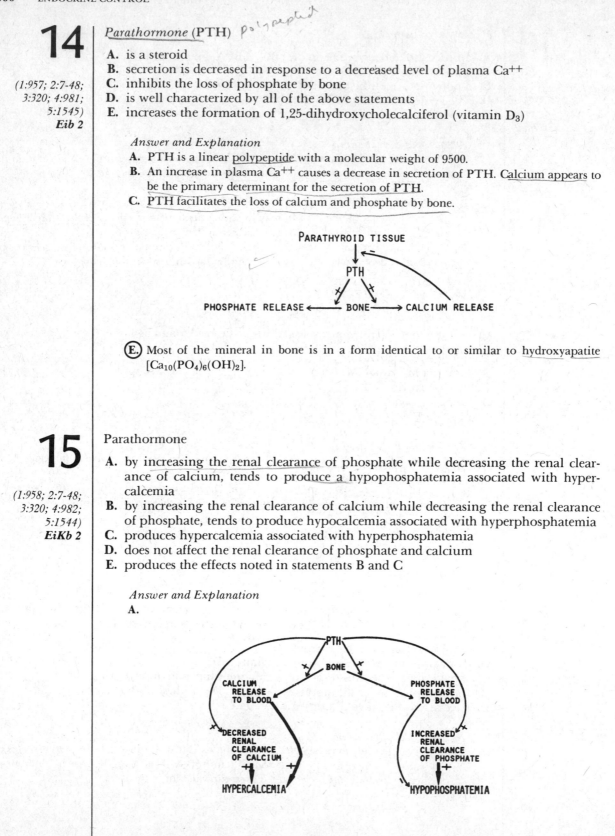

E. Most of the mineral in bone is in a form identical to or similar to hydroxyapatite $[Ca_{10}(PO_4)_6(OH)_2]$.

15

(1:958; 2:7-48;
3:320; 4:982;
5:1544)
EiKb 2

Parathormone

A. by increasing the renal clearance of phosphate while decreasing the renal clearance of calcium, tends to produce a hypophosphatemia associated with hypercalcemia
B. by increasing the renal clearance of calcium while decreasing the renal clearance of phosphate, tends to produce hypocalcemia associated with hyperphosphatemia
C. produces hypercalcemia associated with hyperphosphatemia
D. does not affect the renal clearance of phosphate and calcium
E. produces the effects noted in statements B and C

Answer and Explanation

A.

16

(1:957; 2:7-48; 3:320; 4:981; 5:1544)
EiKb 2

Increases in *parathormone* concentration can (select the one best answer)

A. increase the absorption of Ca^{++} from the intestine
B. increase the excretion of Ca^{++} in the urine
C. increase the release of Ca^{++} into the blood by the osteocytes and osteoclasts
D. cause all of the responses noted above
E. decrease blood Ca^{++}

Answer and Explanation

A. It probably performs this function by facilitating the synthesis of *1,25—dihydroxycholecalciferol* (a metabolite of vitamin D) by the kidney.
B. Frequently, the response to parathormone is a decrease in renal *clearance* of Ca^{++} and an increase in *plasma Ca^{++}* concentration, which collectively cause an increase in *excretion* of Ca^{++} in terms of mg of Ca^{++} per hour.
C. The osteocytes seem to be responsible for the early rise and the osteoclasts the maintained rise in plasma Ca^{++}.
Ⓓ

17

(1:960; 2:7-51; 3:322; 4:984; 5:1548)
Eiba 3

Copp and Davidson, in the early 1960s, perfused the thyroparathyroid apparatus of the dog with hypercalcemic *blood*. They found that the dog decreased its plasma Ca^{++} level in response to this procedure from 10 to 9.5 mg of plasma calcium per 100 ml within 15 minutes. *Parathyroidectomy*, on the other hand, produced the same change in plasma Ca^{++} after 3 hours. What conclusions would you draw from this experiment? How has this experiment changed our thinking concerning the control of plasma Ca^{++} concentration?

Answer and Explanation
Since the hypercalcemic infusion produced a more rapid fall in plasma Ca^{++} than is produced by the loss of PTH secretion, some other mechanism is contributing to the fall in plasma Ca^{++} under these circumstances. It is now recognized that this other mechanism is the release of a hypocalcemic factor, *calcitonin*, by the thyroid gland.

18

(1:961; 2:7-51; 3:320; 4:984; 5:1549)
Eib 2

Which one of the following statements is most correct? *Calcitonin* in mammals

A. is released by the C cells of the thyroid
B. is released in increased quantities in response to increases in plasma Ca^{++} and gastrin
C. inhibits the release of calcium from the bone into the blood
D. inhibits the action of parathormone on the bone
E. is characterized by all of the above statements

Answer and Explanation
E. The role of calcitonin in human beings is not settled. It may play a role in skeletal development in the young and it may protect a mother from demineralization of her bones during pregnancy and lactation. It has proven useful in the treatment of Paget's disease (osteitis deformans).

19

(1:966; 2:7-46;
3:322; 4:985;
5:1519)
Eib 2

List six or more hormones that act directly on bone to modify its *calcium reservoir*. Indicate whether each increases or decreases the calcium content of bone.

Answer and Explanation

- *Parathormone:* causes demineralization of bone by osteocytes and osteoclasts.
- *Calcitonin:* inhibits demineralization.
- *Insulin, growth hormone, thyroid hormone:* increase *collagen* synthesis (since bone acquires Ca by its deposition within collagen an increase in collagen will facilitate the calcification of bone).
- *Glucocorticoids* (cortisol, corticosterone, etc.): decrease collagen synthesis and can therefore cause demineralization.
- *Prostaglandins* PGE_1 and PGE_2: cause demineralization.
- *Estrogens* inhibit the action of parathormone on bone.

20

(1:966; 2:7-46;
3:313; 4:976;
5:1520)
EibN 2

Which one of the following statements is incorrect? The *calcium* in the plasma

A. exists bound to plasma proteins (about 50%) and in the free state (about 50%)
B. becomes more ionized (i.e., less bound) and therefore more active when the pH falls
C. becomes more active when the plasma phosphate level rises
D. has an action on the irritability of the cell that is approximately opposite to that of plasma K^+
E. facilitates a more forceful contraction of the heart and a greater release of acetylcholine, insulin, and salivary amylase

Answer and Explanation

(C) D. Ca^{++} and H^+, within limits, tend to decrease the *irritability* of the cell, whereas K^+ and $HPO_4^=$ tend to increase it.

E. Ca^{++} not only stabilizes many cell membranes but is part of the *activation-contraction coupling* and *activation-secretion coupling* processes in muscle, neurons, endocrine glands, and exocrine glands.

28
Hypothalamus, Pituitary, and Other Endocrine Glands

1

*(1:973; 2:7-17;
3:188; 4:920;
5:1484)*
ENSK 2

The *pituitary gland*

A. consists of a posterior lobe that is largely controlled by releasing hormones
B. is, in part, controlled by nervous connections with the brain
C. is well characterized by both of the above statements
D. synthesizes all of the hormones it releases
E. contains an anterior lobe, the master gland, that is concerned solely with the control of other endocrine glands (the thyroid, ovary, and adrenal cortex, for example)

Answer and Explanation

A, **B.** The *releasing hormones* play an important role in controlling the secretions of the anterior pituitary. The posterior pituitary is controlled largely by nervous connections with the hypothalamus.

D. *Oxytocin* and *antidiuretic hormone* (ADH = vasopressin) are *synthesized* in the *hypothalamus.*

E. *Growth hormone* (GH) facilitates the cellular uptake of amino acids in endocrine and nonendocrine structures by a direct action. It facilitates the release of fatty acids by adipose tissue, through a direct anti-insulin action.

2

*(1:976; 2:7-13;
3:346; 4:1034;
5:1623)*
ESa 2

If the *pituitary* of the rat were *removed* and implanted on the kidney, the plasma concentration of

A. adrenocorticotrophic hormone would increase
B. follicle-stimulating hormone would increase
C. growth hormone would increase
D. prolactin would increase
E. thyrotropic hormone would increase

Answer and Explanation

D. The *hypothalamus* produces *releasing hormones* that control the secretion of hormones from the anterior pituitary. The net effect of each of the releasing hormones, except those controlling *prolactin* secretion is to increase secretion. Therefore, when the anterior pituitary is removed from the influence of the hypothalamus and its PIH (*prolactin inhibitory hormone*), it releases more prolactin and less of its other secre-

(Continued)

prolactin inhibiting hormone

tions. Apparently what has been called PIH is the *dopamine* secreted by the tubero-infundibular neurons into the *portal hypophyseal vessels*. Hormones that increase prolactin secretion include PRH (*prolactin-releasing hormone*), TRH (*thyroid-stimulating hormone-releasing hormone*), and *estrogens*.

3

(1:979; 2:7-13; 3:323)
EKai 1

In human beings, total *hypophysectomy* does not cause

A. a decreased secretion of cortisol
B. a decreased secretion of testosterone
C. cessation of the menstrual cycle
D. a decrease in the hypoglycemic action of insulin
E. sterility in the male

Answer and Explanation

A. The most important action of pituitary ACTH is to stimulate the secretion of *cortisol* and other glucocorticoids.
B. Pituitary LH stimulates the secretion of testosterone.
C. Pituitary FSH and LH regulate the menstrual cycle.
(D.) There will be an exaggerated sensitivity to insulin due to loss of pituitary ACTH and growth hormone.
E. Pituitary FSH stimulates spermatogenesis.

4

(1:1057; 2:7-62; 3:309; 4:945; 5:1588)
EKGa 2

Total *adrenalectomy* is fatal in human beings, but hypophysectomy is not. This is because

A. ADH from the pituitary and aldosterone from adrenals are antagonistic (i.e., hypophysectomy causes a decreased production of both ADH and aldosterone)
B. GH and cortisol are antagonistic
C. aldosterone secretion is not markedly reduced after hypophysectomy
D. cortisol secretion is not markedly reduced after hypophysectomy
E. the adrenal cortex hypertrophies after hypophysectomy

Answer and Explanation

A. Both of these hormones reduce urine volume. They are not antagonistic.
B. GH and cortisol have antagonistic actions on protein metabolism. On the other hand, aldosterone injections are more effective in the prevention of death due to adrenalectomy than cortisol injections.
(C.) D. Hypophysectomy causes a marked decrease in the release of cortisol but little or no change in the secretion of aldosterone.
E. Parts of it atrophy.

5

(1:994; 2:7-20; 3:330; 4:7-156, 5:1482)
Ebl 1

Growth hormone (GH)

A. is not produced in physiologically important quantities in the nonpregnant adult
B. secretion is stimulated by the somatostatin released by the hypothalamus and the somatomedins released by the liver
C. secretion is increased in response to cortisol
D. secretion is decreased during a hypoglycemia caused by fasting
E. secretion is increased during psychologic stresses

Answer and Explanation

A. GH is most important prior to puberty, but also plays an important role in the normal adult. His daily production of GH is about 0.2 to 1.0 mg/day.

B, C, D, (E.) GH secretion is decreased by *somatostatin*, *somatomedins*, and *cortisol*; and increased by *hypoglycemia* and psychologic *stress*.

6

*(1:932; 2:7-148;
3:300; 4:968;
5:1476)*
EGWbl 2

Match each of the hormones listed in Column I with the one most appropriate listing of actions in Column II. Place your answer in the space provided in Column I.

Column I (Hormone)

1. _____ Growth hormone (GH)

2. B _____ Insulin *B*

3. _____ Glucagon *D*

4. _____ Cortisol

↑Glucose
GH, Insulin → +ive N Balance
Cortisol + glucagon   Neg. N. Balance

Column II (Actions)

A. Increases blood sugar
   Increases gluconeogenesis in the liver
   Increases ketone body formation
   Cortisol Increases lipolysis in adipose tissue
   Causes a negative N balance
   And chemically similar hormones are essential for surviving a severe stress
   And chemically similar hormones prevent water intoxication in response to a severe water load

B. Decreases blood sugar
   Decreases gluconeogenesis in the liver
   2. Insulin Decreases ketone body formation
   Decreases lipolysis in adipose tissue
   Increases glycogenesis in the liver
   Causes a positive N balance

C. Increases blood sugar
   Increases gluconeogenesis in the liver
   GH Increases ketone body formation
   Increases lipolysis in adipose tissue
   Causes a positive N balance

D. Increases blood sugar
   Increases gluconeogenesis in the liver
   Glucagon Increases ketone body formation
   Increases lipolysis in adipose tissue
   Causes a negative N balance

Answer and Explanation

1. **C.** *GH* is the only substance listed that increases *blood sugar* and causes a positive *N balance*. While preventing the catabolism of protein and amino acids, GH increases the availability of *glucose*, *lipids*, and *ketone bodies* in the blood.

2. **B.** *Insulin* is the only substance listed that decreases blood sugar. It increases the movement of blood sugar into most cells and in so doing increases CHO catabolism and the formation of *glycogen*. Associated with this increased dependence on CHO as a source of calories, there is a decreased concentration of ketone bodies and *free fatty acids* in the blood and a decreased *protein* catabolism.

3. **D.** *Glucagon* and *cortisol* both increase blood sugar, cause a negative N balance, and are secreted in increased quantities during stress. Glucagon, on the other hand, is not essential for surviving a severe stress or a severe water load. For a comparison between glucagon, epinephrine, and insulin see Chapter 26, question 22.

4. **A.** In the following diagram we note the role of CRF (*corticotropin-releasing* factor or *hormone* = **CRH**) and ACTH (*adrenocorticotropic hormone*) in the control of metabolism.

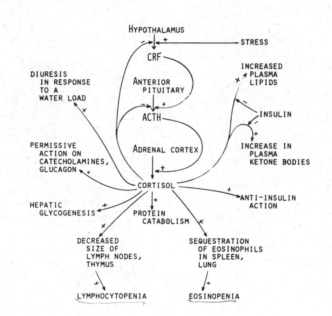

7

(1:1033; 2:7-61; 3:290; 4:953; 5:1558)
EiWa 3

The *adrenal gland* can be divided into four functional units: (1) the capsule, (2) the zona glomerulosa, (3) the zona fasciculata and reticularis, and (4) the medulla. The hypothalamus plays a major role in the control of (select the one best answer)

A. only the zona glomerulosa
B. only the zona fasciculata and reticularis
C. only the medulla
D. the zona glomerulosa and medulla
E. the zona fasciculata and reticularis and the medulla

Answer and Explanation
E. These conclusions are discussed after question 8.

8

(1:1043; 2:7-61; 3:290; 4:953; 5:1561)
EiWa 2

The *pituitary* plays a major role in the control of *adrenal* secretion by (select the one best answer)

A. only the zona glomerulosa
B. only the zona fasciculata and reticularis
C. only the medulla
D. the zona glomerulosa and the medulla
E. the zona fasciculata and reticularis and the medulla

Answer and Explanation

A. All the zones of the cortex release *glucocorticoids* and sex hormones, but the zona glomerulosa is the major source for *aldosterone*. It is the only zone in the cortex that does not atrophy within the first 6 weeks of *hypophysectomy*.

B. These areas show marked atrophy after hypophysectomy. This apparently results from the lack of ACTH secretion by the anterior pituitary. The release of ACTH is controlled by a hormone released by the *hypothalamus*, corticotropin-releasing hormone (CRH).

C. The hypothalamus controls the *adrenal medulla* through internuncial neurons impinging on the preganglionic sympathetic neurons in the thoracolumbar cord. Hypophysectomy produces little or no change in the adrenal medulla.

9

(1:1034; 2:7-54; 3:294; 4:953; 5:1558)
EcWi 2

Which one of the following includes a substance or substances not secreted by the *adrenal cortex?*

A. cortisol and corticosterone
B. aldosterone
C. glucagon
D. deoxycorticosterone and 11-deoxycortisol
E. estradiol, progesterone, and androgens

Answer and Explanation

C. Glucagon is the only substance listed that is not a steroid. Some of the other substances listed play little or no role in the control of body activity but are important in certain pathologic states in which there is a hypersecretion (*feminizing* or *masculinizing*, tumors of the cortex, for example).

10

(1:1047; 2:7-54; 3:295; 4:948; 5:1558)
EWb 1

Which one of the following includes a substance or substances with potent *glucocorticoid* activity?

A. cortisol and corticosterone
B. aldosterone
C. dehydroepiandrosterone
D. deoxycorticosterone and 11-deoxycortisol
E. estradiol and progesterone

Answer and Explanation

A. Cortisol has four times the potency of aldosterone in facilitating *hepatic glycogenesis* and exists at a higher concentration in the blood. It is the body's major glucocorticoid. *Corticosterone* is less potent than *cortisol* and more potent than *aldosterone*. The cortex secretes 7 times as much cortisol per day as corticosterone and 130 times more cortisol than aldosterone.

11

(1:1052; 2:7-56; 3:299; 4:954; 5:1558)
ESc 1

Which one of the following includes a substance or substances with a *masculinizing* action?

A. cortisol and corticosterone
B. aldosterone
C. dehydroepiandrosterone
D. deoxycorticosterone and 11-deoxycortisol
E. estradiol and progesterone

(Continued)

Answer and Explanation

C. This *androgen* is a precursor of *testosterone*, which, in turn, is a precursor of *estradiol*. It is produced in both the male and female *adrenal cortex* and has less potent androgenic activity than testosterone. Its physiologic importance, except as a precursor of other substances, is unknown.

12

(1:1045; 2:7-60; 3:302; 4:950; 5:1567)
EbW 2

Glucocorticoids in the blood

A. are usually at their highest concentration in the evening
B. decrease the total white blood cell count
C. increase growth in lymphoid tissue and facilitate the immune response
D. perform all of the above functions
E. play a role opposite to those mentioned above

Answer and Explanation

A. They usually approach 0 concentration in the evening and maximum concentration in the morning after waking. An individual who moves from the United States to Japan will take about a week to change his intrinsic pattern of secretion of glucocorticoids. It is also of interest to note that glucocorticoid therapy is more potent when administered at midnight than at noon.
B. They increase the concentration of neutrophils, platelets, and erythrocytes and decrease the concentration of eosinophils, basophils, lymphocytes, and plasma cells. The overall effect is to increase the total white cell count.
C. They are used in organ transplants as immunosuppressants.
E.

13

(1:926; 2:7-79; 3:7-80; 4:970; 5:1642)
EWla 2

Three different patients receive the same oral dose of glucose. The following *glucose tolerance curves* are obtained from them:

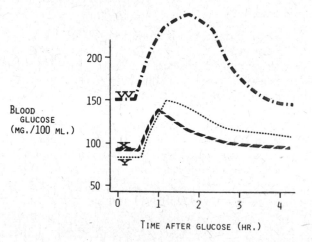

Use graphs W, X, and Y to complete (i.e., fill in the blanks) statements 1 through 3.

1. ___ is a normal response because the blood sugar level _____ .
2. ___ probably has *diabetes mellitus* because the blood sugar level _____ .
3. ___ probably has severe liver damage because the blood sugar level _____ .

Answer and Explanation

1. **X.** The sugar level remains elevated for a short period.
2. **W.** The sugar level initially is high.

3. Y. The sugar level initially is normal, but remains elevated in response to glucose for a prolonged time. The *liver* damage has slowed the rate that *insulin* is destroyed.

14

(1:923; 2:2-58; 3:269; 4:959; 5:1638)
EWba 2

The condition of *diabetes* was described in an Egyptian papyrus in 1500 B.C. In the second century, Aretaeus named it diabetes, meaning to flow through a siphon, and it was described as "a strange disease which consists of the flesh and bones running together in the urine." In 1674, Sir Thomas Willis noted that the urine of diabetes is sweet to the taste and added the term mellitus, or honeyed. In 1889, von Mering and Minkowski noted an unusual number of flies attracted to the urine of pancreatectomized dogs and reported the similarity between diabetes mellitus in human beings and the condition they had produced in dogs. DeMeyer, in 1909, suggested that the hypothetic antidiabetic substance produced by the islets of Langerhans be called insulin. Unfortunately, experimentally-produced diabetes in dogs could not be alleviated by feeding raw pancreas or injecting aqueous solutions of pancreatic extracts. Why is this the case?

Answer and Explanation
The proteolytic enzymes in the *pancreas* destroy the biologic activity of *insulin*.

15

(1:923; 2:7-73; 3:27; 4:959; 5:1638)
EWba 2

In 1921, Frederick *Banting* and Charles *Best* used an important technique and made a monumental discovery. What was this technique and what was the discovery?

Answer and Explanation
They ligated the ducts from the pancreas to the duodenum and waited until the exocrine portion of the pancreas was depleted of enzymes in response to the ligation. They were then able to extract a biologically active insulin from the pancreas. This extract prevents diabetes in response to *pancreatectomy*.

16

(1:932; 2:7-83; 3:279; 4:965; 5:1659)
EWa 2

In 1923, Bernardo A. Houssay noted that the symptoms of *pancreatic diabetes* could be ameliorated by the removal of an organ from the body. In 1936, Long and Lukens noted that the symptoms of diabetes could be ameliorated by the removal of a different organ. What were these organs and what principles do they release that might explain their tendency to produce a diabetes-like condition?

Answer and Explanation
The *Houssay animal* is one in which both the pancreas and the *pituitary* have been removed. The *Long-Lukens animal* is one in which both the pancreas and *adrenals* have been removed. In the latter, *cortisol* is the diabetes-exaggerating principle. It increases plasma lipids and ketone bodies and facilitates protein catabolism, all actions that insulin antagonizes. In the former case, *ACTH* and *GH* are the diabetes-exaggerating principles. GH increases blood sugar, fatty acids, and ketone bodies and facilitates hepatic gluconeogenesis, all actions that insulin antagonizes.

17

(1:1009; 2:5-47; 3:272; 4:928; 5:1638)
EuNa 2

Which one of the following statements is incorrect? In *diabetes insipidus*

A. and diabetes mellitus, we are dealing with conditions in which there is an inadequate production of a sulfur-containing peptide or polypeptide
B. and diabetes mellitus, we are dealing with conditions in which there is polyuria
C. and diabetes mellitus, we are dealing with conditions in which there is glycosuria

(Continued)

D. we are dealing with a condition that can be produced by destruction of the supraoptic nuclei, the hypothalamohypophyseal tract, or the posterior pituitary

E. there may be an insensitivity of the kidney to ADH

Answer and Explanation

A. ADH is a peptide containing nine amino acids and a disulfide bond.

B, C. In diabetes mellitus, the glucose in the urine causes an *osmotic diuresis*. In diabetes insipidus, the urine is virtually devoid of glucose and is markedly hypotonic. It is due, at least in part, to a low permeability of the collecting ducts to water in the renal medulla.

E. This is a rare form of the disease called nephrogenic diabetes insipidus.

18

(1:923; 2:5-85; 3:282; 4:959; 5:1660)
EWbc 2

Some preparations of commercially produced *insulin* have been found to cause increases in blood sugar prior to causing decreases in the *blood sugar*. This was because

A. insulin, like epinephrine, has different actions at different concentrations

B. it is not until the disulfide bridges of insulin are ruptured that insulin is capable of lowering blood sugar levels

C. both of the above statements are operational

D. insulin in some preparations was contaminated with norepinephrine

E. insulin in some preparations was contaminated with glucagon

Answer and Explanation

B. Rupture of the *disulfide bridges* causes a loss of biologic activity.

E. This problem does not occur if one uses synthetic insulin. On the other hand, insulin extracted from the pancreas has been frequently contaminated with *glucagon*.

19

(1:930; 2:7-80; 3:283; 4:967; 5:1663)
EWbc 2

Glucagon

A. levels in the blood decrease within 30 minutes after a meal and increase during fasting

B. concentrations in the blood are elevated in diabetes mellitus

C. is well characterized by statements A and B

D. is a polypeptide which is larger than insulin

E. and insulin both facilitate hepatic glycogenesis

Answer and Explanation

C.

D. Human insulin (molecular weight = 5734) contains 51 and glucagon (molecular weight = 3485) contains 29 amino acids.

E. They have opposite actions on hepatic glycogenesis. See question 22 in chapter 26.

20

(1:937; 2:7-80; 3:282; 4:967; 5:1663)
ERWb 3

In question 19, it was noted that *glucagon* levels increase during *fasting* and during *diabetes mellitus*. In fasting, there is *hypoglycemia*, and in diabetes mellitus, there is *hyperglycemia*. It is frequently stated that glucagon release is initiated by hypoglycemia. Either justify these statements or develop a better alternative. How is the relationship between (1) blood sugar levels and glucagon release similar to the relationship between (2) *plasma* Po_2 and the stimulation of peripheral *chemoreceptors*?

Answer and Explanation

Glucagon release is indirectly related to glucose availability to the alpha cells of the pancreas. In both diabetes mellitus and fasting, there are reduced quantities of glucose available to the pancreatic cells. In the case of diabetes, there is hyperglycemia, but transport of glucose into the cell is reduced. This relationship is similar to histotoxic hypoxia. After the injection of KCN, for example, the plasma Po_2 ranges from normal to elevated, but since this O_2 cannot be utilized by the cell, the cell responds as if there were a decreased Po_2. In other words, in both histotoxic hypoxia and diabetes mellitus, there are reduced quantities of an AVAILABLE nutrient but not the nutrient itself.

21

(3:323; 4:953; 5:1478)
Ea 1

Melanocyte-stimulating hormone (MSH)

A. is produced by the pituitary gland
B. and ACTH produce hyperpigmentation
C. and adrenal insufficiency due to primary adrenal disease produce hyperpigmentation
D. is well characterized by all of the above statements
E. and hypopituitarism produce hyperpigmentation

Answer and Explanation
D. MSH produces increases in pigmentation in the human subject by causing a dispersal of melanophore granules within the cells of the body. In adrenal insufficiency a decreased release of glucocorticoids causes an increased release of ACTH.
E. In hypopituitarism there may be a decreased release of ACTH, which will cause pallor.

22

(1:1003; 3:188; 4:927; 5:1484)
Enc 1

Antidiuretic hormone (ADH) and *oxytocin* are similar in that

A. they are both synthesized and bound in the hypothalamus with a carrier protein called neurophysin
B. they migrate from the hypothalamus to the posterior pituitary in the axoplasm of neurons which have their cell bodies in the hypothalamus
C. they are stored in greatest quantity in the axon terminals in the posterior pituitary
D. they share all of the above characteristics
E. none of the above statements are correct

Answer and Explanation
C. Less than 5% of the total hormone produced by the hypothalamus is found there. On the other hand, the pituitary has enough ADH stored to meet an individual's needs for it for over a week. The posterior pituitary is a good area for storage because it is well equipped for rapid release of hormones into the blood, since, unlike the hypothalamus, it has vessels lacking a *blood-brain barrier*.
D.

23

(1:1007; 2:7-30; 3:190; 4:927; 5:1486)
EKo 1

Under resting conditions, the *osmotic pressure* of the plasma is 290 mOsm. If it decreases to 280 mOsm

A. there will be a reduced stimulation of the supraoptic neurons from the hypothalamus to the posterior pituitary
B. ADH secretion will decrease and, as a result, urine volume will increase
C. the events noted above will both occur
D. ADH secretion will decrease and as a result, systemic peripheral resistance to blood flow will decrease
E. the events noted above will all occur

(Continued)

Answer and Explanation

C.

D. ADH is sometimes called *vasopressin*. In large doses, vasopressin produces vasocon- striction, but it is doubtful that a healthy person ever produces enough to cause changes in peripheral resistance. Its major action is to decrease water loss in the urine.

24

(1:867; 2:4-23;
3:372)
EKib 2

The *kidney* probably releases hormones or hormone-like agents

A. in response to decreases in renal arterial pressure or constriction of renal artery
B. that indirectly facilitate the release of aldosterone
C. that facilitate the production of erythrocytes
D. that are characterized by each of the above statements

Answer and Explanation
D.

25

(1:867; 2:7-62;
3:372; 4:947;
5:1591)
EKp 1

What hormone does the *kidney* release in response to a decrease in renal arterial pressure, which indirectly facilitates the secretion of *aldosterone*?

Answer and Explanation
Renin.

```
        LIVER                           KIDNEY

                                          RENIN
        PLASMA ALPHA-2-GLOBULIN                   ANGIOTENSIN I
                                                       ACE
        ADRENAL CORTEX                            ANGIOTENSIN II
                       +

        ALDOSTERONE
```

26

(1:733; 2:4-23;
3:373; 4:58; 5:1130)
Ebg 1

What hormone-like substance does the kidney release in response to hypoxia, which facilitates the production of red blood cells?

Answer and Explanation
Renal erythropoietic factor (REF = erythrogenin).

```
        LIVER          KIDNEY      ACCESSORY TISSUE
                HYPOXIA    +
        GLOBULIN      ERYTHROPOIETIC FACTOR
                                          ERYTHROPOIETIN

              RED BONE MARROW              FORMATION OF PROERYTHROBLASTS
```

29
Fertilization, Growth, and Maturation

1

Fertilization of the ovum usually occurs

A. in the uterus
B. up to 5 days after ovulation
C. in the uterus up to 5 days after ovulation
D. 6 days before implantation
E. in the uterus up to 5 days after ovulation and 6 days before implantation

(1:1106; 2:7-118; 3:365; 4:1021; 5:1619)
ESG 1

Answer and Explanation
A. It usually occurs in the ampullar portion of the *uterine* (fallopian) *tube*.
B. Fertilization usually must occur 6 to 24 hours after ovulation.
D.

2

Which of the following statements is incorrect? Human *spermatozoa* normally

A. are not motile in the epididymus but are activated by the secretions of the prostate and other accessory glands of the male
B. are produced at an accelerated rate when testicular temperature is raised from 35 to 37°C
C. contain 23 chromosomes
D. contain either an X or a Y chromosome but not both
E. can survive in the female reproductive tract for 1 or 2 days or longer

(1:1081; 2:7-106; 3:347; 4:992; 5:1624)
ESG 1

Answer and Explanation
B. The descent of the *testes* into the *scrotal sac* sends them into an environment about 2°C below the temperature in the abdomen. If they remain in the abdominal cavity or if their temperature in the scrotal sac is changed from 35 to 37°C, they will stop producing spermatozoa, and if this situation persists, degeneration of the tubular walls and sterility will eventually occur.
C. This is half of the *chromosomes* contained in most human cells.
D. If the fertilizing spermatozoan contains an X chromosome, the progeny will be female.
E. *Survivals* up to 6 days have been reported.

3

(1:1081; 2:7-106;
3:349; 4:995;
5:1624)
ESGf 1

Which one of the following statements is incorrect? Human *spermatozoa* normally

A. at a concentration of 1 million/ml of semen in the ejaculate will fertilize the ovum
B. contain enzymes in their head that facilitate the penetration of the ovum
C. once one has penetrated an ovum, initiate a process that prevents other spermatozoa from penetrating that ovum
D. are released in greater numbers in the ejaculate of a 21-year-old after 3 days of abstinence than after 1 day of abstinence
E. take about 74 days to change from a relatively undifferentiated spermatogonium into a mature spermatozoan

Answer and Explanation
A. There are normally 100 million sperm per ml of ejaculate. About half the men with counts between 20 and 40 million per ml are *sterile*. All with counts below 20 million per ml are sterile.

4

(1:1081; 2:7-106;
3:349; 4:994;
5:1619)
ESGf 1

Which one of the following statements is *incorrect?* Human *spermatozoa* normally

A. remain motile in the female reproductive tract for a period of less than 15 minutes after ejaculation during intercourse
B. leave the penis, suspended in a liquid (semen), most of which comes from the prostate gland and seminal vesicles immediately prior to and during ejaculation
C. derive from spermatogonia, cells in the seminiferous tubules that contain the diploid number of chromosomes
D. are stored in the seminiferous tubules, rete testis, epididymis, and vas deferens
E. may remain viable in the epididymus for periods up to 60 days

Answer and Explanation
A. They remain *motile* for periods in excess of 2 hours. It probably takes a spermatozoan over an hour to migrate up to the ovum. One sign of sterility in the male is spermatozoa with short-lived motility.
B. About 80% of the *seminal fluid* comes from the seminal vesicles and *prostate*.

5

(1:1074; 2:7-86;
3:342; 4:1021;
5:1955)
SGa 2

A child with a *genotype* of

A. XX/XY will develop into a true hermaphrodite
B. XY will contain a Barr body in its cells
C. XXX will develop excessively wide hips
D. XO will develop into a normal female
E. XXY will develop into a true hermaphrodite

Answer and Explanation
A. The true *hermaphrodite* has some cells that are female (XX) and some that are male (XY). It may contain both an ovary and a testis.
B. Females (XX) contain *Barr bodies* (usually seen as an extra quantity of chromatin attached to the nuclear membrane in the karyoplasm) in each of their nucleated cells (the cells of mucous membranes, for example).
C. She will develop normally.
D. In this condition (Turner's syndrome), the individual has the *external genitalia* of an immature female and gonads that are either rudimentary or absent (incidence = 0.03%).
E. He will develop the external genitalia of a male. He will have abnormal seminiferous tubules and, since he has two X chromosomes, will be *chromatin-positive* (i.e., have a Barr body). This condition (Kleinfelter's syndrome) has an incidence of 0.26% and is frequently associated with mental retardation.

6

(3:339; 4:1037; 5:1947)
Gbl 1

Which one of the following statements is most correct?

A. 2 months after fertilization, the egg is about 8 cm long and called a fetus
B. in the embryo, most of the blood cells are formed in the yolk sac
C. both of the above statements are correct
D. in the early stages of fetal life, most of the blood cells are formed in the liver and spleen
E. all of the above statements are correct

Answer and Explanation
E.

7

(1:647; 2:6-40; 3:526; 4:1044; 5:1977)
Glba 2

Which one of the following statements is incorrect? A *premature infant* is more likely than a full-term infant to

A. suffer from severe jaundice of hepatic origin
B. lack fontanelles   *soft spot in cranium*
C. suffer from anemia
D. lack surfactant
E. have difficulty regulating body temperature

Answer and Explanation
A. The newborn will frequently be *jaundiced.* This is probably because the glucuronide conjugating system of the *liver* has not fully developed. The jaundice will usually disappear by the 10th day of life. In the premature infant the problem is much more severe. His deficiency in hepatic glucuronyl transferase activity may lead to a sufficiently serious jaundice to produce central nervous system damage (kernicterus).
(B.) The bone that forms the cranium is membranous bone. It starts as a soft connective tissue membrane that becomes infiltrated with osteoblasts. In both the premature and mature newborn infant, calcification of this membrane will be incomplete. In other words, both will have soft spots called *fontanelles* in the cranium. The fontanelles normally have disappeared in the 18-month-old infant.
C. The normal infant is born with a *reserve of iron* that will last him about 6 months. This is important because cow's milk is deficient in iron. The premature infant, on the other hand, is frequently deficient in iron because these stores are laid down late in pregnancy.
D. This can lead to lung collapse (respiratory distress of the newborn).
E. Temperature regulating mechanisms mature in late pregnancy and early infancy.

8

(1:410; 2:4-95; 3:417; 4:76; 5:1133)
GbgK 1

In the *newborn*

A. the ability to manufacture antibodies is poorly developed and will not be comparable to that of the adult until approximately the age of 12 months
B. there is practically no resistance to infection
C. lymphocyte precursors have not yet been changed to T lymphocytes by the thymus
D. the brain cells are less tolerant of a lack of O_2 than in the adult
E. the mechanism for concentrating urine is more efficient than in the adult

Answer and Explanation
(A.)
B. The newborn has a good titer of 7S gamma globulins (IgC), which he has received from the maternal circulation. These will have disappeared by 10 weeks of age, but by this time he is producing his own. At birth, however, he does lack the heavier 19S

(Continued)

gamma globulins (IgM and IgA). Thus, he is susceptible to infection by gram-negative organisms but resistant to many other types of infection.

C. B lymphocytes (B: first noted in birds) produce antibodies. T lymphocytes (T: sensitized in the thymus) promote cellular immunity. Most of the T lymphocytes are produced by the thymus prior to birth and during the several months after birth. If the thymus is removed several months before birth, the body loses most of its ability to reject transplanted organs.

D. They are more tolerant of O_2 *lack*.

E. His *kidney* has one half the concentrating ability of the kidney in the adult.

9

(3:331; 4:988)
GD 1

Which one of the following statements is *incorrect*?

A. the first primary teeth appear at about 6 months and are usually the central and lateral incisors

B. there are a total of 32 primary teeth

C. the first permanent teeth appear at about 8 years and are the central incisors and the first molars

D. there are a total of 32 permanent teeth

E. the third molars appear between the ages of 17 and 21

Answer and Explanation

B. There are a total of 20 primary *teeth*. The last to appear are the second molars at between 16 and 24 months.

10

(1:1030; 2:7-42;
3:331; 4:940;
5:1475)
GSE 1

In the child at birth, the distance from the top of the pubic symphysis to the top of the head is 1.7 times the distance from the top of the pubic symphysis to the soles of the feet. In other words, the ratio between the *upper segment* and the *lower segment* is 1.7. Which one of the following statements is most correct?

A. in the normal adult, this ratio is about 1

B. in the untreated, markedly hypothyroid child, this ratio will remain well above 1 throughout life

C. in the normal adult, this ratio is 1, but in the hypothyroid child, it will remain well above 1

D. in the normal adult, this ratio is 1, but in the hypothyroid child and the eunuch, the ratio will remain well above 1

E. the ratio stays at approximately 1.7 in the normal subject

Answer and Explanation

Ⓒ

D. Loss of testicular function prior to puberty will, if untreated, result in the development of an adult lacking many of the secondary sexual characteristics of the male (muscular development, deep voice, facial hair, characteristic distribution of body fat, etc.) and may prolong the growth period. The upper segment/lower ratio will, however, stabilize at a value slightly less than 1 in later years.

11

(1:993; 2:7-156;
3:335; 4:926;
5:1475)
GEa 1

Acromegaly is a condition caused by

A. an excessive release of growth hormone, which causes, in the adult, enlarged hands and feet and protrusion of the lower jaw

B. an excessive release of growth hormone, which causes, in the adult, an elongation of the humerus, femur, tibia, and fibula

C. an excessive release of thyroid hormone, which causes, in the adult, enlarged hands and feet and a protrusion of the lower jaw

D. an excessive release of thyroid hormone, which causes, in the adult, an elongation of the humerus, femur, tibia, and fibula
E. an inadequate release of thyroid hormone

Answer and Explanation
A. The hypersecretion of GH that produces acromegaly occurs after most of the long bones have ceased to grow. It will not, therefore, cause increases in the length of the femur and humerus. It may increase the length of some bones that have not ceased to grow yet, but it will in all cases increase the size of soft tissue (the tongue and abdominal viscera, for example) and will increase the thickness of a number of bones. This latter effect is particularly obvious in the hands, feet (the subject may have to wear a shoe that is size 14 or larger), the nose (it may be twice normal size), and the mandible.

12

Define *adolescence.*

Answer and Explanation
Adolescence is the period during which the secondary sexual characteristics are developing and complete physical maturity is being obtained.

(1:1088; 2:7-156; 3:343; 4:1001)
GS 1

13

Which one of the following statements is most correct?

A. puberty usually occurs earlier in boys than in girls
B. the testes, seminal vesicles, and prostate, like the nervous system, gradually increase in size during the first 16 years
C. starting at about age 10, there is a gradual decrease in the growth rate until it ceases sometime between ages 14 and 21
D. menarche is an abrupt and prolonged cessation of menstruation
E. all of the above statements are incorrect

(1:1105; 2:7-94; 3:343; 4:1001; 5:1602)
GS 1

Answer and Explanation
A. The first *menstrual period* in girls occurs at about age 13 1/2 years. Boys first ejaculate *spermatozoa* at an average age of 15 years. Pubic hair appears in girls at about age 12 and in boys at about age 13.
B. The nervous system reaches its adult size at about age 10 years. At 1 year, a testis has a volume of 0.7 ml and at 8 years 0.75 ml, but after about age 10 the testes, prostate, and penis in boys, the breasts in girls, and other secondary sexual structures begin to grow rapidly. In the 19-year-old, for example, the testis occupies 16.5 ml.
C. Starting at about 12 years of age, the rate of *growth* rapidly increases and then stops at about 14 to 19 years of age. For example, at age 11 an average boy would weigh 36 kg and be 144 cm tall. At age 14, he would be 50 kg and 170 cm.
D. *Menarche* is the beginning of the menstrual cycle.
(E.)

14

Two years prior to the occurrence of *puberty*

A. the gonads are insensitive to pituitary gonadotropins
B. the pituitary and hypothalamus are devoid of gonadotropins
C. the concentration of adrenal androgens in the blood is markedly lower than in boys and girls at 15 years of age
D. all of the above statements are true
E. none of the above statements is true

(1:1088; 2:7-94; 3:332; 4:1001; 5:1629)
GSEa 1

(Continued)

Answer and Explanation

A, B. Prior to puberty the gonads are sensitive to *gonadotropins* and there are gonadotropins in the pituitary and hypothalamus, but they are not being released. Lesions in
the ventral hypothalamus near the infundibulum can cause precocious puberty in
children. This observation has led to the suggestion that, prior to puberty, a hypothalamic mechanism inhibits the release of gonadotropins.

C. The increase in *androgens* at puberty lead to the development of pubic and axillary
hair and may also contribute to the development of acne.

15

*(1:1088; 2:7-94;
3:343; 4:1001;
5:1481)*
GSEa 2

Puberty before the age of 10 years

A. may be caused by destruction of the pituitary
B. causes a delay in the closure of the epiphyseal plates
C. in over 80% of the cases is caused by pathology
D. is well characterized by the above statements
E. is not well characterized by any of the above statements

Answer and Explanation

A. Removal of the *pituitary* will prevent the onset of puberty. A lesion in the hypothalamus, on the other hand, may cause a premature puberty.

B. There would be a premature growth spurt followed by a premature closure of the
epiphyses. The individual who experiences premature puberty is usually shorter than
the average adult.

C. It has been estimated that puberty prior to 10 years of age is the result of a pathologic
condition in less than 25% of the cases. There is a wide range of normal for the onset of
puberty. Menarche, for example, frequently occurs between the ages of 9 and 17. There
is one case of a child who began menstruating at age 3, became pregnant at 4 years, 10
months, and was delivered by cesarean at 5 years, 7 months.

E.

16

*(1:1105; 2:7-11;
3:343; 4:1015;
5:1613)*
ES 1

During puberty the normal female experiences

A. an increased secretion of estrogen
B. an increased secretion of FSH
C. both A and B
D. a decreased secretion of LH
E. all of the above

Answer and Explanation

C. There is an increased secretion of *estrogen, follicle-stimulating hormone* (FSH), and
luteinizing hormone (LH).

17

*(1:1082; 2:9-60;
3:350; 4:996; 5:901)*
SNCG 1

Erection of the penis

A. occurs in response to a reflex stimulation of parasympathetic neurons that cause
arteriolar dilation in the penis
B. is initiated by afferent neurons from the glans penis or descending fibers in the
cord
C. may occur 5 to 10 years prior to puberty as well as in the elderly
D. is well characterized by all of the above statements
E. is not characterized by any of the above statements

Answer and Explanation

C. Erection has been noted at all ages ranging from 1 day old to 80 years old. In the very young, it is usually due to irritation of the genitourinary tract.

(D.)

18

(1:1083; 2:9-60; 3:351; 4:997; 5:1630)
SNfM 1

In a healthy 21-year-old man,

A. emission is caused by stimulation of sympathetic neurons that initiate the contraction of the smooth muscle of the epididymis, vas deferens, and seminal vesicles
B. ejaculation is caused by a spinal reflex involving both the lumbar and sacral cord, which initiates the contraction of a skeletal muscle, the bulbocavernosus muscle
C. emission is associated with the expulsion of seminal fluid from the prostate, the seminal vesicles, and ejaculatory duct
D. during the last stages of ejaculation, the urethra becomes insensitive, and further ejaculation is usually unnoticed
E. all of the above phenomena are characteristic of ejaculation

Answer and Explanation

(E.)

19

(1:557)
SG 1

Kinsey in 1956 listed the five most common causes of *ejaculation* in healthy 15-year-old boys. List each of these, placing the most frequent source first, the least frequent source last, etc. Define your terms.

Answer and Explanation

- *Masturbation:* self-stimulation of the penis, clitoris, or other structures that can bring about sexual arousal and orgasm.
- *Coitus:* involving the insertion of the penis in the vagina.
- *Homosexual relationships:* involving two or more individuals of the same sex.
- *Nocturnal emissions:* ejaculation while dreaming.
- *Petting:* sexual acts beween a male and female not involving coitus. Kinsey also listed animal intercourse as a cause of ejaculation less common than the above. The incidence of masturbation in a 15-year-old boy averages about 2 per week, but incidences as high as 23 per week are not pathologic. The incidence of nocturnal emissions averages less than once a month, although incidences of 12 per week have been reported.

20

(1:1088; 2:7-91; 3:351; 4:999; 5:1629)
SGE 1

List the changes in the external *genitalia* (i.e., the male structures outside the abdominal cavity with the exception of the ducts leaving the testis) that *testosterone* produces in boys at puberty.

Answer and Explanation

- *Penis:* more than doubles its length and increases its width.
- *Scrotum:* development of *dartos muscle* and its response to changes in temperature, development of pigmentation, and rugal folds.
- *Testes:* double in volume, *spermatogenesis* begins.

21

(1:1086; 2:7-91; 3:351; 4:999; 5:1629)
SGE 1

List the changes in the internal genitalia that *testosterone* produces in boys at puberty.

Answer and Explanation
- *Epididymis:* growth and secretion. It lies adjacent to the testes in the scrotal sac.
- *Seminal vesicles:* growth and secretion. They are evaginations of the vas deferens. ✓
- *Prostate:* growth and secretion.
- *Bulbourethral glands:* growth and secretion.

22

(1:1086; 2:7-91; 3:351; 4:999; 5:1629)
SGEM 1

Testosterone affects six different body systems that bear no direct relationship with erection or ejaculation in the boy at *puberty*. What are these systems and what are the effects that testosterone has on them?

Answer and Explanation
- *Body hair:* male distribution of pubic hair, hair on chest, face, and around the anus (varies with genetic background), recession of hair at temples, axillary hair, baldness (varies with genetic background).
- *Sebaceous glands:* secretion increases and thickens, tendency to develop acne.
- *Larynx:* enlarges, vocal cords thicken and lengthen, voice deepens.
- *Skeleton:* long bones increase in length and then stop growing (closure of epiphyseal plates, pronounced growth of shoulder girdle)
- *Skeletal muscles:* hypertrophy (pronounced in upper torso), positive N balance.
- *Central nervous system:* increase in sex drive (not directed to one sex) and libido.

23

(1:1086; 2:7-91; 3:354; 4:1002; 5:1629)
SGE 2

A boy is *castrated* at 5 years of age. If he remains untreated, at 25 years of age he will have a number of characteristics similar to those of a 25-year-old woman but different from those of a 25-year-old normal man. What are these characteristics?

Answer and Explanation
- *Body hair.*
- *Larynx:* voice does not deepen.
- *Skeleton:* narrow shoulders.
- *Skeletal muscle.*

24

(1:1088; 2:7-102; 3:345; 4:1002; 5:1630)
SGE 2

Six months after castration, an adult man will have

A. a decreased secretion of LH
B. atrophy of the prostate gland
C. abolition of the libido
D. an increased pitch of the voice
E. all of the above

Answer and Explanation
A. Testosterone inhibits the release of *luteinizing hormone-releasing hormone* (LRH).
B.
C. *Libido* may decrease, but it is not abolished.
D. The growth of the *larynx* that occurs during puberty, unlike that of the *prostate*, is permanent.

25

(1:1102; 2:7-121; 3:360; 4:1011; 5:1608)
SGE 1

Estrogen in the *adolescent* girl causes

A. growth and development of the fallopian tubes, uterus, vagina, clitoris, and labiae
B. growth and development of the breasts (development of tubular duct system, pigmentation of nipples)
C. closure of the epiphyses of long bone
D. a deposition of subcutaneous adipose tissue in breasts, buttocks, hips, and thighs
E. all of the above changes

Answer and Explanation
E.

26

(1:1103; 2:7-121; 3:362; 4:1012; 5:1610)
SGE 1

Progesterone, in the absence of estrogen, in the adolescent girl causes

A. growth and development of the fallopian tubes, uterus, vagina, clitoris, and labiae
B. growth and development of the breasts (development of tubular duct system, pigmentation of nipples)
C. closure of the epiphyses of long bone
D. a deposition of subcutaneous adipose tissue in breasts, buttocks, hips, and thighs
E. none of the above changes, because most of its actions require previous or simultaneous exposure to estrogen

Answer and Explanation
E.

27

(1:988; 2:7-96; 3:362; 4:1006; 5:1604)
ESN 1

Follicle-stimulating hormone (FSH)

A. in the presence of LH, facilitates the release of estrogen by the theca interna of the graafian follicle
B. facilitates spermatogenesis
C. performs both of the above functions
D. release is apparently unaffected by impulses impinging on the hypothalamus
E. is well characterized by all of the above statements

Answer and Explanation
A, B, C. FSH plays an important role in both men and women in facilitating the production of mature eggs (i.e., sperm and ova).
D. The hypothalamus secretes a *releasing* hormone that facilitates FSH secretion. Its release is affected by *estrogens*, stress, and possibly other influences.

28

(1:991; 2:7-105; 3:353; 4:1000; 5:1627)
SENa 1

Which one of the following statements is most correct?

A. castration of the adult man causes a prompt increase in the concentration of circulating FSH and LH
B. testosterone decreases the release of LH but not of FSH
C. statements A and B are both correct
D. a substance from the Sertoli cells decreases the release of FSH
E. all of the above are correct

Answer and Explanation
E. It has been suggested that the testes release a substance other than testosterone (possibly inhibin from the Sertoli cells) which inhibits *FSH* (follicle-stimulating hormone) secretion.

29

(1:1109; 2:7-108;
3:346; 4:1033;
5:1477)
ESG 1

In normal adult human subjects, *prolactin*

A. in the presence of ovarian and adrenal hormones, promotes the growth and development of the mammary glands
B. facilitates the secretory activity of the corpus luteum
C. facilitates the development of the Leydig cells of the testis
D. facilitates the release of FSH
E. performs all of the above functions

Answer and Explanation
(A.) It is also one of the important hormones responsible for lactation by the mother after the birth of her child.
B. Prolactin has also been called *luteotropic hormone* (LTH) because it helps to maintain the corpus luteum in the rat. It apparently does not have this function in the human female.
C. It has no well-defined function in the human *male*.

30

(1:1084; 2:7-106;
3:333; 4:992;
5:1629)
SEG 1

Which one of the following statements is incorrect?

A. androgens are in greater concentration in the blood of the male fetus than in the blood of the male child
B. androgens are formed by the seminiferous tubules of the testis
C. androgens are secreted in small quantities in the adult female
D. androgens in men decrease in concentration after the age of 30
E. LH has little effect on the secretion of androgens by the adrenal cortex

Answer and Explanation
A. *Testosterone* is, at least in part, responsible for the growth of the penis and scrotum during prenatal development.
(B.) The *Leydig cells* (interstitial cells adjacent to the seminiferous tubules) secrete androgens.
C. The adult female produces between 0.27 and 0.35 mg of testosterone per day. Some is from the *ovary*, some from the *adrenal cortex*, and some from the conversion of *androstenedione* by other tissues.
D. These decreasing concentrations of androgens are gradual in men, unlike the decrease in concentration of estrogen and progesterone in women after menopause.

31

(1:1084; 2:7-106;
3:351; 4:997;
5:1629)
SEGc 2

Testosterone

A. production in the male increases markedly at puberty because the concentration of LH in the plasma rises markedly
B. production increases at puberty because hypothalamic sensitivity to steroids increases
C. is well characterized by both of the above statements
D. does not normally penetrate the cell membrane
E. is well characterized by all of the above statements

Answer and Explanation
(A.) LRF = LHRH. See question 28.

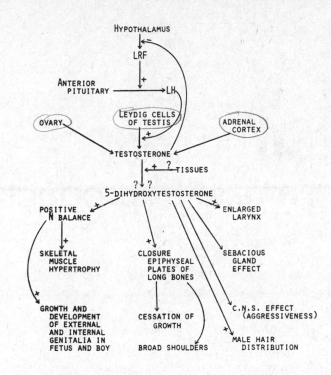

B. The increased testosterone production at puberty is apparently due to a decreased sensitivity of the hypothalamus to steroids such as testosterone.

D. Steroids characteristically penetrate the cell membrane. It has been suggested that testosterone, once in the cell, is changed to 5-dihydroxytestosterone and it is this latter substance that is biologically active.

32

(1:1088; 2:7-94; 3:343; 4:1002; 5:1632)
GSEN 1

Puberty does not occur in

A. children who have been castrated
B. children who have been pituitarectomized
C. children who have been castrated or pituitarectomized
D. girls who have had their uteri removed
E. any of the above cases

Answer and Explanation

A. The *testis* and *ovary* produce the male and female eggs as well as the male and female hormones responsible for the secondary sexual characteristics.

B. The pituitary *gonadotropins* (FSH and LH) are essential for the production of mature eggs and the adequate production of male and female hormones.

Ⓒ

30
Female Reproductive System

1

(1:1094; 2:7-110; 3:354; 4:1006; 5:1612)
Sf 2

In the normal, healthy 25-year-old woman, *menstruation*

A. occurs 1 to 2 days after ovulation
B. occurs approximately every 4 weeks except during or immediately after pregnancy
C. occurs several hours after the formation of a corpus luteum in the ovary
D. is a period of secretion by the endometrium of the uterus
E. occurs 1 to 2 days after the formation of a corpus albicans in the ovary

Answer and Explanation
A. In a 28-day cycle, menstruation begins about 14 days after *ovulation*.
B. Various stressful situations (death in the family, worry, an automobile accident) may cause a missed period.
C. The *corpus luteum*, through the *estrogen* and *progesterone* it produces, prevents menstruation.
D. The secretory phase of the cycle begins on about the 14th day. Menstruation involves the sloughing of about two-thirds of the endometrium of the uterus.
E. In the absence of fertilization, the corpus luteum degenerates into a structure, the *corpus albicans*. Since this structure, unlike the corpus luteum, does not secrete estrogen or progesterone, menstruation will ensue.

2

(1:1094; 2:7-112; 3:356; 4:1013; 5:1612)
Sf 2

In the normal, healthy 25-year-old woman who is not pregnant, menstruation

A. lasts about 1 day during each 28-day cycle
B. occurs 1 or 2 days after an increase in the estrogen and progesterone levels of the blood has begun
C. is associated with a blood loss of about 30 ml
D. is associated with a dilation of the basal segment of the spiral artery of the endometrium of the uterus
E. is associated with all of the above phenomena

Answer and Explanation
A. Menstrual bleeding is generally greatest during the first 3 days and ends about the 5th day.
B. When the corpus luteum stops releasing these hormones, there is a decreased *blood flow* to the endometrium, *necrosis*, and endometrial sloughing.

C. The total volume of menstrual fluid lost per month is about 70 ml. Of this, about 30 ml is liquid *blood*.

D. The basal segment of the artery constricts. This initiates the necrosis and serves to prevent an excessive blood loss.

3

(1:1094; 2:7-113; 3:362; 4:1012; 5:1613)
ES 1

Which of the following hormones exerts little or no control over the *endometrium of the uterus* in its proliferative phase, but during its secretory phase is directly responsible for the changes that occur?

A. progesterone   *secreting phase*
B. follicle-stimulating hormone (FSH)
C. estrogen   *proliferation phase*
D. prolactin
E. luteinizing hormone (LH)

Answer and Explanation
A. Progesterone is released in increased quantities during the second half of the menstrual cycle after the mature follicle has ruptured and part of it has changed into a corpus luteum.
B. FSH exerts its influence on the uterus by controlling the release of hormones from the ovary (follicle and corpus luteum).   *+ testis (for production of Eggs + sperm)*
C. Estrogen controls the proliferative phase of the endometrium.
D. Prolactin probably exerts no influence on the endometrium.
E. LH, like FSH, exerts its influence on the endometrium by controlling the ovary.

4

(1:1092; 2:7-110; 3:355; 4:1006; 5:1602)
SfG 1

Which one of the following statements is incorrect? In a normal, healthy 25-year-old woman who is not pregnant

A. each ovary contains well over 100,000 ova, all of which were formed either prior to or soon after birth.
B. the ova lie adjacent to, but are not a part of, the graafian follicle
C. at the beginning of each cycle, a number of follicles in one ovary start to grow and develop a cavity
D. on about the 6th day of the cycle, most of the developing follicles undergo atresia
E. at ovulation, the follicle ruptures, extrudes its liquor folliculi, but the granulosa and theca cells of the follicle remain in the ovary and fill with blood to form the corpus hemorrhagicum

Answer and Explanation
B. Each *ovum* lies in the middle of the follicle. In the mature follicle, it lies surrounded by cumulus oophorus, which in turn is partially surrounded by the liquor folliculi of the antrum.

5

(1:1094; 2:7-113; 3:358; 4:1006; 5:1614)
ES 1

In a normal, healthy 25-year-old woman

A. the plasma LH is at its lowest concentration during the 2 days prior to ovulation
B. the plasma FSH is at its lowest concentration during the 2 days prior to ovulation
C. ovulation is followed by a decline in plasma estradiol
D. ovulation is followed by a decline in plasma progesterone
E. all of the changes listed above are characteristic

(Continued)

Answer and Explanation

A. It is the high LH concentration that is, in part, responsible for initiating the ovulation.

B. The *FSH* concentration increases prior to ovulation.

Ⓒ. The *estrogen* concentration falls temporarily as the remains of the ruptured follicle reorganize into a corpus luteum, which will also secrete estrogens.

D. The *progesterone* concentration will rise with the formation of the corpus luteum.

6

(1:1100; 2:7-112; 3:358; 4:1011; 5:1613)
ESNf 2

In a normal, healthy 25-year-old woman with a *menstrual cycle* of 28 days

A. the proliferative phase of the uterus is caused by estrogen produced by the graafian follicle

B. menstruation is caused by progesterone from the corpus luteum

C. injections of estrogen and/or progesterone will cause an enlargement of the ovary and an increase in production of mature graafian follicles

D. the concentration of estradiol in the plasma begins to fall at ovulation and continues to decrease until menstruation

E. all of the above statements are characteristic (i.e., true)

Answer and Explanation

Ⓐ.

B. Menstruation occurs when the *corpus luteum* stops producing *progesterone* and *estrogen.*

C. Estrogen acts on the *pituitary* and/or *hypothalamus* to decrease the production of FSH. *Progesterone*, probably through a permissive action, facilitates this inhibition. In other words, these two hormones, by inhibiting FSH release, tend to cause ovarian atrophy and prevent the development of a mature graafian follicle and ovulation. This, apparently, is how they act in *birth control pills*.

D. After ovulation, the corpus luteum is formed. It produces enough estrogen to cause a second rise in plasma estrogen.

7

(1:1096; 2:7-116; 3:363; 4:1006; 5:148)
ES 1

The following diagram explains the changes that occur in the normal 25-year-old woman during the first 10 days of her *menstrual cycle*:

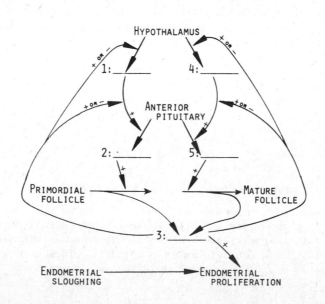

Hormones 1 and 4 are suspected to be released by the hypothalamus and facilitate the release of hormones by the anterior pituitary. What are they?

Answer and Explanation
1. Gonadotropin-releasing hormone (GnRH) or follicle-stimulating-hormone-releasing hormone (FRH).
4. GnRH or luteinizing hormone-releasing hormone (LHRH). Some investigators now believe that a single decapeptide (GnRH or LHRH) controls the release of both FSH and LH.

8

(1:1096; 2:7-116; 3:363; 4:1006; 5:1613)
ES 1

Identify, in the diagram in question 7, hormones 2, 3, and 5. You will note that hormone 3 facilitates endometrial growth, can inhibit the release of hormone 2, and can facilitate the release of hormone 5. In other words, as the concentration of hormone 3 rises, there can be a progressive decline in hormone 2 and a progressive increase in hormone 5.

Answer and Explanation
2. Follicle-stimulating hormone (FSH).
3. Estrogen (estradiol, for example).
5. Luteinizing hormone (LH).

9

(1:1096; 2:7-116; 3:362; 4:1006; 5:1481)
ES 2

During a normal 28-day menstrual cycle, there are marked fluctuations in the plasma concentrations of FSH and LH. These fluctuations are due to the fact that estrogen

A. inhibits the release of FSH and LH during the first 10 days of the follicular phase
B. inhibits the release of FSH and LH during the last 10 days of the luteal phase
C. facilitates the release of FSH and LH prior to, during, and after ovulation
D. does all of the above
E. inhibits the release of FSH and LH prior to, during, and after ovulation

Answer and Explanation
D. When *estrogen* concentration is high (in midcycle), there is a *positive feedback* relationship between estrogen and the *gonadotropins* (FSH and LH). During the rest of the cycle, when estrogen concentrations are lower there is a *negative feedback*. This explains the "+ or −" notation in the figure in question 7.

10

(1:319; 2:7-112; 3:176; 4:720; 5:1615)
ECS 3

How is the control of *FSH* release by *estrogen* similar to the control of peripheral resistance by epinephrine? In both systems the response is

A. initiated by the stimulation of beta receptors
B. initiated by the activation of adenylate cyclase
C. initiated by both of the above
D. concentration dependent
E. blocked by atropine

Answer and Explanation
D. Small doses of *epinephrine* decrease systemic *resistance* to blood flow. Large doses increase resistance.

(Continued)

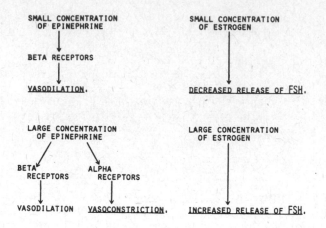

11

(1:1101; 2:7-112; 3:363; 4:1006; 5:1614)
SfEN 1

In a normal 25-year-old woman, 4 days prior to the onset of *menstruation* there are

A. decreasing plasma concentrations of estrogen and progesterone because of a decreased release of prolactin by the pituitary
B. decreasing plasma concentrations of estrogen and progesterone because the corpus luteum, in the absence of pregnancy, has a short life span
C. decreasing plasma concentrations of estrogen and progesterone because of decreasing concentrations of LH
D. increasing concentrations of progesterone
E. increasing concentrations of estrogen and progesterone

Answer and Explanation

A. *Prolactin* from the anterior pituitary is sometimes called luteotropic hormone (LTH) because, in rodents, it facilitates the survival of the corpus luteum and its release of estrogen and progesterone. It apparently does not have this function in the woman.
Ⓑ. The best explanation for the conversion of the corpus luteum into a *corpus albicans* after only 12 days of function is its own intrinsic characteristics.
C. *LH* is capable of prolonging the life of the corpus luteum and facilitating its production of progesterone, but in the absence of pregnancy sufficiently large concentrations of LH do not appear.

12

(1:1080; 2:7-112; 3:350; 4:1019; 5:1630)
SEa 2

Sterility is produced by

A. bilateral ligation and section of the vas deferens ✓
B. destruction of the anterior pituitary of the female ✓
C. both of the above
D. destruction of the posterior pituitary of the female
E. all of the above

Answer and Explanation

A. The *spermatozoa* cannot pass from the testis to the urethra under these conditions. They are resorbed.
B. The *gonadotropins* produced by the anterior pituitary are essential for ovulation.
Ⓒ.
D. The hormones of the *posterior pituitary* are ADH and *oxytocin*. Neither is essential for ovulation, fertilization, or birth.

13

(1:1096; 2:7-112; 3:364; 4:1019; 5:1633)
SEK 1

A group of patients is asked to take oral *contraceptive pills* that contain estrogen and a progesterone-like substance. These pills

A. will, if taken daily throughout the year, prevent menstruation
B. probably depress the release of gonadotropins by the anterior pituitary
C. perform both of the above functions
D. may cause an increased extracellular fluid volume
E. perform all of the above functions

Answer and Explanation

A. The usual way to administer these pills is to either give a few blank pills each month or to stop taking these pills for a few days each month. It is the withdrawal of estrogen and progesterone that causes *menstruation.*
D. Estrogens in large concentrations probably act on the kidney to increase Na and water retention.

Ⓔ

14

(1:1091; 2:7-112; 3:354; 4:1006; 5:1602)
SG 1

Which one of the following statements is incorrect?

A. ovaries are essential for cyclic uterine activity
B. ovaries start to form ova at puberty
C. the cervical mucus becomes less viscous at the time of ovulation
D. a foreign body in the uterus prevents implantation
E. in most women, 14 days after the onset of menstruation is a more fertile period than 22 days after the onset of menstruation

Answer and Explanation

A. The *ovaries* produce estrogen and progesterone.
Ⓑ Apparently no additional ova are formed after birth. At *puberty*, a ripening process begins. This is a situation different from what we find in the male.
C. This facilitates the penetration of spermatozoa into the oviduct.
D. The use of intrauterine devices is a more effective means of *contraception* than condoms. It is less effective than the use of oral contraceptives.
E. At approximately 14 days after the beginning of menstruation, ovulation occurs.

15

(1:1106; 2:7-118; 3:365; 4:1022; 5:1619)
SGE 2

When fertilization occurs, there is a missed *menstrual period.* This first missed period results from the fact that after fertilization

A. the corpus luteum degenerates
B. a trophoblast is formed which secretes estrogen and progesterone
C. a trophoblast is formed which secretes gonadotropins
D. the ovaries decrease their production of estrogen and progesterone
E. all of the above occur

Answer and Explanation

A. The *corpus luteum* hypertrophies and releases increased quantities of estrogen and progesterone.
B, Ⓒ The first missed menstrual period is due to the *human chorionic gonadotropin* (hCG is a glycoprotein with a molecular weight of about 100,000) released by the developing placenta. hCG facilitates the continued growth of the corpus luteum and its release of increasing quantities of estrogen and progesterone. As the *placenta* develops, it will eventually, itself, produce enough steroids to make the corpus luteum unnecessary.
D. The corpus luteum of the ovary increases its production of *estrogen* and *progesterone.*

16

(1:1106; 2:7-118;
3:365; 4:1023;
5:1619)
SE 2

During most of *pregnancy*, the maintenance of human gestation is dependent upon the secretions of the

A. corpus luteum of the ovary
B. anterior pituitary
C. corpus luteum and anterior pituitary
D. placenta
E. adrenal cortex

Answer and Explanation

A. Removal of the *ovaries* after the first trimester of pregnancy does not result in abortion.
B. The *anterior pituitary* has its effect on pregnancy through its control over the secretions of the ovary.
C, (D.) Until the *placenta* has formed and matured, the corpus luteum helps to prevent the sloughing of the endometrium of the uterus. After the placenta has formed, it produces chorionic gonadotropins which control the production of estrogens and progesterone by the corpus luteum. After it has matured, it produces its own estrogens and progesterone and the corpus luteum undergoes degenerative changes.
E. Although the *adrenal cortex* of the fetus, and possibly of the mother, secretes large quantities of hormones during pregnancy, there is no evidence that these play the key role that the hormones of the placenta do.

17

(1:1107; 2:7-118;
3:365; 4:1026;
5:1620)
SGE 1

During pregnancy, *human chorionic gonadotropin* (hCG)

A. slowly increases in concentration throughout pregnancy
B. starts to decrease in concentration at about the middle of the first trimester of pregnancy
C. causes a decrease in concentration of estrogen and progesterone in the blood
D. increases in concentration throughout pregnancy and, while increasing, decreases the concentration of estrogen and progesterone
E. increases in concentration throughout pregnancy and, while increasing, increases the concentration of estrogen and progesterone

Answer and Explanation

B. hCG increases the production of estrogen and progesterone by the *corpus luteum*. The corpus luteum maintains pregnancy until the *placenta* starts producing sufficient quantities of estrogen and progesterone to maintain the pregnancy without the help of the corpus luteum.

18

(1:1107; 2:7-118;
3:365; 4:1026;
5:1619)
SGEu 2

hCG

A. is like LH in that they are both glycoproteins with an alpha unit that has a luteotropic action
B. can be detected in the urine as early as 14 days after fertilization
C. from the urine of an individual who has missed her first menstrual period and is pregnant will cause ovulation in the rabbit and the release of sperm by the frog
D. has all of the above actions
E. from the urine of an individual who has missed five menstrual periods and is pregnant will cause ovulation in the rabbit and the release of sperm by the frog

Answer and Explanation

C. hCG has an LH-like and an FSH-like action. These reactions were the basis of *pregnancy tests*, but now a simple immunologic test done with a sample of the patient's urine on a slide is more common, cheaper, and easier to do.

D.
E. By this time the hCG concentration in the urine is usually so low that it will not be detected by one of the standard pregnancy tests.

19

(1:1107; 2:7-118; 3:366; 4:1026; 5:1619)
GES 1

List the hormones produced by the placenta and their actions.

Answer and Explanation
- *Human chorionic gonadotropin* (hCG): facilitates the release of estrogen and progesterone by the corpus luteum of the ovary.
- *Human chorionic somatomammotropin* (hCS): promotes mammary growth, a positive N balance, and has an action similar to growth hormone.
- *Estrogen* (estradiol): promotes growth and maintenance of the endometrium of the uterus.
- *Progesterone:* promotes growth and maintenance of the endometrium of the uterus.
- *Others:* It also produces substances with TSH (probably due to hCG), renin, ACTH, and relaxin characteristics. *Relaxin* is a polypeptide with a molecular weight of about 9000. It is produced by the ovary and placenta. It decreases the tension on the pubic symphysis and softens and dilates the uterine cervix during pregnancy.

20

(1:1106; 2:3-232; 3:365; 4:1028; 5:1619)
SMRG 1

During *pregnancy,*

A. spontaneous contractions of the uterus are usually absent until the 36th week
B. the breasts enlarge due to the release of prolactin from the anterior pituitary
C. the O_2 tension in the umbilical artery exceeds that in the umbilical vein
D. the blood in the umbilical vein has an O_2 tension similar to or greater than the blood in an average systemic vein of the mother
E. none of the above statements apply

Answer and Explanation
A. *Uterine contractions* occur throughout pregnancy. During the last weeks of pregnancy they become more rhythmic, occurring as frequently as every 1 to 20 minutes.
B. *Estrogen* and *progesterone* prevent the release of *prolactin* from the pituitary.
C. The umbilical vein carries oxygenated blood away from the *placenta* and therefore has more O_2 than the umbilical artery.
D. The blood in the umbilical vein apparently has a Po_2 well below 40 mm Hg. (See Longo, L.D.; Bartels, H. "Respiratory Gas Exchange and Blood Flow in the Placenta." *DHEW Publications.* No. (NIH) 73–361, 1972. Pgs. 272, 359, 406.) Whereas the mother's venous blood has a Po_2 of about 40 mm Hg.
E.

21

(1:1107; 2:7-118; 3:365; 4:1026; 5:1619)
SEGu 1

From the 10th week of *pregnancy* until delivery there is, in the mother, a progressive increase in the concentration of (check each correct answer):

1. plasma estriol
2. plasma human chorionic somatomammotropin (hCS)
3. urinary pregnanediol
4. plasma pituitary growth hormone (GH)
5. plasma chorionic gonadotropin (hCG)

(Continued)

Which of the following best summarizes your conclusions?

A. statements 1, 2, and 3 are correct
B. statements 1, 2, and 4 are correct
C. statements 1, 3, and 5 are correct
D. statements 3, 4, and 5 are correct
E. statements 4 and 5 are correct

Answer and Explanation
A. The placenta produces and releases progressively greater quantities of *estrogen*, hCS, and progesterone from the 8th week throughout the pregnancy. *Pregnanediol* is the principal metabolite of progesterone. GH does not increase during pregnancy. hCG is at its highest concentration on the 8th week of pregnancy.

22

(1:1008; 2:7-31; 3:366; 4:1031; 5:1622)
SEMG 1

The event or series of events that initiate *parturition* (labor) are not well defined for human beings. We do know, however, that during the last day of pregnancy

A. oxytocin is released by the posterior pituitary and decreases the frequency of uterine contractions
B. stretching the uterine cervix is a potent stimulus for oxytocin release
C. prostaglandin formation in the decidua of the uterus is depressed
D. all of the above statements are true
E. progesterone facilitates uterine contractions

Answer and Explanation
A. It increases the frequency of *uterine contractions*.
(B.)
C. Oxytocin, during labor, increases uterine contractions by (1) increasing the production of prostaglandins in the decidua and (2) a direct action on the uterus.
E. *Progesterone* usually inhibits uterine contractions.

23

(1:1112; 3:366; 4:1032; 5:1622)
SMG 1

Which one of the following statements is most correct?

A. during the first stages of labor, uterine contractions occur every 15 minutes and last about 10 to 15 seconds
B. as labor progresses, the contractions occur less frequently but are more forceful
C. the chorion and amnion generally rupture 1 to 2 days prior to labor and, as a result, the amnionic fluid rushes out the vagina
D. in an ideal pregnancy, an average woman should gain less than 15 pounds
E. all of the above statements are correct

Answer and Explanation
(A.)
B. As *labor* progresses, the contractions eventually occur every 2 to 4 minutes and last 45 to 60 seconds each.
C. They generally rupture when the head of the fetus is penetrating the cervix of the uterus (the first stage of labor) but may break days or weeks before the onset of labor or not break until after the child is born.
D. On an average, the woman gains 21 pounds. This gain is considered desirable.

24

(1:1113; 3:367)
SG 1

Which one of the following statements is most correct?

A. in a lactating mother, within 24 hours after the end of parturition the uterus stabilizes within 10% of its weight prior to pregnancy

B. it is usually safe to have coitus during menstruation, the first 8 months of pregnancy, and 1 month after parturition

C. in the postpartum period, most women who nurse their children have less interest in coitus than those who do not nurse

D. there is a reduced interest in sexual intercourse in the woman during the second trimester of pregnancy

E. all of the above statements are true

Answer and Explanation

A. After delivery, the *uterus* weighs about 1 kg. After 4 to 5 weeks, it has involuted and returned to a weight of 50 g.

B.

C. The reverse is true.

D. There is an increased interest in sex. During the third trimester, there is usually a decreased interest in sex.

25

(1:1111; 4:1029;
5:1954)
GS 1

List the causes of a *weight* gain in a woman who is 9 months *pregnant*.

Answer and Explanation

Fetus:	7 pounds
Amniotic fluid and placenta:	4 pounds
Uterus:	2 pounds
Protein retention by mother:	3 pounds
Increased body *fat* by mother:	2 pounds
Increased *water* retention by mother:	3 pounds
Total:	21 pounds

26

(1:1112; 2:7-19;
3:368; 4:1031;
5:1623)
SENG 2

Which one of the following statements is incorrect? In a 24-year-old woman

A. the breasts enlarge during pregnancy because of the high circulating levels of estrogen and progesterone

B. the breasts enlarge during pregnancy because of the synergistic actions of estrogen and prolactin

C. after parturition an abrupt decrease in estrogen concentration initiates lactation

D. after parturition the progesterone concentration increases abruptly

E. suckling facilitates the release of oxytocin from the posterior pituitary

Answer and Explanation

C, D. The expulsion of the placenta during parturition eliminates the major source of estrogen and progesterone during pregnancy.

E. *Suckling* inhibits the release of PIH (prolactin-inhibiting factor). In other words, suckling causes the release of *prolactin* by withdrawing the inhibition of PIH secretion.

27

(1:1112; 2:7-31;
3:367; 4:1034;
5:1623)
SENG 2

In the *lactating* woman

A. oxytocin causes the contraction of the myoepithelial cells

B. prolactin causes the secretion of milk

C. nursing inhibits the secretion of FSH and LH

D. oxytocin is essential for milk ejection

E. all of the above relationships exist

(Continued)

Answer and Explanation

C. About half of all nursing mothers do not *menstruate* until their child is weaned. Mothers who do not nurse have their first menstrual period about 6 weeks after delivery.

D. This is true for women but not for all species of mammals.

Ⓔ

28

(1:1000; 2:7-13; 3:368; 4:1034; 5:1623)
SENG 2

The following diagram illustrates some of the probable relationships that exist in the *lactating* woman who is nursing a child (**PRF:** prolactin-releasing factor or hormone):

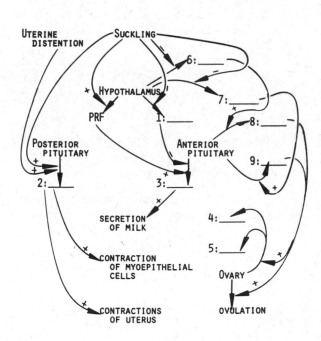

Identify agents 1 (inhibits the release of a substance from the anterior *pituitary*, which facilitates the secretion of milk by the *mammary glands*), 2 (facilitates the contraction of the myoepithelial cells of the mammary glands), and 3 (facilitates the secretion of milk).

Answer and Explanation
1. Prolactin-inhibiting hormone (PIH)
2. Oxytocin
3. Prolactin

29

(1:1090; 2:7-113; 3:368; 4:1034; 5:1623)
SENG 1

Identify, in question 28, hormones 4 and 5 (released by the ovary). Both of these hormones facilitate the development of the breasts during pregnancy. It is the abrupt decrease in the concentration of hormone 4 that initiates lactation.

Answer and Explanation
4. Estrogen (estradiol)
5. Progesterone

30

*(1:1096; 2:7-21;
3:368; 4:1034;
5:1623)*
SENG 2

Identify, in question 28, agents 6 and 7. Their release by the *hypothalamus* is inhibited by *suckling*. They both, through their action on the *pituitary*, facilitate *ovulation*. Agent 6 indirectly facilitates spermatogenesis in the male, and agent 7 facilitates the secretion of testosterone in the male.

Answer and Explanation
  6. Follicle-stimulating-hormone releasing hormone (FRH). FRH and LHRH may be the same substance.
  7. Luteinizing hormone-releasing hormone (LHRH). Its secretion is decreased by prolactin.

31

*(1:1094; 2:7-21;
3:368; 4:1035;
5:1623)*
SENG 2

Identify, in question 28, agents 8 and 9. Both are released by the anterior pituitary in response to LHRH. Increases in agent 8 at the end of menstruation precede increases in agent 9.

Answer and Explanation
  8. Follicle-stimulating hormone (FSH)
  9. Luteinizing hormone (LH)

32

*(1:1106; 3:345;
4:1016; 5:1612)*
SEcs 1

A 28-year-old woman differs from a woman 2 years after menopause in that the postmenopausal woman has

  A. an elevation of gonadotropins caused by a decreased secretion of ovarian steroids
  B. a tendency toward hot flashes as a result of increased secretion of estrogen
  C. elevations in plasma concentrations of gonadotropins and ovarian steroids
  D. an incapacity to achieve orgasm
  E. a decreased concentration of gonadotropins in the plasma

Answer and Explanation
  A. The decreased secretion of the ovarian steroid estrogen results in (1) an elevation in the plasma concentration of gonadotropins and (2) a tendency toward hot flashes.
  D. Aging is the culprit, not menopause. The average woman 1 to 2 years postmenopause experiences three orgasms per month.

PART NINE
SENSATION AND INTEGRATION

31
Vision

1

(1:132; 2:8-61;
3:125; 4:741)
sN 1

Electromagnetic waves with *wavelengths* shorter than those for light are (select the one best answer)

A. gamma rays
B. x-rays
C. ultraviolet waves
D. all of the above
E. infrared light waves

Answer and Explanation
   D. *Light* has a wavelength between 3970 and 7230 A ($39.7 - 72.3 \times 10^{-6}$cm).

2

(1:132; 2:8-61;
3:125; 4:741; 5:512)
sN 1

Light with the longest *wavelength* is

A. blue
B. red
C. yellow
D. green
E. violet

Answer and Explanation
   B.

3

(1:132; 2:8-61;
3:125; 4:741; 5:512)
sN 1

Light with the greatest *frequency* is

A. blue
B. green
C. red
D. violet
E. yellow

Answer and Explanation
   D. Frequency and wavelength are inversely related:

$$\text{Frequency} = \frac{\text{velocity}}{\text{wave length}}.$$

4

*(1:100; 2:8-62;
3:114; 4:724; 5:484)*
sN 1

The *refractive index* of the *lens* of the eye is

A. greater than that for the aqueous humor of the eye because light travels faster in the lens

B. greater than that for the aqueous humor of the eye because light travels slower in the lens

C. less than that for the aqueous humor of the eye because light travels faster in the lens

D. less than that for the aqueous humor of the eye because light travels slower in the lens

E. not explained by any of the above statements

Answer and Explanation

B.
$$\text{Refractive index} = \frac{\text{velocity of light in air}}{\text{velocity of light in other medium}} .$$

The refractive index for the lens is between 1.386 and 1.406 and that for the aqueous humor is 1.336.

5

*(1:99; 2:8-62;
3:114; 4:725; 5:488)*
sN 2

A lens that brings parallel rays of light to focus 2 cm from its center has a focal length of 0.02 m and a *refractive power* of

A. less than 1 diopter

B. between 1 and 10 diopters

C. between 11 and 25 diopters

D. between 26 and 46 diopters

E. greater than 47 diopters

Answer and Explanation

E.
$$\text{Refractive power} = \frac{1}{\text{focal length in meters}} = \frac{1}{0.02 \text{ m}} = 50 \text{ diopters}.$$

6

*(1:100; 2:8-62;
3:114; 4:727; 5:488)*
sN 1

Most of the *refraction* that occurs in the eye occurs at the

A. anterior surface of the cornea

B. posterior surface of the cornea

C. anterior surface of the lens

D. posterior surface of the lens

E. anterior surface of the cornea and lens (i.e., they both have refractory powers within 1 diopter of the other)

Answer and Explanation

A. The anterior surface of the *cornea* has a refractory power of 48.2 diopters. The other surfaces mentioned have refractory powers below 9 diopters. The importance of the lens is that its refractory power can be modified by the stimulation of parasympathetic neurons.

7

*(1:102; 2:8-65;
3:114; 4:730; 5:484)*
sN 2

An individual with *emmetropic eyes*, in order to see an object distinctly (select the one *incorrect* statement)

A. does not have to accommodate if the object is 200 ft away

B. does not have to accommodate if the object is 20 ft away

C. must stimulate parasympathetic neurons to his ciliary muscle if the object is 5 ft away

D. must increase the tension on the suspensory ligament of his eyes if the object is 5 ft away

E. must increase the refractory power of his lens if the object is 5 ft away

Answer and Explanation

A, B. Light rays from a single point 20 or more feet away strike the cornea essentially parallel to one another and in an emmetropic individual come to focus on a single point on the retina. Because of this relationship in optics objects 20 or more feet away are said to be at *infinity*.

C, D, E. In an emmetropic individual objects closer than 20 ft away come to focus "behind the retina" (i.e., they cause a blurred image). The response of an individual to a blurred image is called *accommodation*:

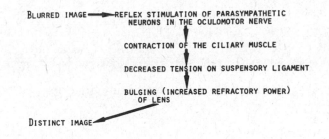

8

(1:102, 2:8-111, 3:114; 4:728; 5:491)
sMN 2

In *accommodation* to a near object (select the *incorrect* statement)

A. the lens becomes more convex

B. there is a convergence of the two eyes owing to the stimulation of somatic efferent neurons

C. there is a dilation of the pupils owing to the stimulation of sympathetic neurons

D. there is a response that can be blocked by atropine

E. there is a contraction of medial rectus muscle of the eye

Answer and Explanation

C. There is a constriction of the *pupils* as a result of a decreased tension on the *suspensory ligament*. This causes a decreased *field of vision* when one is focused on a near object.

D. *Atropine* prevents the contraction of the ciliary muscle and therefore prevents accommodation.

E. This causes the *convergence* of the eyes.

9

(1:102; 2:8-111; 3:124; 4:728; 5:494)
sMNa 2

In question 8, it was noted that during accommodation there is a *convergence* of the eyes. What happens if there is no convergence during accommodation? How does convergence prevent this problem?

Answer and Explanation

The failure of the eyes to converge during accommodation to a near object causes the sensation of *blurring* or a double image. Convergence prevents this by reorienting the eyes so that light rays from a single point fall on *corresponding points* on each retina. Corresponding points on the two retinas are two spots which, when stimulated simultaneously, give the sensation of a single stimulus. *Strabismus* or squint is a condition in which light from a single point source does not fall on corresponding points on the two retinas. It is

(Continued)

340   SENSATION AND INTEGRATION

sometimes treated by surgical shortening of some of the extrinsic eye muscles or corrective lenses. If untreated, the patient may reject impulses coming from one eye. This eliminates the blurring but may cause other problems.

10

(1:104; 2:8-99; 3:123; 4:733; 5:489)
sa 2

Visual acuity is the ability of the eye to detect the separation between two points. Which one of the following lists starts with the best visual acuity and ends with the worst acuity?

A. 20/16, 20/20, 20/30
B. 20/30, 20/20, 20/16
C. 20/20, 20/16, 20/30
D. 20/20, 20/30, 20/16
E. 20/16, 20/30, 20/20

Answer and Explanation

A. Visual acuity = $\dfrac{\text{distance of object in feet}}{\text{rated distance for object in feet}}$.

In other words, a visual acuity greater than 1 (20/20) is better than average for a young, healthy individual.

11

(1:104; 2:8-65; 3:116; 4:730; 5:486)
sa 2

A subject is given a cycloplegic (homatropine) and asked to read a chart 20 ft away. He complains of blurred vision. The physician, by use of corrective lenses, concludes the image is blurred because it is being focused not on the retina but behind it. What conclusions can you draw from these observations? The patient is suffering from

A. myopia
B. hyperopia
C. presbyopia
D. either myopia or hyperopia
E. either myopia or presbyopia

Answer and Explanation

A. In *myopia*, the light from distant objects comes to focus in front of the retina.
B.
C. *Presbyopia* is a failure to accommodate due to a deterioration of the elastic characteristics of the lens.

12

(1:104; 2:8-65; 3:116; 4:731; 5:486)
sa 3

If, in the study in question 11, a cycloplegic is not used, the *hyperopia* might not be identified. Why?

Answer and Explanation

In most cases of hyperopia, the subject's distant visual acuity is excellent because he can *compensate* for his hyperopia by contracting his ciliary muscle. The person with emmetropia, on the other hand, does not have to contract his ciliary muscle for excellent visual acuity in distant vision. The cycloplegic prevents the contraction of the ciliary muscle.

13

(1:104; 2:8-65;
3:116; 4:731; 5:486)
sa 2

A bus driver complains of headaches and eye strain after each day's work. She is given a pair of glasses that have a prescription for each lens of +3 diopters (spherical) and is told to wear them while driving the bus. The results are that her headaches and *eye strain* disappear. The reason is that the patient is suffering from

A. myopia, and the lenses correct for this
B. hyperopia, and the lenses correct for this
C. emmetropia, and the lenses correct for this
D. astigmatism, and the lenses correct for this
E. strabismus

   Answer and Explanation
   A. *Myopia* is corrected for by concave lenses (i.e., lenses with a minus (−) diopter rating: diverging lenses).
   (B.)
   C. This subject is suffering from healthy vision (emmetropia).
   D. *Astigmatism* is not corrected for by spherical lenses.
   E. This prescription does not correct for *strabismus* (squint).

14

(1:104; 2:8-65;
3:116; 4:731; 5:488)
sa 3

In the study in question 13, a patient was told to wear glasses while using her eyes for distant vision (i.e., driving a bus). Her condition was *hyperopia*, commonly called farsightedness. Why should this subject wear glasses for distant vision when her visual acuity for distant objects is already excellent?

   Answer and Explanation
   Her glasses (converging lenses) were prescribed to prevent eye strain due to the excessive use of the ciliary muscle, not to improve visual acuity.

15

(1:104; 3:116;
4:732; 5:485)
sa 3

Why would a patient be prescribed glasses with a −2 diopter *spherical lens* and a +3 diopter *cylindrical lens* at an axis of 180°?

   Answer and Explanation
   The −2 diopter lens is a correction for *myopia*. The cylindrical lens at 180° is a correction for *astigmatism*.

16

(1:103; 2:8-64;
3:116; 4:729; 5:488)
Gsa 3

What optical defect in the eye necessitates the wearing of *bifocal* lenses? What is the most common cause of this defect?

   Answer and Explanation
   When there is a failure to accommodate (i.e., *presbyopia*), bifocal or trifocal glasses are usually recommended. In this condition, the *lens* has lost much or all of its *elastic recoil*. Therefore, when the ciliary muscle contracts and the suspensory ligament decreases its tension on the lens, the lens DOES NOT change its convexity. In other words, there is either an inadequate or no accommodation. The major cause of presbyopia is *aging*. During the aging process, there is a progressive loss of elasticity throughout the body [i.e., in the lens of the eye (presbyopia), in the lungs (emphysema), in the arteries (arteriosclerosis), and in the skin (wrinkles and bags under the eyes)].

17

(1:106; 2:8-65;
3:112)
sN 2

Which one of the following statements is incorrect? The *fovea centralis* is an area in the retina in which there

A. is the greatest concentration of cones
B. is the greatest visual acuity
C. are no rods
D. is a greater sensitivity to light than any other area in the retina
E. are no blood vessels overlying the light receptors

Answer and Explanation
A, B. The high concentration of cones and the fact that each sensory unit consists of but one cone contributes to the great visual acuity in this area.
D. Areas peripheral to the fovea centralis are more sensitive to light than the fovea centralis.

18

(1:107; 2:8-76;
3:122)
sN 3

A farmer is out in his fields on a very dark night. He sees in the distance a post that he has recently painted red. He notes that the post is blurred and appears to be black. When he looks directly at the post, it seems to disappear. What is the physiologic basis of this?

Answer and Explanation
The rods are more sensitive to light than the cones. In the dark, one may be using exclusively *rod* vision. As a result, the visual world is converted from reds, blues, whites, etc. (*photopic vision*) to blacks, grays, and whites (*scotopic vision*). In other words, the farmer sees his red post as black because he, in dim light, lacks functional receptors for red, and since there are no functional receptors, in dim light, in the *fovea centralis*, he must use his more *peripheral receptors* (rods). *Visual acuity* is markedly reduced peripheral to the fovea, and therefore, peripheral vision is blurred vision.

19

(1:106; 2:8-68;
3:109)
sN 1

Which one of the following statements is incorrect?

A. the blind spot or optic disk in the retina is an area where the optic nerve leaves and blood vessels enter the eye
B. the rods and cones are apposed to the sclera of the eye
C. light, in passing to the rods and cones, must pass through a layer of ganglion cells and a layer of bipolar cells
D. the macula lutea is a yellowish spot at the posterior pole of the eye that contains the fovea centralis
E. the iris and ciliary muscles are smooth muscle that constitute the intrinsic muscles of the eye

Answer and Explanation
B. The *rods* and *cones* are apposed to the *choroid layer* of the eye.

20

(2:8-62; 3:115;
4:736; 5:488)
sGMN 2

Which of the following statements is incorrect?

A. the near point is the closest distance an object can be brought to the eye and still be seen distinctly.
B. in aging, the near point gets closer to the eye beginning at age 20 or before
C. dark adaptation takes about 20 minutes or longer
D. wearing red goggles for a half hour before going out into the dark gives a more rapid dark adaptation than wearing blue goggles
E. rods are more likely to become nonfunctional in bright light than cones

Answer and Explanation

B. Starting in the teens, the *near point* begins to recede (i.e., the ability of the lens to increase its convexity decreases). At age 10, the average near point is about 8 cm, and at age 70, it is about 100 cm (i.e., over 3 ft).

D. Since the *rods* are less *sensitive* to red light than to yellow, blue, or violet light, one can permit the rods to dark adapt while using cone vision to work in a moderate intensity of red light.

21

(2:8-76; 3:116; 4:739; 5:520)
scN 1

Which of the following statements is most correct?

A. the rods, but not the cones, are capable of dark adaptation
B. avitaminosis A causes a malfunction of the rods but not of the cones
C. in response to light, the rods and cones produce retinene, and the stimulation of bipolar neurons
D. all of the above statements are correct
E. the rods differ from the cones in that they depend on a photopsin-containing pigment called iodopsin for their function, whereas the cones depend on a scotopsin-containing pigment called rhodopsin for their function

Answer and Explanation

A. Both rods and cones increase their sensitivity to light in dim illumination. *Dark adaptation* by the cones is less marked and more rapid than with the rods.

B. *Avitaminosis A* causes night blindness as a result of impaired rod function and a deterioration of the cones that can lead to total blindness.

C, E.

RHODOPSIN IN RODS $\xrightarrow{\text{LIGHT}}$ PRELUMIRHODOPSIN

RETINENE $_1$ + SCOTOPSIN

IODOPSIN IN CONES $\xrightarrow{\text{LIGHT}}$

RETINENE $_1$ + PHOTOPSIN OR OTHER PROTEIN

22

(1:132; 2:8-105; 3:125; 4:741; 5:512)
sN 2

Approximately 0.5% of all women and 8% of all men have a deficiency in their color perception called *color blindness*. The classification of the types of color blindness is based on the Young-Helmholtz theory of color vision. What is the *Young-Helmholtz theory*? What are the types of color blindness?

Answer and Explanation

The Young-Helmholtz trichromatic theory of color vision holds that there are three types of color receptors and that all of the images we perceive by stimulation of the fovea centralis are due to the stimulation of a combination of these receptors. In other words, the whites, grays, blues, oranges, yellows, reds, and violets we experience in bright light are due to the stimulation of (1) blue receptors (maximum sensitivity for a wavelength of 450 millimicra), (2) green receptors (550 millimicra), (3) red receptors (580 millimicra), or (4) some combination. Color blindness is divided into:

1. anomalous trichromats: weak in one of the three primary colors
2. dicromats: absence of one of the three primary colors (red blind, green blind, blue blind)

(Continued)

3. monochromats: either lack cones (blind in the fovea centralis) or lack two of the primary color receptors. These individuals apparently see only black, white, and shades of gray.

23

(1:102; 2:8-97; 3:115; 4:760; 5:492)
sNM 2

When *light* enters one eye
A. the contralateral pupil constricts
B. there is an ipsilateral reflex stimulation of sympathetic neurons to the iris
C. there will be a change in the size of the pupil that will occur even if the second cranial nerve has been sectioned bilaterally
D. each of the above events occurs
E. none of the above events occur

Answer and Explanation
A. In human beings, unlike some animals, light in one eye causes a reflex *miosis* in both eyes.
B. There is a reflex stimulation of parasympathetic neurons and a depression of sympathetic neurons to the iris.
C. The *optic nerve* is essential for this reflex.

24

(1:120; 2:8-72; 3:111; 4:748; 5:544)
sN 2

Light rays from an object to the left of the visual axis (i.e., from the *left temporal field*)
A. form an inverted, reversed image on the side of each retina to the right of the fovea centralis
B. generate impulses that are carried in the right optic tract
C. stimulate neurons which synapse in the right lateral geniculate nucleus
D. perform all of the above functions
E. generate impulses which all cross over in the optic chiasma

Answer and Explanation
D.
E. Primarily, the neurons from the left eye would carry these signals through the *optic chiasma* to the contralateral side. Neurons stimulated in the right eye would not cross over in the chiasma.

25

(1:120; 2:8-72; 3:111; 4:748; 5:552)
sNa 3

A patient with a total transection of the right *optic tract* would have which of the following symptoms?
A. an ipsilateral loss of the nasal field and a contralateral loss of the temporal field of vision
B. an ipsilateral and contralateral loss of the nasal fields
C. an ipsilateral and contralateral loss of the temporal fields
D. an ipsilateral loss of the temporal field and a contralateral loss of the nasal field
E. an ipsilateral loss of vision

Answer and Explanation
A. The right optic tract contains neurons from the lateral part of the ipsilateral retina (stimulated by light from the nasal field) and from the medial part of the contralateral retina (stimulated by light from the temporal field).

26

*(1:120; 2:8-72;
3:111; 4:748; 5:552)
sNa 3*

A patient with a total transection of the *optic chiasma* would have which of the following symptoms?

A. a loss of the nasal field of the right eye and a loss of the temporal field of vision of the left eye
B. a loss of the nasal fields of both eyes
C. a loss of the temporal fields of both eyes
D. a loss of the temporal field of the right eye and a loss of the nasal field of the left eye
E. a loss of vision in the right eye

Answer and Explanation
C.

27

*(1:120; 2:8-72;
3:111; 4:748; 5:552)
sNa 2*

A patient with a total transection of the right *optic nerve* would have which of the following symptoms?

A. an ipsilateral loss of the nasal field and a contralateral loss of the temporal field of vision
B. an ipsilateral and contralateral loss of the nasal fields
C. an ipsilateral and contralateral loss of the temporal fields
D. an ipsilateral loss of the temporal field and a contralateral loss of the nasal field
E. an ipsilateral loss of vision

Answer and Explanation
E.

28

*(1:102; 2:8-67;
3:176; 4:760; 5:492)
sNMa 1*

Which of the following statements is most correct?

A. miosis is produced by narcotics such as morphine
B. mydriasis is produced by parasympatholytic agents such as atropine
C. mydriasis is produced by brain ischemia
D. all of the above statements are true
E. none of the above statements are true

Answer and Explanation
D.

29

*(1:102; 2:8-71;
3:115; 4:761; 5:493)
sNMa 2*

A patient is found to have lost the *light reflex* (miosis in response to light) but to have otherwise normal vision. He can, for example, see distinctly both near and distant objects. What lesion might explain this set of symptoms?

A. a transection through the geniculocalcarine tract
B. a transection through the superior part of the sympathetic chain
C. a transection through the third cranial nerve
D. a lesion in the tectal region of the midbrain
E. none of the above

Answer and Explanation
A. This will cause a partial blindness and not eliminate the light reflex.
B. Since, in the light reflex, miosis is brought about by a reflex stimulation of *parasympathetic neurons*, destruction of sympathetic neurons will not eliminate the light reflex.
C. This will eliminate the light reflex but will also eliminate accommodation to a near object.
D. The phenomenon described above is called an *Argyll Robertson pupil*. It is seen in neurosyphilis.

32
Other Senses

1

The threshold of *audibility* for a normal young adult is about 2×10^{-4} dynes/cm² and occurs in the middle of the audible range (i.e., at about 2000 cycles/sec). If a sound at this frequency produced 1000 times this pressure, it would have an intensity rating of

(1:168; 2:8-12; 3:133; 4:769; 5:430) svN 2

A. less than 50 decibels (db)
B. between 50 and 70 db
C. between 71 and 90 db
D. between 91 and 110 db
E. greater than 110 db

Answer and Explanation

B. Intensity in db $= 20 \left(\log \dfrac{\text{measured pressure}}{\text{standard measure}} \right) = 20(\log 1000) = 20(3) = 60$ db.

2

Which one of the following statements is most correct?

(1:179; 2:8-12; 3:133; 4:773) svGN 2

A. frequencies of 10^2 and 10^4 cycles/sec have a higher threshold measured in decibels than sounds with a frequency of 2000 cycles/sec
B. the 20-year-old individual differs from the 40-year-old in that the younger individual has a threshold for frequencies of 10^4 hertz (hz) that is 20 db lower than for the older individual
C. both of the above statements are correct
D. the 20-year-old individual differs from the 40-year-old in that the younger individual has a threshold for frequencies of 10^2 hz that is 20 db lower than for the older individual
E. all of the above statements are correct

Answer and Explanation

C. In *aging*, the major hearing loss is in the *high-frequency* range.

3

(1:169; 2:8-12;
3:128; 4:763; 5:457)
sv 1

The ear consists of (select the one incorrect statement)

A. a tympanic membrane which separates two air-filled chambers
B. an auditory tube that connects the inner ear and the middle ear
C. a stapes that is attached to the oval window
D. a malleus that is attached to the tympanic membrane
E. a round window that separates the scala tympani from the middle ear

Answer and Explanation

B. The *auditory tube* or eustacian tube connects the middle ear with the pharynx.

4

(1:169; 2:8-14;
3:128; 4:764; 5:433)
svM 1

Which of the following statements is most correct?

A. obstruction of the auditory tube will, in time, usually cause an outward bulging of the tympanic membrane
B. the tympanic membrane continues vibrating several seconds after sound waves stop
C. the tensor tympani and stapedius are skeletal muscles inserted on the malleus and stapes that contract in response to loud sounds
D. high-pitched sounds generate waves with a maximum amplitude near the apex of the cochlea, whereas low-pitched sounds produce their maximum amplitude near the base of the cochlea
E. all of the above statements are correct

Answer and Explanation

A. Obstruction causes an inward bulging of the tympanic membrane because air is slowly removed from the middle ear by the blood.
B. The tympanic membrane is very nearly critically damped.
C. This protects the auditory receptors from excessive stimulation.
D. The reverse is true. Low-pitched sounds generate their maximum amplitude wave near the apex of the *cochlea.*

5

(1:170; 2:8-12;
3:134; 4:764; 5:433)
svMN 1

Which of the following statements is most correct?

A. sound waves with an intensity of 0 decibels are inaudible
B. loud sounds cause a reflex dampening of the movement of the auditory ossicles
C. the ossicles are the only way of conducting sound waves across the middle ear
D. the helicotrema is at the base of the cochlea
E. deafness due to nerve damage is more improved by a hearing aid than deafness due to ossicular damage

Answer and Explanation

A. Zero *decibels* represents a sound intensity of 2×10^{-4} dynes/cm^2. Most normal, young adults can hear a frequency of 2000 hz at this intensity.
B. This protects the cochlea from damage.
C. *Loud sounds* can set up vibrations through either the *oval or round window* in the absence of a functional *tympanic membrane* or functional *ossicles.*
D. The *helicotrema* is at the apex of the cochlea and connects the perilymph of the scala vestibuli with that of the scala tympani. Because of this path, a bulging of the oval window inward will cause a bulging of the round window outward.
E. Hearing is impossible in the absence of the *auditory nerves.* A hearing aid merely amplifies the signal delivered to the *organ of Corti.*

6

(1:180; 2:8-32;
3:134; 4:773; 5:433)
svNa 3

Two patients are tested as follows. A tuning fork is set into vibration. *Subject 1* reports he can initially hear the tuning fork distinctly but when its sound disappears, if it is placed on his temporal bone, he hears it again. *Subject 2* reports she can hear the tuning fork only when it is placed on her temporal bone. If you assume that subject 1 has normal hearing, what conclusions can you draw concerning subject 2?

A. there is probably damage to the cochlear branch of the eighth cranial nerve
B. there is probably damage to the organ of Corti
C. there is probably damage to either the auditory nerve or the organ of Corti
D. there is probably damage to either the tympanic membrane, ossicles, or oval window
E. none of the above statements are true

Answer and Explanation

A, B, C. The tuning fork against the temporal bone is delivering vibrations to the peri-lymph, endolymph, and *organ of Corti*, which, in turn, is stimulating neurons in the *auditory nerve*. In short, the major defect here is apparently in the delivery of impulses to the inner ear.

(D.) This is the system that delivers impulses to the inner ear. Occlusion of the *auditory tube* can produce sufficient impairment of *tympanic* function to produce the above symptoms, as can otitis media (impaired function of the *ossicles*) and otosclerosis (im-mobilization of the stapes and *oval window*).

7

(1:180; 2:8-25;
3:130; 4:771; 5:466)
svNn 1

Which of the following statements is incorrect? Impulses in the right *auditory nerve*

A. are initiated by distortion of hair cells in the organ of Corti
B. are carried to the cochlear nuclei of the medulla (here first-order neurons synapse with second-order neurons)
C. result in the stimulation of fourth-order neurons in the medial geniculate body of the thalamus
D. result in the stimulation of the auditory area of the cerebral cortex (i.e., the occipital lobe)
E. strongly stimulate both the ipsilateral and contralateral auditory areas of the cortex

Answer and Explanation
D. The auditory area of the *cortex* is the superior temporal gyrus.

8

(1:188; 2:8-17;
3:135; 4:768;
5:1244)
svi 2

Endolymph

A. lies in the bony labyrinth
B. has Na^+ and K^+ concentration resembling intercellular fluid
C. bathes the organ of Corti and the cupulae of the semicircular canals
D. when one is traveling at a constant velocity produces a displacement of the cupula of the semicircular canal
E. is well characterized by each of the above statements

Answer and Explanation
A. It lies in the *membranous labyrinth*.
B. Its concentration of Na^+ and K^+ resembles that of intracellular fluid.
(C.)
D. Endolymph produces a displacement of the *cupula* during *acceleration* and decelera-tion but not during a constant velocity.

9

*(1:191; 2:8-55;
3:138; 4:645; 5:498)*
shNa 2

Which one of the following statements is incorrect?

A. the semicircular canals do not lose their function in the weightless condition
B. when cold water is infused over the eardrum, nystagmus results
C. nystagmus may result from damage to the semicircular canals or cerebellum
D. nystagmus is a quick movement of the eyes and is preceded by a slow movement
E. nystagmus may be caused by damage to the cochlea, retina, or one of the first six cranial nerves

Answer and Explanation
A. They continue to be stimulated by *acceleration* and deceleration.
B. This caloric test is used clinically to assess unilateral *semicircular canal* function.
Ⓔ None of these systems when damaged produce *nystagmus.*

10

*(1:193; 2:8-46;
3:130; 4:601; 5:353)*
sNn 1

Which one of the following statements is incorrect?

A. sensory neurons from the utricle, saccule, and semicircular neurons travel in the 8th cranial nerve
B. first-order neurons from the semicircular canals synapse in the vestibular nuclei of the brain stem or in the cerebellum
C. second-order neurons from the vestibular nuclei synapse with motor neurons in the cranial nerves III, IV, and VI
D. second-order neurons from the vestibular nuclei synapse in the contralateral thalamus
E. second-order neurons from the vestibular nuclei descend in the spinal cord in the dorsal columns

Answer and Explanation
C. It is apparently through these synapses that *nystagmus* may be produced.
D. These impulses will be relayed to the cerebral cortex.
Ⓔ They will descend in the vestibulospinal tract in the lateral and ventral columns. The *dorsal columns* contain ascending axons.

11

*(1:188; 2:8-128;
3:141; 4:781; 5:594)*
sNn 1

Which of the following *sensations* is served by sensory neurons with no primary relay in the *thalamus* and no projection area on the *neocortex* (select the one best answer)?

A. touch
B. acceleration
C. taste
D. smell
E. taste and smell

Answer and Explanation
D. Impulses pass from the *olfactory epithelium* to the prepyriform cortex and the peri-amygdaloid area without passing through the thalamus. From these areas, impulses may then be relayed to the thalamus and hypothalamus.

12

*(1:202; 2:8-131;
3:142; 4:780; 5:594)*
sNn 1

Which one of the following statements is most correct?

A. the olfactory epithelium is in a part of the nasal cavity that is well ventilated by inspired air
B. the olfactory epithelium adapts less readily than stretch receptors and to about the same degree as pain neurons

(Continued)

C. adaptation to one type of odor causes adaptation to all types of odors
D. all of the above statements are incorrect
E. all of the above statements are correct

Answer and Explanation
A. Most of the air that is inspired does not pass over the olfactory epithelium. A *sniff* (rapid inspiratory air flow), on the other hand, will cause sufficient turbulence in the nasal cavity to increase the quantity of odoriferous material reaching the *olfactory epithelium.*
B. The olfactory epithelium *adapts* more readily than does a pain fiber.
Ⓓ

13

(1:202; 2:8-128; 3:141; 4:706; 5:594)
sNn 1

Which one of the following statements is incorrect?

A. the receptor neuron in the olfactory epithelium synapses in the olfactory bulb
B. some of the axons in the olfactory tract pass via the anterior commissure to the contralateral olfactory bulb
C. the rhinencephalon is almost totally concerned with olfactory functions
D. the rhinencephalon is sometimes called the limbic system
E. the so-called odor of peppermint, menthol, and chlorine is initiated by neurons in the trigeminal nerve

Answer and Explanation
Ⓒ In human beings, only a small part of the rhinencephalon is concerned with olfactory function. It is much more concerned with sexual behavior, the emotions, and motivation.
D. Now that it is realized that only a small part of this system in human beings is concerned with olfaction, it is more common to call it the *limbic system* than the "smell brain."
E. These substances as well as a number of *nasal irritants* act by stimulating free nerve endings that enter the *fifth cranial nerve.* These endings may also initiate *sneezing,* lacrimation, and respiratory inhibition.

14

(1:202; 2:8-118; 3:143; 4:778; 5:586)
sN 1

Which one of the following statements is most correct?

A. most, or possibly all, taste buds are sensitive to all of the gustatory modalities
B. there are over ten well-defined distinct gustatory modalities
C. the taste projection area is at the foot of the postcentral gyrus
D. all of the above statements are correct
E. none of the above statements are correct

Answer and Explanation
A, B. There are only four gustatory *modalities.* They are sweet, sour, bitter, and salt. There is no one taste bud that responds to all four. Some respond to only one. Others may respond to two different modalities.
Ⓒ

15

(1:198; 2:8-115; 3:144; 4:776; 5:586)
sNn 1

Which one of the following statements is incorrect?

A. the taste buds are confined to the tongue
B. the sensory neurons subserving the sense of taste are in the seventh, ninth, and tenth cranial nerves
C. first-order neurons from the taste buds synapse in the nucleus solitarius

D. second-order neurons from the nucleus solitarius send axons up the contralateral medial lemniscus

E. if all of the sensory neurons from a taste bud are destroyed, the taste bud will degenerate

Answer and Explanation

A. *Taste buds* are found on the tongue, epiglottis, and pharynx.

B. The *facial* and *glossopharyngeal* nerves innervate the tongue and the *vagus* innervates the epiglottis and pharynx.

16

(1:146; 2:8-142; 3:99; 4:602; 5:353)
spNn 2

First-order neurons from

A. touch receptors pass up the spinal cord in the ipsilateral dorsal columns to the medulla

B. pressure receptors synapse in the dorsal horn with second-order neurons, which pass up the contralateral ventral spinothalamic tract to the thalamus

C. touch and pressure receptors are not segregated from one another in the spinal cord

D. touch and pressure receptors are well characterized by all of the above statements

E. sensory neurons are not segregated in the spinal cord by function

Answer and Explanation

D. First-order neurons from *touch and pressure* receptors both ascend in the ipsilateral *dorsal columns* of the spinal cord. Second-order neurons for touch and pressure both ascend in the contralateral *ventral spinothalamic tract.*

E. Neither the dorsal columns nor the ventral spinothalamic tract is important in initiating the sensation of pain, heat, or cold. Signals initiating these sensations are segregated in the *lateral spinothalamic* tract.

17

(1:148; 2:8-142; 3:99; 4:602; 5:402)
sNpn 2

In the cervical cord, axons originating from the sacral cord lie lateral to axons originating from the lumbar cord in

A. the spinothalamic tracts

B. the dorsal columns

C. both the spinothalamic tracts and the dorsal columns

D. neither the spinothalamic tracts nor the dorsal columns

E. neither the spinothalamic tracts nor the dorsal columns because there is no segregation by source in these areas

Answer and Explanation

A. If you trace the *spinothalamic tracts* from the caudal cord to the medulla, you will note that as you ascend, second-order neurons from the contralateral side are added to the medial part of these tracts. This *segregation* is characteristic of these contralateral pain, temperature, pressure, and touch tracts.

B. The *dorsal columns* contain axons from the ipsilateral side. As you ascend this tract, the axons originating from the more cephalad parts lie lateral to the axons entering the cord from a more caudal area.

18

(1:149; 2:9-113; 3:101; 4:603; 5:360)
sN 2

During an operation on the brain, a surgeon exposes a part of the lateral surface of the *postcentral gyrus* halfway between the medial longitudinal fissure and the temporal lobe. She then stimulates this area with fine electrodes and asks the patient what he feels. The patient

A. might reply that he feels as if something were touching his index finger
B. might reply that he feels heat in his foot
C. might reply that he sees light
D. might reply that he feels a tingling in his head
E. would not reply because brain surgery cannot be performed on a conscious subject

Answer and Explanation

Ⓐ *Sensations* for the hand and finger are located in this area. Other sensations that might also be felt are numbness, tingling, movement, warmth, cold, and pain.
B. The area for the foot is adjacent to the superior sagittal sinus.
C. The postcentral gyrus is in the parietal lobe of the cortex. Sensations of light are caused by stimulation of the *occipital lobe*.
D. Although the stimulus is being applied to the brain the sensation is being projected to the finger in much the same way a blind person *projects* his sense of feeling to the tip of his cane.
E. Brain surgery on a conscious subject is sometimes necessary when the patient's response to the stimulation of certain areas in the brain is needed. In this technique, sensation in the scalp is eliminated by a local anesthetic. Since there are no pain fibers in the brain (just internuncial neurons), this procedure can be relatively painless.

19

(1:147; 2:8-141; 3:102; 4:601; 5:352)
sNMa 2

A patient is asked to stand erect with his feet close together. With his eyes open, he seems quite normal, but when he closes his eyes, there is a pronounced body sway, and the patient sometimes loses his *balance*. Other signs of malfunction are an elevated *touch threshold*, an impaired ability to localize sensation, and an impaired ability to recognize objects such as keys by touching them (*astereognosis*). Which one of the following is most likely to be the cause of this condition?

A. bilateral destruction of the ventral spinothalamic tract
B. bilateral destruction of the dorsal columns
C. a cerebellar lesion
D. bilateral destruction of the corticospinal tract
E. a transection of the midbrain

Answer and Explanation

A. This tract carries touch and pressure fibers, and its destruction will cause an elevated touch threshold but will not produce the other symptoms listed.
Ⓑ The dorsal columns are more important than other ascending tracts in the detailed localization of touch sensation, the determination of form, and in balance with the eyes closed.
C, E. In both a *cerebellar lesion* that produced problems with balance and a *midbrain* transection, the balance problem would exist with both eyes open and closed.
D. In this condition there would be a loss of voluntary control of skeletal muscle. Stereognosis would be normal.

20

(1:147; 2:8-142; 3:99; 4:601; 5:350)
sNMa 2

List the symptoms of a bilateral loss of the *dorsal columns*. Include in your list the signs mentioned in question 19, as well as others.

Answer and Explanation

• Positive Romberg sign: swaying motion or loss of *balance* when standing erect with the feet close together.

- Ataxia: a loss of muscle *coordination* and a resultant abnormal gait.
- Inability to recognize limb *position* with eyes closed.
- Touch threshold elevation.
- Decreased number of touch receptors on skin.
- Loss of *two-point discrimination*: cannot distinguish between a single touch stimulus and two stimuli applied simultaneously.
- *Astereognosis*: inability to recognize common objects by touch.
- Loss of *vibratory sense*: patient cannot distinguish between a vibrating tuning fork and a silent one by touching them.

21

(2:2-16; 3:105; 4:614; 5:413)
sNn 2

Most of the *pain fibers* from the

A. ureter, intestine, liver, stomach, and heart travel with sympathetic neurons and enter the dorsal roots at T-1 through L-1.
B. rectum, prostate, urethra, trigone of the bladder, cervix of the uterus, and upper vagina enter the cord at S-2, 3, and 4.
C. pharynx, esophagus, and trachea travel up the vagus and enter the central nervous system at the medulla
D. viscera travel via the routes indicated above
E. viscera travel in a manner different from any of those listed above

Answer and Explanation
A. This is why *sympathectomy* so effectively abolishes pain from a damaged heart, stomach, or intestine.
Ⓓ.

22

(1:155; 2:2-16; 3:102; 4:619; 5:414)
sN 2

Which one of the following statements is most correct?

A. all pain fibers in the dorsal roots are unmyelinated
B. phantom pain is the sensation of pain in the absence of the stimulation of pain neurons
C. an example of referred pain that is common is a coronary ischemia that is associated with a pain in the left arm
D. the theory that some types of pain are caused by the release of kinins or other tissue substances is no longer held by most competent investigators
E. all of the above statements are correct

Answer and Explanation
A. The unmyelinated fibers are responsible for *dull pain*. The myelinated fibers are responsible for *sharp* pain.
B. *Phantom pain* is the projection of the sensation of pain to a part of the body that has been lost (i.e., removed surgically). It is due to the stimulation of pain fibers that previously innervated the lost part.
Ⓒ. In many cases, the irritation of the viscera (diaphragm, ureter, etc.) is interpreted as pain from the skin.
D. Many investigators hold that most stimuli that induce pain do so by causing the release of chemical agents which act on pain fibers. This is consistent with the observation that the sensation of pain frequently outlasts the noxious stimulus that caused it.

23

*(1:159; 2:2-16;
3:106; 4:617; 5:414)*
sNa 2

It has been suggested that one mechanism responsible for *referred pain* is *convergence*. Explain or diagram how this might work.

Answer and Explanation

If you assume that the localization of pain depends on which area or areas of the brain are stimulated, then you can explain referred pain on the basis that pain fibers from the heart and pain fibers from the arm, for example, have an identical final common pathway:

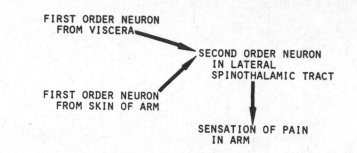

24

*(1:159; 2:2-16;
3:106; 4:620; 5:414)*
sNna 2

Irritation of the peritoneum covering the diaphragm will produce a pain that is usually referred to the

A. testes
B. ventral surface of the abdomen
C. skin between the 11th and 12th rib
D. shoulder or neck
E. head

Answer and Explanation

D. When *pain is referred*, it is usually projected to a structure that developed from the same *dermatome* (i.e., the same embryonic segment) as the irritated structure. Thus, pain caused by an irritation (1) of the diaphragm is referred to the neck or shoulder, (2) of the heart to the chest, axilla, or inside of the left arm, (3) of the ureter to the groin or testes, (4) of the stomach to between the scapulae, and (5) of the appendix to the umbilical area.

25

*(1:65; 3:107;
4:616; 5:411)*
Ns 1

Our understanding of the *enkephalins* is incomplete. At present they are generally believed to (check each correct answer)

1. be released only by neurons scattered throughout the brain
2. increase intestinal motility
3. increase the release of substance P by first-order pain fibers
4. compete with morphine for receptor sites
5. function as synaptic transmitters
6. depress pain perception, produce a state of euphoria, and cause pupillary constriction

Answer and Explanation

1. They are also released in the *substantia gelitinosa* of the cord, the retina, and the gastrointestinal tract.
2. They decrease *GI motility*.
3. They apparently act by *presynaptic inhibition* to decrease the release of *substance P* by first-order pain fibers.
(4.) (5.) (6.) It has been suggested that *acupuncture* has its analgesic effect by causing these *opioid peptides* to be released. They, like *morphine*, may have the effects listed in statement 6.

26

(1:65; 3:107; 4:616;
5:413)
Ns 1

Beta-endorphin is generally believed to

A. be secreted by neurons in the thalamus, hypothalamus, brain stem, and retina
B. be released in response to stimulation of the periaquaductal gray and to relieve intractable pain
C. well characterized by both of the above statements
D. be secreted by the same cells that secrete enkephalin
E. well characterized by all of the above statements

Answer and Explanation

Ⓒ

D. Both are *opioid peptides*, but are secreted by different cells.

27

(1:295; 2:9-123;
3:103; 4:690; 5:407)
sNa 2

Lesions in the central nervous system are frequently produced surgically to treat intractable *pain* and sometimes develop from a pathologic process. Some of the following are procedures or processes that affect the sensation of pain. Select the one statement that is *incorrect*:

A. prefrontal lobotomy: the white matter of the frontal lobe is incised; a markedly increased threshold to the sensation of pain results
B. unilateral anterolateral cordotomy: a lateral spinothalamic tract is incised; a loss of the pain and temperature sense in the skin of the contralateral side below the lesion results
C. unilateral lesion in the postcentral gyrus: little change in the threshold for pain on either side of the body results
D. syringomyelia in the cervical enlargement of the cord: destruction of fibers crossing the cord below the central canal; a loss of the pain and temperature sense restricted to the upper extremities results
E. tic douloureux: attacks of severe pain in the face, lips, or tongue due to stimulation of the trigeminal nerve

Answer and Explanation

Ⓐ *Prefrontal lobotomy* has little effect on the threshold to pain. It does cause a marked change in attitudes, however. After this operation, pain usually is no longer an unpleasant sensation.

B. Bilateral *anterolateral cordotomy* is presently the most common neurosurgical procedure for the relief of pain in terminal cancer patients. It does not produce the marked personality changes that result from lobotomy.

C. Lesions in the *postcentral gyrus* may cause agnosias (inability to recognize certain stimuli), loss of fine touch discrimination, and disorders of spatial discrimination, but they produce only mild changes in the threshold for pain. Pain apparently can be recognized at the subcortical level.

28

(1:148; 2:9-117;
3:102; 4:615; 5:402)
sNa 2

Which one of the following statements is incorrect?

A. proprioception and fine touch are affected by a cortical lesion to approximately the same degree as pain sensation
B. the size of the cortical receiving area on the postcentral gyrus for a part of the body is proportionate to the number of receptors in that part
C. an increase in the frequency of impulses carried by a neuron or an increase in number of neurons stimulated are both signals to the central nervous system that are interpreted as an increase in intensity of sensation
D. an individual who, because of disease or heredity, has no functional, myelinated

(Continued)

pain fibers can still feel pain but will have a slower reflex withdrawal in response to intense heat than a normal subject
E. most visceral sensation travels the same pathway as somatic sensation in the spino-thalamic tracts and the thalamic radiations

Answer and Explanation
(A.) *Proprioception* and *fine touch* are more affected by cortical lesions than *pain*.
D. Unmyelinated and *myelinated* fibers carry the sensation of pain. The former conduct more slowly than the latter. In the absense of myelinated pain fibers, a small child is much more likely to be injured than a normal child.
E. The major exceptions to this rule are the few sensory fibers that travel to the brain in the vagus nerve.

29

(1:149; 2:8-136; 3:101; 4:603; 5:350)
sNa 2

Which one of the following statements is most correct?

A. a 10°C piece of steel feels colder than a 10°C piece of wood
B. sympathectomy abolishes most or all pain from the heart, stomach, and intestine
C. touch receptors are more numerous on the fingers and lips than on the trunk and thighs
D. all of the above statements are correct
E. two-point discrimination requires only one sensory neuron with multiple axon branches

Answer and Explanation
A. Since steel is a better conductor than wood, it will conduct heat more rapidly from the body and therefore, seem cooler.
B. Visceral pain fibers pass through the *sympathetic trunk* and into the cord via the dorsal root.
(D.)
E. Two-point *discrimination* requires at least two sensory neurons.

33
Central Nervous System

1

(1:73; 2:9-31;
3:544; 4:640; 5:787)
N 1

Check each of the following that is a part of the *brain stem*

1. cerebellum
2. cerebral hemispheres
3. cervical cord
4. diencephalon
5. medulla oblongata
6. midbrain
7. pons

Answer and Explanation
The brain stem includes the medulla oblongata, pons, and midbrain, and some investigators also include the diencephalon (i.e., **4, 5, 6,** and **7**).

2

(1:235; 2:9-19;
3:180; 4:560;
5:1237)
N 1

Check each of the following structures which is found in the *medulla oblongata*

1. cerebral aqueduct
2. fourth ventricle
3. amygdaloid nucleus
4. hypoglossal nucleus
5. spinal nucleus of the trigeminal nerve
6. nucleus gracilis
7. nucleus cuneatus
8. hypothalamus
9. inferior olivary nucleus
10. pyramidal decussation
11. medial geniculate bodies
12. restiform body (inferior cerebellar peduncle)
13. red nucleus
14. medial and spinal vestibular nuclei
15. reticular formation

Answer and Explanation
The medulla contains structures **2, 4, 5, 6, 7, 9, 10, 12, 14,** and **15.**

3

(1:148; 2:9-19;
3:540; 4:560;
5:1070)
N 1

Check each of the following structures that is found in the *pons*

1. fourth ventricle
2. caudate nucleus
3. motor nucleus of the facial nerve
4. nucleus of the abducent nerve
5. lateral geniculate body
6. pineal body
7. superior olivary nucleus
8. lateral and superior vestibular nuclei
9. reticular formation
10. corpus callosum
11. middle cerebellar penduncle
12. dentate nucleus
13. motor and superior sensory nuclei of the trigeminal nerve

Answer and Explanation
The pons contain structures 1, 3, 4, 7, 8, 9, 11, 12, and 13.

4

(1:181; 2:9-19;
3:164; 4:560; 5:790)
N 1

Check each of the following structures that is found in the *midbrain*

1. medial geniculate bodies
2. substantia nigra
3. fornix
4. habenular nuclei
5. corpora quadrigemina
6. mamillary body
7. red nucleus
8. nucleus of trochlear nerve
9. hippocampus
10. decussation of superior cerebellar peduncle
11. cerebral aqueduct
12. reticular formation

Answer and Explanation
The midbrain contains structures 1, 2, 5, 7, 8, 10, 11, and 12.

5

(1:181; 2:9-35;
3:181; 4:701; 5:923)
N 1

Check each of the following structures that is found in the *diencephalon*

1. basal ganglia
2. mamillary bodies
3. pineal body
4. subthalamus
5. nucleus cuneatus
6. supraoptic nucleus
7. cerebellum
8. arcuate nucleus
9. fornix
10. lateral geniculate bodies
11. lateral lemniscus

Answer and Explanation
The diencephalon contains structures 2, 3, 4, 6, 8, 9, 10, and 11.

6

(1:275; 2:9-111; 3:156; 4:657; 5:883)
Na 2

A unilateral destruction of *extrapyramidal fibers* in the medulla causes

A. spastic paralysis contralateral to the lesion
B. muscle fasiculations
C. both of the above
D. ipsilateral hypotonicity
E. flexion of the leg

Answer and Explanation

A. The spasticity is due to the loss of inhibition of the stretch reflexes.
B. This is characteristic of a *lower motor neuron lesion.*
D. Contralateral *hypertonicity* is found after these lesions. On the other hand, certain lesions restricted to the internal capsule (i.e., pyramidal fibers) produce a contralateral hypotonicity and muscle weakness.
E. When there is hypertonicity, the leg is usually extended.

7

(1:244; 2:7-162; 3:156; 4:136; 5:90)
NMa 2

A *lower motor neuron* lesion is characterized by

A. pronounced skeletal muscle atrophy
B. flaccidity
C. both of the above
D. hyperreflexia
E. none of the above

Answer and Explanation

C. The *atrophy* produced by a lower motor neuron lesion is more severe than that produced by an upper motor neuron lesion.

8

(1:244; 2:9-19; 3:99; 4:610; 5:762)
sNna 2

A patient presents the following history. One month ago, he had begun to experience *pain* in the ventral surface of the right thigh and leg. On examination, the physician noted an impaired *vibratory sense* and *tactile discrimination* in the leg. There was no *Babinski* sign on either side. The upper extremities and the contralateral leg appeared normal. What conclusions can you draw? There may be a tumor (check each correct statement)

1. at S-4
2. at T-12
3. at C-2
4. compressing the dorsal root
5. compressing the ventral root

Answer and Explanation

1, **4.** The thighs, legs, and feet are innervated by neurons entering the cord at L-2 through S-2. The sensory nature of the problem suggests the involvement of the sensory root (i.e., the dorsal root).

9

(1:147; 2:9-21; 3:108; 4:602; 5:383)
sNna 2

If the signs in the case in question 8 are being caused by a tumor, the tumor apparently has caused a disruption of function in the

A. lateral corticospinal tract
B. the ipsilateral dorsal column
C. the contralateral dorsal column

(Continued)

D. the lateral spinothalamic tract

E. the dorsal column and the corticospinal tract

Answer and Explanation

A. The lack of a *Babinski* sign as well as other signs is consistent with the view that the tumor is having little or no effect on the lateral column.

(B.) The impaired *vibratory sense* and *tactile discrimination* are consistent with a diagnosis of ipsilateral dorsal column malfunction.

C. The dorsal column carries impulses from its ipsilateral side.

D. The fibers in this tract are stimulated by *pain and temperature* fibers from the contralateral side of the body and do not affect tactile discrimination and the vibratory sense.

10

(1:244; 2:9-83; 3:159; 4:657; 5:885)
Na 1

What is a Babinski sign? How does one test for it?

Answer and Explanation

Normally, in the adult, if the lateral border of the sole of the foot is stroked with a blunt instrument, there is a reflex plantar flexion of the toes. If there is *dorsiflexion* of the great toe and fanning of the other toes instead of the plantar flexion, this is called a Babinski sign and is strongly indicative of a disorder in *pyramidal function.*

11

(1:245; 2:9-22; 3:156; 4:653; 5:885)
N 2

Which one of the following statements is most correct?

A. over 80% of the axons in the pyramidal system originate from the precentral gyrus (area 4)

B. the internal capsule contains neurons that are a part of the frontopontine, thalmocortical, corticospinal, and corticopontine tracts, as well as part of the auditory and optic radiations

C. the lateral corticospinal tract contains between 30% and 50% of the pyramidal fibers in the cord

D. the lateral and anterior corticospinal tracts are formed almost entirely of neurons from the contralateral cerebral cortex

E. neurons leaving the pyramidal tract above its decussation, for the most part, innervate the ipsilateral somatic efferent neurons in cranial nerves IV, V, VI, VII, IX, X, XI, and XII

Answer and Explanation

A. The neurons in the *pyramidal system* originate from (1) the precentral gyrus (area 4), (2) the premotor area (area 6), and (3) the parietal lobe (primarily areas 1, 2, and 3). Possibly, as few as 35% of the fibers originate from area 4.

(B.) The *internal capsule* contains both ascending and descending fibers.

C. The *lateral corticospinal tract* contains about 80% of the pyramidal fibers in the cord.

D. The *anterior corticospinal tract* contains fibers from the ipsilateral cortex.

E. They innervate the contralateral somatic efferent neurons either directly or through internuncial neurons.

12

(1:147; 2:9-25; 3:99; 4:600; 5:885)
sNMa 3

A patient had the following clinical signs:

1. motor *paralysis*, hypertonicity, and *hyperreflexia* on the left side of the body below the neck with the exception of the diaphragm, shoulders, and parts of the arm, forearm, and hand

2. loss of the *position sense*, tactile discrimination sense, and vibratory sense in the same areas

3. *numbness* in the left thumb that was noted nowhere else in the body
4. loss of *pain* and temperature sense below the neck on the right side of the body with the exception of the shoulders and parts of the arm, forearm, and hand.

The cause of these signs is probably a lesion in

A. the left side of the pons
B. the right side of the pons
C. the left side of the spinal cord
D. the right side of the spinal cord
E. none of the above areas

Answer and Explanation

A, B. If a unilateral lesion were in the pons, clinical signs 1, 2, and 4 would be on the contralateral side. Since the loss of pain is on one side of the body and the motor paralysis on the other, a unilateral pontine lesion is unlikely.

C. The lesion is on the same side as the motor paralysis and the proprioceptive deficit. The motor paralysis is caused by the destruction of descending fibers which have crossed over in the brain stem. The proprioceptive deficit is caused by the destruction of ascending fibers in the dorsal columns. They do not cross over in the spinal cord. The loss of the pain sense is contralateral to the lesion because pain fibers *cross-over* almost as soon as they enter the cord.

13

(1:146; 2:8-143; 3:99; 4:600; 5:885)
sNMa 3

The symptoms in the case in question 12 are consistent with a diagnosis of a lesion restricted to the left

A. dorsal and lateral funiculi
B. dorsal funiculus
C. ventral funiculus
D. lateral and ventral funiculi
E. rubrospinal tract

Answer and Explanation

A. The clinical picture is the result of a destruction of the *dorsal funiculus* (symptom 2) and *lateral funiculus* (symptom 1 is due to the destruction of the lateral corticospinal tract; symptom 4 is due to the loss of the lateral spinothalamic tract).

14

(1:73; 2:9-19; 3:99; 4:610; 5:473)
sNMa 2

The clinical signs in the case in question 12 are consistent with a diagnosis of a *hemisection of the spinal cord*. It probably occurred at

A. C-2
B. C-6
C. T-2
D. T-5
E. L-1

Answer and Explanation

A. A lesion here would cause a motor paralysis of the *diaphragm* and an involvement of the scalp and neck, as well as all of the shoulder, arm, and forearm.

B. This lesion usually leaves the innervation of the diaphragm and shoulder and parts of the arm and forearm intact.

C. A lesion at T-2 would usually leave the upper appendage intact.

15

(1:145; 2:8-148;
3:99; 4:600; 5:383)
sNna 3

Why, in the case in question 12, is there a *numbness* of the left thumb but no complete loss of the sensation of touch anywhere else? The lesion

A. destroyed the left dorsal root but left intact the right ventral spinothalamic tract
B. destroyed the left dorsal root but left intact the right dorsal funiculus
C. destroyed the ventral horn but left intact the right ventral spinothalamic tract
D. destroyed the ventral horn but left intact the right dorsal funiculus
E. left intact the right lateral spinothalamic tract

Answer and Explanation

Ⓐ Destruction of the left *dorsal root* at C-6 usually causes a complete sensory loss in the left thumb. Touch fibers from the left side of the body (1) ascend the cord in the left *dorsal funiculus* (column) and (2) stimulate neurons which ascend the cord in the right *ventral spinothalamic tract*. Thus, a unilateral lesion in the cord will not cause a complete loss of the sensation of touch on either side.

C, D. Destruction of the ventral horn will cause motor paralysis but no sensory deficit.

E. The *lateral spinothalamic tract* carries pain and temperature signals but no touch signals.

16

(1:148; 2:8-137;
3:99; 4:602; 5:383)
sNa 3

A *unilateral transection* of the spinal cord causes below the lesion

A. an ipsilateral loss of the vibration sense and tactile discrimination and a contralateral increase in the touch threshold
B. an ipsilateral loss of the vibration sense and tactile discrimination and a contralateral decrease in the touch threshold
C. a contralateral loss of the vibration sense and tactile discrimination and a contralateral decrease in the touch threshold
D. a contralateral loss of the vibration sense and tactile discrimination and a contralateral increase in the touch threshold
E. a contralateral spastic paralysis

Answer and Explanation

A. The loss of the vibration sense results from the destruction of the dorsal columns (uncrossed ascending axons), and the increased touch threshold results, at least in part, from the destruction of the ventral spinothalamic tract (crossed ascending axons).

17

(1:300; 2:9-127;
3:216; 4:697; 5:639)
sNa 3

A patient who is naturally righthanded has his *corpus callosum* severed. In a clinical test, he is shown a series of pictures and told to say the word "horse" every time he sees a picture of a horse. When both of his eyes are open, he performs this task without error. If his *optic chiasma* is

A. also severed, he probably could not perform this task if his right eye is closed
B. also severed, he probably could not perform this task if his left eye is closed
C. also severed, he probably could not perform this task with either eye closed
D. either intact or severed, he probably could not perform this task if his right eye is closed
E. either intact or severed, he probably could not perform this task if his left eye is closed

Answer and Explanation

A, Ⓑ C. The left cerebral hemisphere is the part of the forebrain where most of the *language* functions are centered in righthanded individuals and over 70% in lefthanded individuals. Since the destruction of the optic chiasma prevents impulses from passing from the retina of one eye to the contralateral cortex, and the destruction of the corpus callosum prevents most impulses from passing from one cortex to the contralateral brain, answer **B** is the only possibility. Some of the other commissural tracts in the brain include the anterior commissure, the suprachiasmatic commissure, the posterior commissure, and the habenular commissure. They apparently cannot prevent the visual receptive *aphasia* noted in this case.

D, E. Even if there is visual input into only one eye, that eye has connections via the optic chiasma to the contralateral cortex.

18

(1:300; 2:9-127; 3:156; 4:697; 5:639)
sNMa 3

The *corpus callosum* has been completely severed in more than 20 human subjects. Prior to this operation the subjects were suffering from frequent *epileptic* seizures that could not be controlled by drugs. After the operation, there was diminution in the frequency of the seizures. A number of tests were performed on these subjects. In one, the subject covers his left eye and sits at a table containing a number of objects (a key, a nut, a penny, etc.). The visual field in the right eye that sends impulses to the left brain (i.e., the right temporal visual field) is occluded. In other words, the subject is using only his right brain for vision. The subject is now told to pick up whatever object is named on a lantern slide. The word "nut" is shone on the screen. The subject will

A. not pick up the nut
B. pick up the nut with his right hand but be unable to tell you what he has done
C. pick up the nut with his left hand but be unable to tell you what he has done
D. pick up the nut with his right hand and be able to tell you what he has done
E. pick up the nut with his left hand and be able to tell you what he has done

Answer and Explanation
B. The left cerebrum plays an essential role in initiating motor activity in the right arm and leg in response to olfactory, visual, and auditory stimuli. Since the stimulus is being delivered to the right brain and the major connecting link between the right and left sensory areas has been destroyed, the right arm will not pick up the nut.
C. He will not be able to tell you what he has done because an oral or written description requires sensory input into the left cerebrum. The right cerebrum is capable of simple language comprehension, but not the production of language.

19

(1:300; 2:9-127; 3:216; 4:697; 5:639)
sNMa 2

In a subject with the *corpus callosum* destroyed, the word "hat" was seen only by the right brain and the word "band" only by the left brain. If he were asked what he saw he would say the word

A. "band" and write with the left hand the word "hat"
B. "hat" and write with the left hand the word "band"
C. "band" and write with the left hand the word "band"
D. "hat" and write with the left hand the word "hat"
E. "hat" and write nothing with the left hand

Answer and Explanation
A. If the subject were asked what kind of band it was, he would just as likely say "jazz band" as "hat band." Sperry received the Nobel Prize in Medicine in 1981 for much of the work on which questions 16 through 18 are based.

20

(1:307; 2:9-127; 3:216; 4:687; 5:647)
sNMa 2

The *right hemisphere* is better than the left hemisphere in its

A. sensitivity to signals from the right cochlea
B. sensitivity to signals from the left olfactory epithelium
C. sensitivity to signals from the left cochlea and the left olfactory epithelium
D. sensitivity to signals from the right cochlea and right olfactory epithelium
E. nonverbal ideation (i.e., constructing with colored blocks a mosaic that matches a colored picture, or most other types of copying)

Answer and Explanation

A, B, C, D. There is an ipsilateral projection of *smell* and a predominantly contralateral projection of *hearing*.

Ⓔ The left hemisphere has been characterized as "almost illiterate in respect to pictorial and pattern sense, at least, as displayed by its copying disability" (see Eccles JC, *The Understanding of the Brain*, New York, McGraw-Hill, 1973). It has been suggested, for example, that the removal of the right temporal lobe seriously limits musical ability.

21

(1:244; 2:9-82; 3:156; 4:652; 5:885)
NMa 1

Which one of the following statements is most correct?

A. hemiplegia may be caused by a contralateral lesion in the internal capsule
B. paraplegia may be caused by a bilateral lesion at C-5
C. monoplegia may be caused by a unilateral lesion at C-5
D. hemiplegia may be caused by a contralateral lesion at C-5
E. all of the above statements are correct

Answer and Explanation

Ⓐ Most of the corticospinal fibers cross at the *pyramidal decussation*.

B. This causes *quadriplegia*. Paraplegia is caused by a bilateral lesion in the middle or caudal thoracic cord.

C. This causes an ipsilateral hemiplegia.

D. It is caused by an ipsilateral lesion at C-5.

22

(1:244; 2:9-82; 3:160; 4:638; 5:912)
NMsa 1

Immediately following the *transection of the spinal cord* in human beings, there is

A. a period of spinal shock that rarely lasts more than 24 hours
B. a general increase in skeletal muscle tone
C. a retention of urine and feces
D. a retention of urine and feces associated with an increase in skeletal muscle tone
E. all of the above statements are correct

Answer and Explanation

A. In human beings, *spinal shock* may last only 2 weeks or for several months.

B. There is a decrease in muscle tone (*flaccid paralysis*) and a loss of reflexes during the period of spinal shock.

Ⓒ The tendency to retain urine can result in a GU tract infection.

23

(1:275; 2:9-88; 3:162; 4:626; 5:790)
NMa 2

A patient 3 weeks after an accident has (1) an increased *tone* in his antigravity muscles in the four limbs, (2) bilaterally hyperactive knee jerks, and (3) an inability to maintain his body *temperature*. In addition, it is noted that (4) muscle tone could be modified by turning the patient's head. The accident has apparently produced a

A. hemidecortication
B. decortication with an intact brain stem
C. midcollicular transection

D. bilateral section of the cord

E. either a midcollicular transection or a bilateral section of the cord

Answer and Explanation

A, B. In *decortication*, you obtain extension of the hind limbs but a moderate flexion of the arms. This is called decorticate rigidity.

Ⓒ The signs of this lesion are *decerebrate rigidity* (extension of the arms and legs). The temperature regulation problem is due to the lesion separating most of the body from descending hypothalamic neurons.

D. In this condition, the *tonic neck reflexes* have no effect below the lesion.

24

(1:265; 2:9-97; 3:166; 4:650; 5:808)
NMa 2

A patient exhibits *athetoid* movements. She probably has a

A. cerebellar lesion

B. lesion in the basal ganglia

C. thalamic lesion

D. hypothalamic lesion

E. lesion in the cerebral cortex

Answer and Explanation

A. In athetosis, there is a continuous series of writhing, worm-like movements in the extremities. In a *cerebellar lesion*, you will generally find few problems in the resting individual. Here the major problems concern walking, grabbing, talking, and other types of willed activity.

Ⓑ Lesions in the *basal ganglia* produce involuntary contractions in the otherwise inactive individual. Athetosis may be produced by a lesion in the *lenticular nucleus*.

25

(1:262; 2:9-46; 3:166; 4:781; 5:805)
NM 1

Which one of the following statements is incorrect? The *lenticular (lentiform) nucleus*

A. lies between the lateral side of the internal capsule and the claustrum

B. contains the putamen and globus pallidus

C. is part of the basal ganglia

D. receives its major input from the olfactory tract

E. sends messages to the motor cortex via the thalamus

Answer and Explanation

D. It receives a much greater and more important input from the caudate nucleus, substantia nigra, and cortical area 6.

26

(1:262; 2:9-46; 3:165; 4:648; 5:805)
NM 1

List the structures which form the *basal ganglia*.

Answer and Explanation

Some classification systems include all of the following as basal ganglia. Some systems include only the caudate nucleus and lenticular nucleus.

1. The major group of *extrapyramidal nuclei*:
   the *caudate nucleus*
   the lenticular nucleus (*putamen* and *globus pallidus*).

2. Structures that do not belong to the extrapyramidal system (i.e., functionally they are not basal ganglia):
   the *amygdaloid nucleus*

(Continued)

the *claustrum*

the *internal capsule.*

3. Structures that belong to the extrapyramidal system, but were *not* initially considered basal ganglia:

the *substantia nigra*

the *subthalamic nucleus*

the *red nucleus.*

27

(1:265; 2:9-100; 3:166; 4:649; 5:807)
NM 2

A child demonstrates irregular, spasmodic, *involuntary movements* of the limbs and facial muscles. He most likely has a lesion in the

A. caudate nucleus
B. precentral gyrus of the cortex
C. postcentral gyrus of the cortex
D. rubrospinal tract
E. midline of the cerebellum

Answer and Explanation

(A.) *Hyperkinetic syndromes* such as the one discussed above (chorea or St. Vitus' dance) are usually caused by a malfunction of the basal ganglia.

D. This lesion is most likely to cause spasticity.

E. This lesion is most likely to cause difficulty in maintaining an upright stance and is associated with a *staggering gait.*

28

(1:265; 2:9-100; 3:208; 4:650; 5:809)
NcMa 1

Which one of the following is present in high concentration in the *putamen* and *caudate nucleus* and has been shown to be at a reduced concentration in *Parkinson's disease*?

A. acetylcholine
B. dopamine (3, 4-dihydroxyphenylethylamine)
C. histamine
D. gamma-aminobutyric acid (GABA)
E. melatonin

Answer and Explanation

B. It has been suggested that in parkinsonism the dopaminergic neurons in the substantia nigra have been destroyed and that this causes a loss of inhibition to the caudate nucleus and putamen. The administration of dopamine has proven successful in the treatment of parkinsonism.

29

(1:255; 2:9-94; 3:171; 4:666; 5:845)
NM 2

A lesion in the right *cerebellum* is likely to cause

A. dysmetria on the left side (i.e., an inability to stop a finger or other body part at a desired point; pastpointing, for example)
B. inability to rapidly pronate and supinate the left hand
C. both of the above problems
D. dysmetria on the right
E. dysmetria on the right and an inability to rapidly pronate and supinate the right hand

Answer and Explanation

E. The cerebellum, unlike the cerebral cortex, usually responds to a unilateral lesion with an ipsilateral symptom.

30

(1:298; 2:9-110;
3:148; 4:675; 5:284)
eN 1

The *electroencephalogram* (EEG) from a normal adult human who has her eyes closed and is letting her mind wander has as its predominant pattern

A. an alpha rhythm (i.e., a pattern with a lower frequency than a beta rhythm and a higher frequency than a theta or delta rhythm)
B. a rhythm that is unaffected by changes in the blood
C. a rhythm that is the same in the child under similar circumstances
D. a rhythm that is not characterized by any of the above statements
E. a rhythm that is well characterized by all of the above statements

Answer and Explanation
(A.) During an alpha rhythm the wave frequency is 8 to 12/sec.
B. Decreases in blood sugar, glucocorticoids, body temperature, or increases in arterial P_{O_2} decrease the frequency of alpha waves.
C. The predominant pattern prior to adolescence is the theta rhythm (4 to 7 per second).

31

(1:299; 2:9-110;
3:149; 4:680; 5:318)
esNM 1

Dreaming usually is associated with

A. the period of sleep when the EEG contains slow waves (less than 8 per second)
B. rapid eye movements (REM)
C. an increased muscle tone
D. a decreased threshold to extrinsic stimuli
E. all of the above

Answer and Explanation
A. Dreaming is associated with an *EEG* containing patterns similar to those in an alert subject (high frequency, low voltage).
(B.)
C. There is a decreased *muscle tone*.
D. The reverse is true. Dreaming is a period when arousal is difficult.

32

(1:299; 2:9-146;
3:153; 4:680; 5:316)
GNS 1

Which one of the following statements is incorrect?

A. dreaming in men is frequently associated with an erection of the penis
B. the newborn sleeps about 16 hours, the adolescent about 8 hours, the 40-year-old about 7 hours, and the 70-year-old about 6 hours per day
C. dreaming in men after the age of 40 is rare
D. REM sleep and paradoxic sleep are the same
E. at 6 months of age, 70% of the sleep is NREM sleep, and at 15 years of age, 80% of the sleep is NREM sleep

Answer and Explanation
(C.) After the age of 3 years, between 18% and 23% of *sleep* is occupied with *dreaming*. If a person is awakened during or immediately after a dream, he will remember it. On the other hand, if he is awakened well after he has finished dreaming, he will probably have forgotten his dream. Most dreams last approximately 20 minutes and, in some cases, are associated with teeth grinding.
E. NREM sleep is sleep when there is no rapid eye movement.

33

(1:213; 2:9-147;
3:146; 4:628; 5:311)
Ns 2

The ascending, midbrain reticular formation

A. is most active during wakefulness
B. is most active when there is a minimal stimulation of the extero- and interoceptors of the body
C. sends all or almost all of its impulses to the thalamus
D. sends all or almost all of its impulses via the thalamus to the precentral gyrus
E. is well characterized by all of the above statements

Answer and Explanation

A.

B, C, D. Most of the neurons carrying sensory messages to the cortex give off *collaterals*, which stimulate the *reticular-activating system* (RAS) in the reticular formation. As a result, many signals are sent from the RAS that bypass the thalamus and project diffusely to the cortex. These signals serve to keep the subject alert.

34

(1:213; 2:9-147;
3:162; 4:628; 5:311)
Ns 2

The stimulation of the (check each correct statement)

1. facilatory reticular substance in the midbrain facilitates the stimulation of gamma efferent neurons
2. facilatory reticular substance in the midbrain will not produce an increased muscle tone if all of the dorsal roots are cut
3. lateral corticospinal tract will not produce an increased muscle tone if all of the dorsal roots are cut
4. vestibulospinal tract will not produce an increased muscle tone if all of the dorsal roots are cut
5. Renshaw cell facilitates the stimulation of alpha efferent neurons
6. motor cortex, basal ganglia, or *cerebellum* can inhibit the stretch reflex

Answer and Explanation

1, 2. The stimulation causes (a) a stimulation of *gamma efferent neurons*, (b) contraction of *intrafusal fibers*, (c) a stimulation of afferent neurons (in the dorsal root) from the muscle spindle, (d) a stimulation of *alpha efferent neurons*, (e) contraction of extrafusal fibers and an increase in *muscle tone* (see Chapter 2, questions 19 through 23).

3, 4. These 2 tracts act directly on the alpha efferent neuron and therefore produce an increased muscle tone even if the dorsal roots are cut.

5. The *Renshaw cell* is an interneuron that feeds back on the motor neuron to depress its activity. If its activity is prevented hypertonicity results.

6. It is the loss of *inhibitory signals* from the *cortex* and *basal ganglia* to the gamma efferent neurons that probably causes *hypertonicity* after a high pontine *transection*.

35

(1:193; 2:8-48;
3:132; 4:645; 5:819)
Ns 2

Displacement of endolymph in the *semicircular canals* causes the excitation of an ampulla and a reflex eye rotation. This reflex is due to the following sequence of events. The stimulation of

A. ipsilateral *cochlear nuclei* and the contralateral 2nd *cranial nerve*
B. ipsilateral *vestibular nuclei* and the contralateral 2nd cranial nerve
C. ipsilateral cochlear nuclei and the contralateral 3rd and 6th cranial nerves
D. ipsilateral vestibular nuclei and the contralateral 3rd and 6th cranial nerves
E. none of the above

Answer and Explanation

D. These two nerves are motor to the extrinsic and intrinsic muscles of the eye.

Items 36 to 39 concern the ascending, central, *auditory pathway* for messages to the auditory cortex.

36

(1:181; 2:8-26; 3:131; 4:770; 5:462)
Ns 1

These messages are carried from the inner ear in the

A. third and fifth cranial nerve (C.N.)
B. fifth cranial nerve
C. sixth cranial nerve
D. seventh cranial nerve
E. eighth cranial nerve

Answer and Explanation
E. C.N. III, IV, and V carry messages to and from the extrinsic and intrinsic muscles of the eye. C.N. VIII carries messages from the cochlea, semicircular canals, and vestibule.

37

(1:181; 2:8-26; 3:131; 4:770; 5:462)
Ns 1

These messages then pass (on their way to the cortex) via the

A. caudate nucleus or cochlear nuclei
B. caudate nucleus or red nucleus
C. superior olivary nuclei or caudate nucleus
D. superior olivary nuclei or cochlear nuclei
E. red nucleus or cochlear nuclei

Answer and Explanation
D. The caudate nucleus and red nucleus are *basal ganglia*. The basal ganglia, through an extensive system of interconnections with the cortex and the brain stem, serve to program skeletal muscle contraction. They do not serve as a major relay station for sensory messages to areas of consciousness.

38

(1:181; 2:8-26; 3:131; 4:770; 5:462)
Ns 1

These messages then pass (on their way to the cortex) via the

A. lateral lemniscus that is either ipsilateral or contralateral to their source
B. lateral lemniscus that is ipsilateral to their source
C. tractus solitarius that is either ipsilateral or contralateral to their source
D. tractus solitarius that is ipsilateral to their source
E. tractus solitarius that is contralateral to their source

Answer and Explanation
A. The *tractus solitarius* lies in the medulla and cervical cord. It carries viseral afferent fibers and fibers from the taste buds. Impulses from one cochlea travel in both the right and left *lateral lemniscus*.

39

(1:181; 2:8-26; 3:131; 4:770; 5:462)
Ns 1

These messages than pass (on their way to the cortex) via the

A. inferior colliculus to the medial geniculate nucleus
B. inferior colliculus to the nucleus cuneatus
C. inferior colliculus to the hippocampus
D. fornix to the medial geniculate nucleus
E. fornix to the nucleus cuneatus

(Continued)

Answer and Explanation

A. The *nucleus cuneatus* is an area of synapse in the medulla for sensory fibers in the fasciculus cuneatus. The fornix connects the hippocampus to the hypothalamus. The *hippocampus* is part of the limbic system.

Answer questions 40 and 41 by selecting one of the following possibilities for each question

A. hearing
B. smell
C. taste
D. vision
E. smell and taste

40

(1:200; 2:9-113; 3:143; 4:603; 5:362)
Ns 2

What modality is represented on the *postcentral gyrus* and does not have a separate cortical projection area?

Answer and Explanation
C.

41

(1:202; 2:128; 3:200; 4:781; 5:594)
Ns 1

What modality has little or no representation on the neocortex?

Answer and Explanation
B. The *neocortex* is the laminated part of the cerebral cortex. The *olefactory epithelium* sends neurons to the olefactory bulb, which relays impulses to the anterior olefactory nucleus, olefactory tubercle, parts of the hypothalamus, parts of the amygdaloid nucleus, the paleocortex, the pyriform, and entorhinal areas.

Questions 42 through 46 consist of two main parts: a statement and a reason for the statement. Choose the best answer for each of these questions from the options A through E below and mark your choice in the space provided:

A. both the statement and the reason are true and are related as cause and effect (i.e., as indicated)
B. both the statement and the reason are true but are not related as cause and effect
C. the statement is true but the reason is false
D. the statement is false but the reason is an accepted fact or principle
E. both the statement and the reason are false

42

(1:352; 2:9-124; 3:214; 4:689; 5:661)
sNa 2

Statement

_____ Patients with a frontal lobe ablation are unable to distinguish between a meatball and a rubber ball because

Reason

the frontal lobe plays an important role in memory (i.e., delayed response performance).

Answer and Explanation
D. Removal of the *frontal lobe* seems to have a more pronounced effect on recent *memory* (i.e., events that occurred 2 minutes earlier) than long-term memory (i.e., events that ocurred 2 days earlier). Some maintain this is because the ablation causes the subject to

be more easily distracted. The ventral hippocampus also plays an important role in short-term memory.

43

(1:352; 2:9-124; 3:214; 4:685; 5:661)
Na 2

Statement

_____ A tumor in the temporal lobe of the forebrain can, through stimulation, cause hallucinations because

Reason

the temporal lobe is apparently the area where long-term memories are stored.

Answer and Explanation

C. One can elicit memories by stimulating the *temporal lobe* or stimulating subcortical areas after the temporal lobe is removed. It is more likely that memories are stored in subcortical areas.

44

(1:332; 2:9-15; 3:204; 4:704; 5:943)
Na 2

Statement

_____ A cat, several months after its telencephalon was destroyed, will respond to having its paw pinched by extending its claws, hissing, snapping, increasing its heart rate, and dilating its pupils because

Reason

it still has its hypothalamus intact.

Answer and Explanation

A. The *hypothalamus* is essential for the expression of *rage*. The absence of the cerebral cortex means that the rage will be poorly directed.

45

(1:300; 2:9-127; 3:216; 4:697; 5:639)
Na 2

Statement

_____ A dog has its optic chiasma cut at the midline. It is then trained with its left eye opened and its right eye closed to bark every time it sees a triangle. If now its left eye is closed and its right eye is opened, it will still bark when it sees the triangle because

Reason

its corpus callosum and anterior commissure are still intact.

Answer and Explanation

A. If these two *cross-over tracts* had also been destroyed before training, the dog would only bark at a triangle when its left eye is open.

46

(1:300; 2:9-127; 3:156; 4:687; 5:651)
Na 3

Statement

_____ A patient with his corpus callosum severed is told to pick up a coin with his left hand. He will be able to perform this task because

Reason

his language center is in the left brain and his control over fine movement on the left side of the body centers in the right brain.

Answer and Explanation

D. Since his *language center* cannot communicate with the appropritate *motor centers*, he cannot perform the task. He can, however, pick up the coin with his right hand.

Index

This index is referenced by Chapter and question number (i.e., 14:11−18 refers to Chapter 14, questions 11 through 18).

754- 1917

Delaware - Spache

1-800-621-8506

$1050 / Twin Bed
couch